© Dennis Gaboury

About the Author

The author of eight previous books, ELINOR BURKETT, who holds a doctorate in history, has reported for the *Miami Herald*, written for many of America's leading magazines, chaired the journalism department at the University of Alaska-Fairbanks, and trained reporters in Central Asia and Africa. She divides her time between the Catskill Mountains of New York and Bulawayo, Zimbabwe.

Golda

ALSO BY ELINOR BURKETT

Gospel of Shame: Children, Sexual Abuse, and the
Catholic Church (with Frank Bruni)

The Gravest Show on Earth: America in the Age of AIDS

Representative Mom: Balancing Budget, Bill, and Baby in the
U.S. Congress (with Susan Molinari)

The Right Women: A Journey Through Conservative America

The Baby Boon: How Family-Friendly America Cheats the Childless

Another Planet: A Year in the Life of a Suburban High School

So Many Enemies, So Little Time: An American
Woman in All the Wrong Places

GOLDA

ELINOR BURKETT

HARPER  PERENNIAL

NEW YORK • LONDON • TORONTO • SYDNEY • NEW DELHI • AUCKLAND

HARPER ● PERENNIAL

A hardcover edition of this book was published in 2008 by HarperCollins Publishers.

GOLDA. Copyright © 2008 by Elinor Burkett. All rights reserved. Printed in the United States of America. No part of this book may be used or reproduced in any manner whatsoever without written permission except in the case of brief quotations embodied in critical articles and reviews. For information, address HarperCollins Publishers, 195 Broadway, New York, NY 10007.

HarperCollins books may be purchased for educational, business, or sales promotional use. For information, please e-mail the Special Markets Department at SPsales@harpercollins.com.

FIRST HARPER PERENNIAL EDITION PUBLISHED 2009.

The Library of Congress has catalogued the hardcover edition as follows:

Burkett, Elinor.
  Golda / Elinor Burkett—1st ed.
      p.   cm.
  ISBN 978-0-06-078665-6
  1. Meir, Golda, 1898–1978. 2. Prime ministers—Israel—Biography.
3. Israel—Biography.
  DS126.6.M42 B87   2008
  956.9405—dc22

ISBN 978-0-06-078666-3 (pbk.)

HB 12.12.2023

In memory of Nathan Cohen, my first guide through Israel

CONTENTS

Golda

INTRODUCTION

One cannot and must not try to erase the past
merely because it does not fit the present.

A woman of steely self-discipline, Golda Meir was rarely known to flinch, even more rarely to allow any emotion to slip beyond her control. But as the U.S. Marine Corps helicopter lifted off from Philadelphia on a gray Thursday morning in September 1969, she clenched her signature black purse and whispered her apprehension to Lou Kaddar, her longtime aide.

Philadelphia had been a triumphal love fest for Israel's prime minister, thirty thousand people crowded into Independence Hall Park singing "Hatikvah" and waving WE DIG GOLDA posters. Both the mayor and the governor of Pennsylvania had shown up to honor the dowdy matron of seventy-one. But Golda knew better than to expect any such outpouring at the White House.

Richard Nixon was an enigma, a Republican who owed nothing to the Jewish vote. For months, his secretary of state, William Rogers, had been pressuring Israel to cooperate with a U.S.-Soviet peace initiative and withdraw from the territories occupied in the 1967 war in exchange for

vaguely worded international security guarantees. The plan was everything Golda hated about Big Power arrogance, so she'd angled for an invitation to Washington both to stop Rogers' momentum and to convince
Nixon to fill her shopping basket with state-of-the-war Phantom jet fighters, A-4 Skyhawk attack bombers, ground-to-air missiles, and a $200 mil-
lion low-interest loan to pay for them.

She might have felt calmer if she'd understood that the White House
was equally anxious. In a secret cable, the U.S. ambassador in Israel had
advised against inviting her at all. She'll try to persuade the president to
abandon his negotiations with the Russians and "presumably her efforts
will be futile," wrote Walworth Barbour. "If such pleas were limited to
private discussions little harm might be done although conversations
could be unpleasant in view of her emotionalism. But if they [are] repeated, as we would have to fear, on *Face the Nation, Meet the Press,* and
before National Press Club, we might face a major internal storm."

Golda had been prime minister of Israel for only six months, but it was
already clear that she was not just another head of state, or just another
prime minister of Israel. The protocol office at the State Department had
been swamped with requests for invitations to the state dinner Nixon and
his wife, Pat, were hosting on the night of her arrival, hundreds of donors
and congressmen, Jewish leaders, governors, activists, and entertainers
hoping to share the White House chef's Sole Veronique with her. In Los
Angeles, Nelson Riddle was busy composing special music for her entrance to the gala Hollywood was throwing in her honor.

In New York, her pending visit had sparked a political brouhaha between Mayor John Lindsay and his opponents in a tough election campaign over who would stand next to her at the airport. Then the *New York
Times* accused the mayor of using Golda's visit for political advantage by
organizing the most lavish dinner in the city's history for her. More than
1,100 guests had been invited, at $25 a person. Three years earlier, when
the city threw a reception marking the opening of the United Nations
General Assembly, they had confined the guest list to 800.

On her first trip to the United States since taking office, Golda had been invited for coffee by the Senate Foreign Relations Committee and breakfast with a group from the House of Representatives. She'd been booked on *Meet the Press* and the *Today* show and for meetings with the editors of the *New York Times*, the *Los Angeles Times*, *Time*, *Life*, *Fortune*, and *Newsweek*. The AFL-CIO had asked her to deliver a speech to their biennial convention in Atlantic City. And a who's who of luminati, from former vice president Hubert Humphrey, Democratic Party heavyweight Averell Harriman, and New York governor Nelson Rockefeller to the Reverend Billy Graham, had appealed for a bit of her time.

Before she left home, her staff had worried obsessively about the two evening dresses and the hats she'd bought, and whether she'd agree to wear the gloves they were convinced diplomatic niceties demanded. But all she could think about was how to explain to Nixon why Israel wouldn't withdraw "a single soldier from a single inch" until the Arabs signed a peace agreement.

Nixon quickly laid Golda's fears to rest. Despite reams of memoranda from his staff warning him not to succumb to the Israeli prime minister's charms, alone with the grandmotherly woman for more than an hour and a half, Nixon couldn't resist her.

"I remember so well when we sat down in the chairs in the Oval Office, and the photographers came in . . . and we were shaking hands, and she was smiling and making the right friendly comments," Nixon recalled. "Then the photographers left the room. She crossed her legs, lit a cigarette and said, 'Now, Mr. President, what are you going to do about those planes that we want and we need very much?' . . . Golda Meir acted like a man and wanted to be treated like a man. There is no question that she was a very strong, intelligent leader in her own right."

That night, at the glittering State dinner, where Leonard Bernstein and Isaac Stern provided the entertainment, he toasted his guest of honor. "This is the first time . . . we have had the honor to receive the head of government of another state who is also a woman. I can only say . . . that I am reminded of the fact that David Ben-Gurion, in referring to our very distinguished guest here this evening, referred to her as the best man in his cabinet.

"I also recall the old Jewish proverb to the effect that man was made out of soft earth and woman was made out of a hard rib," Nixon continued. Then he compared her to the biblical prophet Deborah, who "loved her people and served them well," noting that she made "peace in the land for forty years."

To modern ears those words sound as patronizing as they did to Golda, who flinched when she heard them, having always abhorred that quote from Ben-Gurion and the sentiments behind it. But she came long before Margaret Thatcher, Benazir Bhutto, Corazon Aquino, and Angela Merkel, long before Hillary Clinton commenced a bid for the American presidency that was shadowed less by her gender than by her messy history with the voters she was trying to woo. Golda was a trailblazer and an anomaly in the corridors of power she entered, and she walked through them as she had to, holding her tongue.

When she departed that evening, Pat Nixon kissed her good-bye. The president beamed. And Golda raced back to Blair House, where she and Lou stayed up late giggling about the flowers, the paintings, and the elegant clothing the women had worn.

For a full week, as she traveled from Washington to New York, Los Angeles, and her old hometown of Milwaukee, Golda kept doing it right, catapulting Israel's view of the Middle East conflict onto the front page of the nation's newspapers, charming political and media leaders, and wracking up the sort of crowds that those in power could not ignore. At the National Press Club, she beguiled journalistic heavyweights with her caustic wit. Would Israel consider employing nuclear weapons if the Arabs attacked again? one reporter asked. "We haven't done so bad with conventional weapons," she responded. When the club president asked for her gefilte fish recipe, having heard of its delights from her grandson, she promised to cook lunch for the entire group on her next visit.

In more serious interviews, she proved equally droll. Responding to a question from a *New York Times* reporter about the occupied territories, she answered, "We're not so fortunate that the quarrel between us and the Arab countries . . . is a question of territory. . . . The Arab countries are in lack of a little more sand? . . . If that were the question, somehow during the last 21 years I suppose we could have found some solution—a little sand for him, a little sand for us. Come on, are you dealing with adults, with heads of states?"

In Los Angeles, Governor Ronald Reagan fawned over her, while Gregory Peck and Jack Benny competed for her attention at the Hollywood gala. She took Henry Kissinger along on a sentimental stop at her old elementary school in Milwaukee, where 243 African-American pupils presented her with a scrapbook of little Goldie Mabovitch's old report cards and photographs of a serious young woman in a high-necked blouse.

From the presidential suite at the Waldorf-Astoria in New York, she basked in her unlikely star power. Almost 3,000 Jewish leaders flew into town for a sold-out dinner organized in her honor. And more than 15,000 people sporting SHALOM GOLDA buttons welcomed her at City Hall, where vendors did a brisk business selling miniature Israeli flags for seventy-five cents each. "You don't have to be Jewish to love Golda," proclaimed mayoral candidate John Marchi. Lindsay spoke of the "inseparable bond of spirit" between the city and Israel, declaring, "Madame Prime Minister, New York is yours."

Golda responded with her coy trademark public humility. "I know you are not here to greet me but, through me, the people of Israel, the mothers and fathers, the young boys in the trenches on the hills on the Suez and on the Jordan, the young widows, orphaned children and, yet, a people whose spirit is high, whose determination is unwavering," she said.

She was, at best, half right. Israel might have been the subtext, but the story of Golda's visit was Golda herself. "To think that that woman with such a sad face played such a role in building a homeland for the Jewish people," said one old man who'd traveled across town to hear her speak at City Hall Park.

It was an extraordinary reception for a woman everyone assumed was nothing more than a caretaker prime minister. David Ben-Gurion, Israel's

feisty founding father, had met with Harry Truman, Dwight D. Eisenhower, and John Kennedy but never managed to wrangle an official
invitation to Washington. Golda's predecessor, Levi Eshkol, was hosted
by Lyndon Baines Johnson, but with little hoopla, and when he visited
New York several months before his death in February 1969, city officials
barely acknowledged Eshkol's presence.

But the sassy nicotine-stained grandmother who wore baggy suits and
orthopedic shoes, spoke with an accent in every language but Yiddish,
and led one of the smallest countries in the world had become an international luminary. She topped the Most Admired Woman charts in the
United States, Great Britain, and the Netherlands, packed auditoriums
from Milwaukee to Boston, and was celebrated not only in Hollywood
but also in halls of power from London to Lagos. Her autobiography became an international best seller; both Anne Bancroft and Ingrid Berg-
man would rush to portray her onstage and on-screen.

Golda wore the sheen of a triumphant Israel in the days before that
triumph became fashionably suspect and radiated moral certainty at a
time when to be sure of one's morals was still honorable. Perhaps because
of her age, her sarcastic wit charmed when it might have offended coming out of the ruby red lips of a woman in a lithe young body. The package was perfect, from her sensible shoes, frumpy dresses, and swollen
ankles to her old-fashioned handbag and omnipresent Chesterfields.

She managed to translate her warts—prosaic warts with which so
many could identify—into reassuring virtues. She made people forget,
then, that the same grandmother who heaps food on you also shamelessly
manipulates you into obedience. That the line between dogged idealism
and willful blindness is, at best, fine. Or that moral certainty can easily
slip over the edge into intolerance. In an era when we needed our Goldas
unsullied, she made it easy to overlook the obvious.

Most of the stars in the political and cultural firmament are larger-
than-life creatures, more beautiful, more brilliant, richer than those of us
who lead prosaic lives. Objects of veneration, they nonetheless are *other*.
Golda's very ordinariness—her world-weary mien, the undisguised wrin-

kles that mapped her face, her straightforward language and blunt manner—allowed her to surmount her extraordinary role and to seem accessibly plebian. Her genius lay in her ability to use her ordinariness, to become a canvas onto which others could project their hopes, their political morality, and their personal fantasies. Unlike most celebrities, she reflected not the baser instincts of her minions but their deepest idealism.

Immensely savvy in reading an audience, Golda did so, often with considerable calculation, never hesitating to exploit her personal history or her experience with oppression. She writ large her Russian-Jewish-American-Israeli Cinderella story and, while deprecating the movement, still managed to tug at feminist heartstrings. At a moment of rampant political cynicism, with her candor and indifference to political correctness, she successfully peddled herself as unpackaged, devoid of any political slick, the slickest political trick of them all.

Almost four decades later—thirty years after her death—Golda still retains remarkable star wattage. In 2003, *Golda's Balcony* opened and played to packed houses on Broadway for fifteen months, making it the longest-running one-woman show in Broadway history. At the White House celebration of the fiftieth anniversary of the establishment of Israel, it was a portion of Golda's autobiography that was read during the ceremony on the South Lawn presided over by President Bill Clinton. T-shirt companies do a brisk business selling garb emblazoned with Golda's pithiest quotes. Collectors still compete for Golda memorabilia—crystal paperweights, vintage photographs, magazine covers emblazoned with her image, and recordings of her speeches. Even Americans who can't conjure up the name of the current prime minister of Israel recognize her visage.

Yet the Golda of legend is bereft of the complexity and subtlety that define a full human being. The least introspective of women, she left behind no diaries and few letters that might provide a path through the mythology. Even with her closest friends, she was discreet to the point of obsessiveness about her personal life. And no one—from her son to her worst enemies—seems able to measure her against anything but her towering reputation, which people maintain by alternately protecting or railing against it.

For Americans, Golda is frozen in time, fixed in memories of youth and the idealism of another era, an age of different sensibilities. In the collective consciousness, she remains a selfless old lady who worked into her eighties to defend the Jewish homeland from a hostile world, the larger-than-life vice president of the Socialist International who nonetheless managed to win Richard Nixon's heart, the first female head of state to rule in a Western nation. Even those embarrassed that David became a nuclear-armed Goliath patrolling the West Bank can't resist admiring her as a tough negotiator who also found time to fix tea and cake for Henry Kissinger.

Especially for women who grew up in her shadow, she remains part Superwoman, a dash of Emma Goldman, a smidgen of Nelson Mandela, all wrapped up in the warmth of our grandmothers.

Israelis once shared the American affection for her, and for the same reasons. They too reveled in her sarcasm, her staunch certitude, her strength. She echoed their swagger, softening it with her gray bun and flirtatiousness. But when that smugness was deflated by the Yom Kippur War, they turned all their resentment over the vanishing of their illusion of invincibility on the grandmother they had long adored.

For them, the arc of the Golda story became the tale of a political shark who came out of retirement and ran Israel with an iron fist worthy of a Bolshevik, concentrating power in a small cabal operating out of her kitchen. Leading the nation during the critical years after the occupation of the West Bank, Gaza, and the Sinai, she steered Israel into obduracy, ignored peace signals from Arab capitals, and led her people into the exuberant arrogance that sparked the disastrous Yom Kippur War.

Bitter that Golda didn't shield them—from their own conceit or the wrath of their neighbors—they ignored the zeitgeist of cocky euphoria she inherited, and the outraged public cries that arose at the mere mention of giving up Hebron. After all, blaming Golda is less painful than introspection about the national intransigence she mirrored or a desiccated political culture that roped a sick old woman into staying long beyond her time.

It's more convenient than tarnishing an old warhorse like Moshe Dayan, the champion of permanent occupation, or asking why, if Golda was the architect of Israel's continued belligerence, peace still does not reign in the Middle East. Just as Americans project their idealism onto Golda, Israelis use her as a foil for absolving themselves.

Like most conventional wisdom, the notion that the truth lies somewhere in between is a cheap way of evading the complicated nature of reality. Both the American Golda and her Israeli doppelgänger contain some elements of truth. But neither offers enough of it to be accurate in capturing a complex woman who teetered between idealism and paranoia during a career that spanned more than half a century. In the end, both versions are heavily burnished by contemporary events and evolving mores, and muddled by a dozen colliding political agendas layered onto her memory. Perceived truth, after all, rarely remains static; it shifts, perforce, in space and time. In History 101, no one teaches us that within a decade or two, history will have moved on, casting doubt on much of what we learn. We discover hidden details, new dimensions, and unforeseen consequences. We become more, or less, forgiving of foibles or frailties we once exalted. In the process, we rip the past out of its context and edit it according to our own needs and values, superimposing the present where it cannot belong.

Christopher Columbus, once an intrepid explorer, turns into a racist plunderer five centuries later. Harry Truman, whose presidential approval ratings sank as low as 24 percent, has now found a steady place among the top ten American presidents of all time. And despite his reputation elsewhere, Genghis Khan remains a hero in Ulan Bator. Well-meaning policies that once seemed so right now seem dreadfully, terrifyingly wrong. As our experiences, our priorities, and our understandings are transformed, so too is our view of the past.

But distance doesn't necessarily make a sharper lens; it simply adjusts our focus. If the stubbornness that once seemed so admirable in Golda now appears wanton and her toughness shortsighted, it's not that she has changed. We have. We know what she could not possibly have divined. And as captive of our present as she was of her past, we see her against a backdrop of an age that has teetered from idealism to terror, from optimism to cynicism, while she lived and worked in the shadow of pogroms and holocausts, precariously balancing herself between hope and dread.

Exploring Golda in her own context is more than a journey into the life of a woman who turned the ordinary into the historic, who reshaped the Middle East, forged an alliance between the United States and Israel that we now take for granted, and raised the first international voice that offered a prescient warning about terrorism. It is also an opportunity to delve into this moment through the lens of the past, to reexamine who we have become in light of who Golda was.

*   *   *

During her last weekend in New York in October 1969, Golda turned up at a small private gathering of Jewish intellectuals at the Central Park apartment of writer Elie Wiesel. It was a tense time in America. A month earlier, the Woodstock Nation had erupted, and Henry Kissinger had held his first meeting with the North Vietnamese. Charles Manson and his ersatz family had recently completed their killing spree, William Calley had been indicted for premeditated murder for the deaths of 109 Vietnamese civilians, and the trial of the Chicago 8 had begun, presaging the Days of Rage.

Golda paid no attention to any of the events reshaping America, too wrapped up in her mounting sense that Israel was losing its international sheen and the plight of the Jewish people its power of suasion as new world horrors captured the international imagination. An old lady still burning with images of Russian pogroms and the helplessness she felt during the Holocaust, she was utterly unprepared for the moment when the world began to fall out of love with Israel. Everywhere she traveled across the globe, she begged and pleaded for understanding, although such leaps of imagination often eluded her.

That night at Wiesel's, she poured out her heart, providing a glimpse into the inner Golda. "The world still does not accept us," she told the guests, who had overflowed the chairs onto the floor, sitting rapt from 9 P.M. until well after midnight. "The non-Jewish world has been in two groups—those that killed us and those that pitied us. Before the war, the decent non-Jewish people were with us. Now we're alive and we're certainly not pitied. If we have to have a choice between being dead and pitied and being alive with a bad image, we'd rather be alive and have a bad image."

CHAPTER ONE

*Pessimism is a luxury that a Jew
can never allow himself.*

By the time the bells of Kishinev's churches tolled noon on Easter Sunday, April 19, 1903, the 50,000 Jews of the Bessarabian capital had been huddled in terror for more than six hours. For them, it was the ninth day of Nisan, 5663, the seventh day of Passover, and they should have been celebrating the exquisite memory of freedom and deliverance from Pharaonic enslavement. The year before, they'd barely escaped the threats and bloodlust set off by the ancient calumny that Jews use the blood of young Christians in the ritual preparation of matzo. But this year, they feared they might not be so lucky.

In early February, the battered body of a Gentile boy had been discovered in a Kishinev neighborhood, and then a Christian girl committed suicide in the Jewish hospital. Suddenly, the city was awash in sinister rumors about Jewish plots. *Bessarabetz,* the city's only newspaper, stoked the fire with tales of Jews exploiting the toils of hapless Christians, of Jewish plots to undermine the power of the tsar, and of the international Jewish conspiracy to seize and rule the whole world.

Finally, in late March, an anonymous handbill circulated in the city's taverns and teahouses. "Brothers, in the name of our Savoir, who shed his blood for us; in the name of our Father the Czar, who cares for his people and grants them alleviating manifests, let us exclaim in the forthcoming great day: Down with Zjids! Beat these mean degenerates, blood suckers drunk with Russian blood!"

The leaders of the panic-stricken community had pleaded with Governor Von Raben to arrange police protection, to no avail. So on *Noch Velikoi Soobboty*, the Night of the Great Saturday in the Russian Orthodox calendar, Christians chalked prominent crosses on their homes and stores, and the Jews of Kishinev boarded up their houses, nailed their shutters closed, and hunkered down for the inevitable.

It was a strange phenomenon, the majority petrified of the minority. But in that odd corner of Eastern Europe called the Pale of Settlement, Christians had been tormenting Jews ever since Catherine the Great set her mind on protecting the masses from "evil influences" by banishing Russia's Jews to the farthest and least economically significant reaches of her empire, to pieces of Russian Poland and the Ukraine, to Lithuania, Belorussia, the Crimea, and Bessarabia, until 94 percent of Russia's Jews were confined to the Pale.

There, the Jews coped by insulating themselves in their own world. Signs in Yiddish advertised Jewish-owned groceries and factories, barbers expertly twisted Hasidic side curls, and tinsmiths refurbished dented tea samovars and cooking pots for a population so poor that most Jewish communities were nothing more than warrens of hovels leaning together in dank courtyards.

They lived girded against outbreaks of anti-Jewish violence that felt like natural phenomena, as unpredictable and unavoidable as the eruption of sleeping volcanoes. For more than two decades, the fury had ignited almost regularly—in 30 Ukrainian towns in 1881, on Christmas day that same year in Warsaw, in a total of 166 towns and villages from one end of the Pale to the other. Then again, in 1882 and 1893.

That Easter Sunday in Kishinev, the violence began innocuously enough: just a knot of teenage boys heckling and shoving the few Jews who dared cross Chuflinskii Square. Then the adults came, men who'd been celebrating the resurrection of Jesus with vodka or tuica, the local plum brandy, and fixed on capping it off with the first pogrom of the twentieth century.

By midafternoon, more than two dozen mini-gangs of two to three dozen men had fanned out along the winding alleys of the old Jewish quarter, smashing windows, kicking in doors, raiding small shops. Gradually, the crowd—the same sort of crowd that would soon turn on the tsar himself—was joined by a new and more lethal wave of rioters. Hearing a rumor that Tsar Nicolas II had given permission for a three-day orgy of Jew-beating, students and seminarians streamed out of the Royal School and area religious colleges wielding iron bars and axes.

At 13 Asiasky Street, two families trembled inside a low-roofed outhouse as the rampage swept into their courtyard. They thought they'd found safety. But the crowd found them, stabbing one man, a glazier, in his neck. When he still showed signs of life, they beat him to death with sticks and truncheons. Panicked, the others in the outhouse made a rush for a small attic, but the crowd spotted and followed them. So they tore their way through the roof, seeking protection in the sunlight and the gaze of a policeman, a Russian priest and scores of neighbors below. Five minutes later, their battered bodies became part of the dross of the morning's bloody frenzy.

*　　*　　*

In New Bazaar, 150 Jews organized in self-defense—but they were arrested. A grocer, blind in one eye, offered his attackers all his money, the not-so-princely sum of 50 rubles, for his life and livelihood. They took the cash, then destroyed his shop and gouged out his other eye with a sharpened stick, yelling, "You will never again look upon a Christian child."

Joseph Shainovitch's head was bashed in by a gang of drunks who then diverted themselves by driving nails through his mother-in-law's eyes. A two-year-old boy's tongue was cut out while he was still alive. Bodies were hacked in half or gutted and stuffed with chicken feathers as the city's bourgeoisie sauntered the streets in calm indifference.

For two days, the murderers, looters, and rapists ran free in the streets. When the frenzy finally died down, pages of scripture floated in the wind, and everywhere blood mixed with brick and glass, mortar and mud. The toll was 49 dead, 587 injured. More than seven hundred houses and 588 shops were destroyed. Two thousand families were left homeless.

That year, 76,203 "Hebrew" immigrants arrived in the United States from Eastern Europe, among them a failed carpenter named Moshe Yitzhak Mabovitch.

$*$ $*$ $*$

The second of his three daughters, Goldie, saw none of that carnage, knew none of its victims. But long before the Jews of Kishinev were forced to bury their dead, she was aware of what it meant to belong to the most vilified of Russia's outsiders. One afternoon, when she was playing with a friend on the dusty streets of her hometown of Pinsk, a drunken peasant grabbed the two girls and banged their heads together. "That's what we'll do with the Jews," he laughed. A few weeks later, Goldie was building mud castles with her friends in the alley when a troop of Cossacks rode by on their horses, slashing at the air with sabers and whips. Rather than veer around the girls, the men jumped over them yelling, "Death to the Jews."

Goldie raced inside, bolted the heavy door, and cried frantically for her mother. But she already knew that there was nothing Bluma could do, that Jews were helpless in the face of such hatred, a lesson she'd absorbed when rumors of a pogrom spread through Pinsk. Goldie didn't quite understand what a pogrom was beyond adult whispers about *goyim* brandishing knives and clubs screaming "Christ Killers." But her father, Moshe, and the other men in their apartment building frantically boarded up windows and doors, hoping to barricade their families against the danger.

"That pogrom never materialized," Golda recalled years later. "But to this day I remember how scared I was and how angry I was that all my father could do to protect me was to nail a few planks together while we waited for the hooligans to come."

The Pale of Settlement was a harsh teacher, and terror was just one of its lessons. Normalcy was hunger, the day's porridge carefully doled out in small spoonfuls. Golda's older sister, Sheyna, regularly fainted at school because she left home in the morning with her stomach empty, and her younger sister, Tzipka, wailed nonstop. To welcome the Sabbath on Friday night, the family was lucky to have a bit of dried fish to eat with their potatoes.

Goldie learned about her lot as a woman when Sheyna, suffused with both the idealism of a teenager and her mother's fiery temper, refused to succumb to the despondency and added the misery of incessant fighting to the dreary mix. "I want to go to school," Sheyna insisted. What for? Bluma practically spat. Girls don't need an education; they need to be prepared for marriage,

Things were unlikely to change, or so her high-strung mother predicted regularly. Sheyna, her regal but mule-headed eldest, was bound for trouble with her fancy ideas about education and her prattlings about liberation. Bluma didn't worry about Goldie, her favorite, who was eight years younger than Sheyna. But the rest of the family found Golda's feisty self-absorption so unbearable that one aunt declared that she had been possessed by a dybbuk, a malicious spirit. And Bluma couldn't imagine how little Tzipka, the youngest by four years, was going to grow when she could never afford enough porridge or potatoes to feed her.

Mired in a perpetual cloud of doom, Bluma prophesized on a daily—sometimes hourly—basis that nothing good would ever happen: Moshe would never earn enough money to get them out of an unending series of dank rooms in shtetl hovels, she lamented, with the utter resignation of a stolid peasant babushka who found the strength to cope but never the means to escape. Life would be a perpetual struggle. We'll starve. The goyim will kill us. And the future? *Feh!* What future?

Goldie's father, Moshe, had tried to defy the odds. When he failed to

make a go of things as a cabinetmaker in Pinsk, he'd hauled the family
off to Kiev, a city so anti-Semitic that Jews needed special permits to live
there. But unlike Bluma, who had the stubborn shrewdness of a tough
businesswoman, Moshe had a fondness for ideas and a gullible naïveté
that made him a patsy for thieves and charlatans. He found few buyers for
his furniture, and was rarely paid when he did. Within a year, the rent
was overdue, he was in debt to moneylenders, and the family celebration
for Shavuot, the holiday marking the day the Torah was given to Moses,
was bread, beans, and potatoes.

The only solution Moshe could divine was to sell everything and send
the family back to Bluma's father in Pinsk while he joined the long line of
more than two million Russian Jews making their way to the *goldina medina,* the golden shores of America.

                        *   *   *

It was a blessedly quiet Sabbath morning when Sheyna taught Goldie the
final lesson that would define her life. Bluma had taken Tzipka, her youngest daughter, with her to synagogue. Sheyna was off with her friends.
And Golda was curled up in her favorite hiding spot, on the warming
shelf above the massive black iron stove built into the wall of the kitchen.
Suddenly, a whisper interrupted her reverie, and Golda peered down
from her perch to see Sheyna sneaking in with a group of other teenagers.
While two young men stood watch at the window, the others hunkered
on the floor and listened raptly to a handsome young man with long,
flowing hair rail against the evils of the tsar and the necessity for Jewish
self-defense. Goldie sunk back into the shadows, struggling to absorb
every word.

After an hour or so, the group disbanded, one by one sidling into the
alley. Golda climbed down and confronted her sister.

"If you annoy me, I'll tell Maxim [the neighborhood policeman] what
I heard," she threatened.

"Tell him what?" Sheyna asked angrily.

"That you and your friends shouted, 'Down with the tsar!'"

Sheyna turned stern and sober. "If you do, I'll be flogged with a whip, maybe sent to Siberia or killed."

Sheyna appeased her sister with a lesson on the evils of tsarist Russia, the glories of socialism, and the distant dream of a Jewish homeland. Pulling out a newspaper she'd hidden away, Sheyna pointed excitedly to the photograph of a man with dark hair and a heavy beard. "This is Theodor Herzl," she explained, regaling Goldie with stories about his meetings with the sultan of Turkey, the kaiser of Germany, and the prime minister of England, all to save the Jews.

The name Herzl was buzzing throughout Jewish Pinsk, whispered in synagogues and in alleys, and scores of Jews saved their pennies for his new association, the World Zionist Organization, formed to create a Jewish homeland in Palestine. A secularized Viennese Jew galvanized by the 1894 trial of Captain Alfred Dreyfus, Herzl had concluded that Jews would never be assimilated into European society and limned a utopian vision of a Jewish state where Jews could be freed from life as an unwanted and vilified minority. His ideas had been widely ridiculed by Jewish leaders, but his fantasy of a Jewish state struck a deep chord among the Jewish masses.

Russian peasants and workers were rising against the tsar and Jews had enlisted in the revolution by the thousands, particularly joining the Bund, the popular name for the General Jewish Labor Union of Lithuania, Poland, and Russia, which offered socialism as the answer to Jewish misery. But while sympathy for socialism became almost as widespread as loathing for the tsar, memories of the viciousness of Russian workers ran too deep for most Jews to rush toward them in brotherly embrace.

Young people like Sheyna and her friends instead teetered between Zionism and Socialism, suspicious both of non-Jewish socialists and of the political Zionism of Herzl, which put the fate of Jews in the hands of the world powers. Gripped by a powerful need to redeem centuries of helplessness with bold action, they developed a hybrid philosophy, Labor Zionism, which married class struggle and activism to aspiration for a Jewish homeland run by and for Jewish workers.

Sheyna's group did little more than talk. But as the government clamped down on public meetings and political discussions, talk, too, became dangerous. Rather than meet in homes, they began gathering in the woods or in synagogue after services.

Bluma reacted with the empty gesture of forbidding her daughter to participate in political activity lest they all be exiled or killed, but Sheyna baited her mother mercilessly, reporting on all of her escapades and near run-ins with the police. After months of door slamming, threats, and screaming, Bluma locked Sheyna out of the house.

The expulsion didn't last long, but Bluma laid down a firm proscription against Sheyna's political guru, the long-haired young man Goldie had spied from her perch in the kitchen. The grandson of a prominent Torah scholar, Shamai Korngold was the author of the pamphlets Sheyna distributed, the principal speaker at their gatherings, the fairy-tale revolutionary. She was already hopelessly enamored of her romantic rabble-rouser.

Bluma grew to tolerate Sheyna's political activities until she realized that her eldest daughter was infecting Goldie, who worshipped her older sister. Golda followed Sheyna everywhere, nagging her for reading lessons, for help with her numbers, for her to admire her image in the family's faded mirror. The more Goldie demanded, the more Sheyna disapproved, and the harder Golda worked to live up to her sister's standards, the beginning of a lifelong pattern.

When Labor Zionists organized a fast to mark the anniversary of the 1903 Kishinev pogrom, Goldie vowed to prove to Sheyna that she was worthy by also forgoing food.

"Fasting is only for grown-ups," Bluma told her.

"You fast for the grown-ups," said Goldie, already quick with a rejoinder. "I'll fast for the little children."

Bluma sensed what was coming. An austere perfectionist and the severest of taskmasters, Sheyna was Goldie's heroine. There was nothing she wouldn't do to win her sister's approval.

"It doesn't matter if you've saved enough money or not," Bluma wrote her husband, whose last letter, received almost a year earlier, was from a city with the strange name of Milwaukee.

"Believe me, we must come. Now."

CHAPTER TWO

Don't be so humble, you're not that great.

In Russia, Moshe Mabovitch had sported a well-combed mustache and worn a black suit with a high-collared white shirt on special occasions. But the spare-looking man who tried to pull Goldie and her sisters into his arms at the Milwaukee train station was clean-shaven, wore working-man's clothing, and had renamed himself Morris. Only one thing hadn't changed: he still wasn't making a living.

Within a week of her arrival, undaunted by the fact that she spoke not a word of English and had never run a business, Bluma rented a small store with living quarters behind it and announced that she was opening a *kreml*, a grocery store. With her own special brand of certainty that she knew what was best for everyone, she had everything planned: While Goldie and Tzipka went to school, Sheyna would help run the store. Moshe would forget his part-time job as a railroad carpenter and begin a new career as a building contractor. And they would forget the misery of Russia.

Sheyna was no more amenable to Bluma's ideas than she'd been back home. "If I wanted to work in a *kreml*, I could have stayed in Russia," she thundered haughtily. "Shopkeepers are social parasites."

Still wearing black in mourning for the death of Herzl, Sheyna hated everything about America. But Goldie was enthralled—by her first soda pop, her first ice cream, her first trip to a five-story skyscraper. At Fourth Street School, she blossomed among the heavily immigrant Jewish student body. That winter, she learned to sled in Lapham Park. When she and her girlfriends could wheedle nickels out of their parents, they lined up at the Rose Theater to watch movies and traveling vaudeville troupes.

Mostly, they hung out on Walnut Street, the heart of Jewish Milwaukee, or stopped by Settlement House, where choral groups and literary societies offered free entertainment for the Yiddish-speaking immigrants. When they grew weary of the cold, they stopped by the *shvitz*, the steam room, where immigrant women without hot water at home gathered to soak off the grime and relax over a *glessele* of tea.

Golda's world was a tiny slice of an old-fashioned shtetl washed onto American shores by the flood of Eastern European Jews who arrived in the waning years of the nineteenth century and the early years of the twentieth. Before they arrived, the city's tiny Jewish community—German-speakers who dominated the city's textile and footwear industries—had aggressively avoided calling attention to themselves as "other." Their synagogues held regular Thanksgiving Day services and ostentatiously celebrated Washington's birthday.

When the Hebrew Immigrant Aid Society and the Industrial Removal Office began redirecting Russian and Polish immigrants away from the large East Coast cities into the heartland of America, those established Jewish families winced at the prospect of being identified with semiliterate Russians babbling in Yiddish. As the new Jewish immigrants swelled the community from 2,559 in 1880 to more than 10,000 within a decade, the doyens of Milwaukee Jewish society endeavored to pull the newcomers up by their bootstraps and into the mainstream of America with sewing, cooking, and language classes, and Boy Scout and Girl Scout troops for the children.

Bluma despised the German Jewish ladies with their hats and gloves, flaunting their smug superiority. When they launched a crusade to teach new immigrants the fine points of the American art of cleanliness, she almost spit. "I know how to be clean."

The perimeter of Golda's world was narrowly circumscribed. At school, she never would have mingled with those women's children. Nor did she associate with the "other half," the non-Jewish German kids. Golda didn't really live in America, then, not in the America of Fourth of July picnics, baseball games, and political rallies. She never read the daily Milwaukee newspaper and, in later years, couldn't recall any significant national or local political events that had occurred during her years in the country. In scores of interviews and speeches over the years, she rarely spoke about what America meant to her beyond ritual platitudes about democracy. But in an interview she gave in 1972, she provided a glimpse of precisely how deep the American influence really was:

> In America I was able to rid myself of the terror I had in Pinsk, from Kiev. . . . My father worked and was part of a labor union. . . . On Labor Day he said to my mother, "Today there is a parade. If you come to such and such specific corner, you will see me marching in the parade." My mother took us to see the parade, and while we were waiting for it to begin, police appeared, riding to clear a path for those marching. But my younger sister didn't know this and when she saw the police she began to tremble and cry, "The Cossacks, the Cossacks!" . . . The America that I knew was a place that a man could ride on a horse to protect marching workers: the Russia I knew was a place that men on horses butchered Jews and young socialists.

If for Goldie that was enough, for Sheyna nothing was sufficient. Trapped in her mother's store, she had no independence, no time to go to school or to make friends. Fed up with a dominating mother, a passive father, and family expectations born in a shtetl, she found a job at a men's clothing factory in Chicago and, to the disgrace of her family, moved out.

The absence of Sheyna left an enormous void in Goldie's life. "She was an extraordinary person in every aspect, and served as a shining example for me, my closest friend, and my faithful advisor," Golda wrote long after the mentee had wildly surpassed the mentor. "Sheyna was the only one whose praise, when I was worthy of it—which was not easy, was the most important thing to me."

But within months, Sheyna was back home, moping around the house, trapped between her self-concept as a modern rebel and the reality of her life. Awkwardly, Goldie tried to comfort her, combing her sister's hair, quizzing her endlessly about life as a fiery revolutionary. But Sheyna was too listless to spark any fires. Then, an aunt wrote from Pinsk to report that Shamai Korngold, her old heartthrob, had escaped from Russia and was working in a cigarette factory in New York. Sheyna dashed off a letter, suggesting that he move to Milwaukee. A week later, she and Golda met him at the train station. "Maybe here he will begin to notice me a little," Sheyna confessed to her sister.

Golda naively thought that with Shamai in Milwaukee, Sheyna would be happier. But she hadn't factored Bluma into the equation. To keep her daughter out of the arms of a pie-in-the-sky greenhorn, she summarily barred Shamai from the house. The two might have spent years locked in struggle, but Sheyna collapsed and began coughing up blood. Her parents saw no alternative but to send her off to Denver, to the Jewish Hospital for Consumptives.

Still, Goldie thrived, always at the top of her class in school, although chided by her teachers for talking too much. But it was outside the classroom that she made her mark. When she was eleven years old, Golda and her friends organized the American Young Sisters' Society to raise money for books for needy classmates. Golda talked the owner of Packen Hall into donating the room for a fund-raising gala. So just three years off the boat, she stood before her first audience to deliver an extemporaneous speech about the plight of immigrant children without books.

The local newspaper picked up the story, made more prominent by the photograph it ran, President Goldie Mabovitch singled out as the girl in tight braids in the top row. "A score of little children gave their play-time and scant pennies to charity, a charity organized by their own initiative too," the caption read.

"Golda had fire," said her best friend, Regina Hamburger Medzini. Bluma put it differently. She called Goldie a *kochleffl*, a stirring spoon.

One day at school a Christian boy threw a penny at one of Goldie's friends and ordered her to pick it up. When the girl complied, he mocked her, yelling, "A dirty Jew will pick up every penny." That night, Golda organized her first demonstration, against anti-Semitism, in front of the boy's house.

But Golda didn't have much time for demonstrating or speechifying. Every morning, while Bluma shopped for the day's produce, she opened the store, standing on a wooden crate behind the counter to measure out sugar by the cup or herring by the piece. When Bluma lost track of time, Golda wept at the prospect of being late to school yet again. What's the big deal? Bluma mocked her. "So it will take you a little longer to become a rebbetzin, a bluestocking. We have to live, don't we?"

Although Sheyna and her parents barely communicated, Golda kept up a steady correspondence with her sister, using Regina's house as their mail drop. Shamai had followed Sheyna to Denver, and after her release from the hospital, they had married. The young couple was barely surviving by washing dishes and shoveling snow. Golda sent them money, a dollar or two here or there, some change from her school money or a bit "borrowed," as Golda called it, from her mother's money box.

Golda's letters to Sheyna were the prattling of a young girl and a steady stream of bad news about the situation at home. "I can tell you that Pa does not work yet and the store is not very busy," she wrote in one. In another, she confided that Bluma had moved them again, to a deli on Tenth Street, and that Moshe was working there as a butcher.

Curiously, she didn't complain about her own life, although a storm was brewing with her graduation from elementary school. No one in Golda's extended family had ever finished school and Moshe's eyes teared up as he watched his daughter receive her diploma and stand at the podium in a special white dress to deliver the valedictory address on the importance of being socially useful.

Milestones, after all, have a way of opening the door to disaster, and the ink was barely dry on Golda's newly minted diploma before Golda and her mother were at each other's throats. Golda had mapped out her life: She'd go to high school, continue on to teachers' college, and find a job in a school.

Bluma had another vision for Golda's future: She would find a job as a clerk downtown, then, as Golda later put it, "marry, marry, marry." Golda was a *dervaksene sheine meidle,* a fine upstanding young girl, who'd finished school and spoke English flawlessly. She was a prize, a good catch who could marry a fine professional man who'd support her with none of the heartache or indignity that Bluma herself had suffered. "You could be a very good housekeeper," her mother counseled. "But a very clever woman you'll never be."

Golda was neither quiet nor modest, however, and her dreams for the future weren't forged in a Russian ghetto. "She was so beautiful and everyone spoiled her," Sheyna said. We're both stubborn, she admitted, but "Goldie even more so. She didn't like to admit she was wrong. If she retreated, she was very mad at herself."

Golda, then, was not about to give in. Nor was Bluma. "You want to be an old maid?" she ranted.

Moshe agreed with Bluma. "Men don't like smart women," he taught his daughters. But he urged Bluma to offer Golda a compromise, secretarial school. Unlike teachers, who could not marry, secretaries could do what they wanted in their private lives. But Golda refused to cooperate. "I never can explain to myself why I had such a horror of working in an office," she said. "I wouldn't hear of it. That's where the big clash came."

In the fall of 1912, Golda prevailed and enrolled in North Division High School, to Sheyna's hearty delight. "Tell me the truth," she wrote that October, "how many times did you shout and how many times were you about to commit suicide until you won the battle, I mean the fight for going to school?"

That was only round one in the Golda-Bluma feuds. Round two was sparked two months later, when Bluma announced that she had found the perfect husband, a prosperous man who worked in real estate. Golda refused to consider marrying a man who was twice her age, no matter how many times Bluma predicted that she'd wind up a bitter old maid. Bluma didn't know that she had crossed the line, the invisible line that would become infamous among both Golda's admirers and detractors, where Golda's natural stubbornness transmogrified into steel.

Within days, Golda poured out her heart to Sheyna and Shamai, who responded with a tantalizing offer: "Get ready and come to us."

Golda had a mantra that she repeated often to Regina: "Only those who dare, who have the courage to dream, can really accomplish something. People who are forever asking themselves, 'Is it realistic? Can it be accomplished? Is it worth trying?' accomplish nothing. . . . What's realistic? A stone? Something that's already in existence? That's not realism. That's death."

So late one cold February night, while her parents slept, Goldie packed a small suitcase, tied a rope to the handle, and lowered it out of her second-floor bedroom to her friend Regina, who was waiting below. The next morning, acting as if nothing unusual was happening, she ate her breakfast, tucked her books under her arm, and said good-bye to her family. Instead of walking to school, however, she headed for Regina's, where she picked up her meager belongings. With a quick hug to the girl who'd been her best friend from the first week she'd landed in America, Golda took a trolley to the railroad station and bought a ticket for Denver, spending every dime she'd saved, every cent Sheyna could send her, and small contributions from her friends.

*   *   *

Ever the dogmatist, Sheyna referred to Golda's arrival in Denver as her liberation from "the tyranny and oppression" of their parents. But Golda's was liberation light, involving little struggle, no real hardship, and absolutely no risk. When she got off the train, dressed in an ankle-length black skirt and high-necked white blouse, Sheyna was waiting on the platform with Shamai and their two-year-daughter, Judy. That evening, Golda slept in a comfortable bed in their tidy brick bungalow. Within days, her life assumed a routine remarkably parallel to the one she'd just escaped.

In the morning, she walked twenty blocks to North Side High School. After classes, she worked for Shamai at his new dry cleaning business. At the end of that long day, she made her way home to help Sheyna with dinner and the dishes before settling down to work on her algebra and Latin.

But on weekends, Sheyna's house vibrated with young people arguing socialism and anarchism, Hegelian philosophy and Zionism in beautiful Yiddish and execrable English—and they didn't see Golda as a little girl undeserving of a place at the table. The only person who hesitated to welcome Golda was Sheyna, who tried, unsuccessfully, to shoo her off to her books or her bed.

Golda was riveted as they debated change: Could it come without the class struggle advocated by Lenin? What would the split between the Bolsheviks and the Mensheviks mean for Jews? What role should the emancipation of women play in the revolution? Should they join the Socialist Party? Become pacifists? The debaters weren't serious intellectuals. They were lung-sick bachelors who'd been in the hospital with Sheyna or were the sons of Denver's Jewish consumptives, barbers who thought of themselves as philosophers, janitors who described themselves as poets, wannabe revolutionaries, and kooks.

Golda knew that socialism meant democracy and the right of workers to a decent life. But she'd never heard of Peter Kropotkin or Georg Wilhelm Friedrich Hegel. So she sat quietly and concentrated, going over the conversations in her head, honing what became, in later years, her greatest strength, her ability to listen for hours.

Golda had little inclination to the abstruse or abstract. Although she didn't read any of the books her new friends debated, she knew what she thought, relying, as she always would, on her gut. Kropotkin's anarchist vision of a society free of central government hardly seemed practical. Hegel was impossible to decipher. How could freedom inevitably lead to terror? She hated Schopenhauer, the German father of pessimism. If life is as futile and necessarily full of suffering as he taught, what was the point?

Emma Goldman, the fiery Russian-American rebel, was more her style. But Goldman seemed too focused on America. What about the Jews? Who could worry about birth control and American wages when her brothers and sisters were being slaughtered in Eastern Europe? Even then, Golda was a narrow nationalist.

Inevitably, Zionism captured her imagination. And among the dozen factions in the Zionist movement, she was quickly drawn to the Labor Zionism of Aharon David Gordon, which had all the necessary elements of romance, social justice, and Jewish pride.

A farm manager and intellectual, Gordon had left Russia for Palestine in 1903 and taken up a hoe, determined to till the soil himself. Working the land, he believed, would cleanse Jews of centuries of enforced parasitism.

> The Land of Israel is acquired through labor, not through fire and not through blood. . . . We must create a new people, a human people whose attitude toward other peoples is informed with the sense of human brotherhood. . . . All the forces of our history, all the pain that has accumulated in our national soul, seem to impel us in that direction. . . . We are engaged in a creative endeavor the like of which is itself not to be found in the whole history of mankind: the rebirth and rehabilitation of a people that has been uprooted and scattered to the winds.

Inflamed by Gordon's concept, Golda devoured every issue of magazines like *The Young Maccabean* and *New Judea* for tales of the pioneers who'd return to Palestine. When she could carve out free time, she shook the blue-and-white collection box of the Jewish National Fund to raise money to buy them land.

Gradually, she built a life for herself among Sheyna's gang, spending weekends with them at picnics, concerts, and lectures, much to Sheyna's chagrin. You need to get more sleep, Sheyna nagged her. You need to be careful that men don't think that you're loose. In Golda's account of that time, Sheyna started "watching me like a hawk." Shamai tried to reason with Sheyna that Golda was anything but flighty. But while Sheyna was sparing of words, she was never short of opinions. She looked at her younger sister, who was only sixteen years old, and reminded her that she was "blessed with a lot of good attributes, also faults worth watching."

Still craving Sheyna's approval, Golda was both mortified and furious, the two emotions tugging her in opposite directions. Intoxicated by her first taste of independence, she rebelled. Neither sister could ever recall what the final spark was, but over dinner one night, Sheyna berated Golda one time too many and Golda exploded. "I'm not a baby," she fumed, jaw clenched. "You have no right to boss me around as though I were. . . . I'm leaving."

Sheyna refused to back down. "Go ahead," she snapped.

Golda stormed out of the house. Ten minutes later, she realized she had nowhere to live, no way to support herself, no books, and no change of clothes. She couldn't go home to Milwaukee; her father wouldn't allow her name to be spoken in the house. She was faced with an unambiguous choice: Apologize to Sheyna or drop out of school. As an adult, Golda was notorious for refusing to compromise, for shrinking from owning up to her errors. Demonstrating that same propensity as a teenager, Golda left her books behind, rented an apartment, and struck out on her own.

*   *   *

It's impossible to know how much attention Golda would have paid to Morris Meyerson if she hadn't been "almost as lonely as independent," slaving away at a degrading job and living in a tiny rented room. A Lithuanian immigrant, Morris was the least garrulous of the crowd who hung out at Sheyna's. Not quite twenty-one years old, with thinning hair and steel-rimmed glasses, he was a private, introspective man who worked sporadically as a sign painter and seemed more comfortable with Chekhov and Mozart than with political badinage.

But most of the men Golda had met at Sheyna's avoided her, lest they incur her sister's notorious wrath. So Golda couldn't resist when Morris invited her to a free concert in the park. After the last note was played, they strolled hand in hand, Morris analyzing the symphony, Golda awed by his knowledge. "Oh, now I wish I could hear that music all over again," she said. Morris hesitated, and then offered, "There's a concert downtown next Saturday night. If you'd like, I'll get tickets."

That spring and summer, Golda and Morris spent every Sunday at a free concert in one of Denver's parks. During the week, they attended lectures on science, philosophy, and psychology at the Workmen's Circle, a Jewish immigrant aid society, or went rowing on Sloan's Lake, where Morris read her Byron, Shelley, and *The Rubáiyát of Omar Khayyám*.

Morris did more than ease Golda's loneliness. He punctured the wall she'd erected to shield her vulnerability from a tough mother and an even tougher sister. Bluma and Sheyna both hid their worry behind sarcasm, their love beneath a mountain of hectoring. To survive, Golda had developed a thick hide. Other men had offered her strident lectures on Marxism; Morris lavished her with the gentle gift of poetry, flowers he could ill afford, reproductions of paintings cut out of magazines framed for the walls of her tiny room. He became her teacher, her mentor, her brother, the first person to love Golda the woman.

Morris is not particularly good-looking, but "he has a beautiful soul," she admitted in a postcard to Regina.

Those months in Denver were an intimate interlude in Golda's life, an unusual moment of tenderness. Her time wasn't spent on causes, on organizing or mounting a bully pulpit, but on developing a soft side of herself that she had never seen. In turn, Golda infused Morris with her own vivacity and dynamism, ripping through his cocoon of reticence.

Then, out of the blue, Golda received a letter from her father beseeching her to come home for her mother's sake. It was the first time she'd heard from Moshe in almost two years, and she knew what it cost him to beg her—just as she knew how expensive her sojourn in Denver had become. She was in love with Morris, but she was working in a department store instead of studying. Her father promised that he and Bluma would support her decision to go to high school and become a teacher, if only she'd come back.

The sole breadwinner in his family, Morris couldn't follow her. Nonetheless, he encouraged her to go home, to heal the rift with her family and to study, promising that he'd join her as soon as possible so they could marry.

* * *

At the age of seventeen, Golda had already tamed her first unruly crew, her parents. When she reenrolled in high school, they offered not a murmur of protest. When she talked about teaching, Bluma stayed mum. The only battle they fought was over the identity of the author of the letters arriving from Denver, two, three, or four times a week. "Somebody," Golda responded flipply when Bluma pressed for information. "So who's this Mr. Somebody?"

Frustrated by Golda's secrecy, Bluma steamed opened two letters and told Tzipka, renamed Clara in America, to translate them. That night, Clara confessed her betrayal and Golda stopped talking to her mother for several days. Golda's soon-to-be infamous wrath took its toll. Bluma never defied her again.

Back home, Golda read through the long book list Morris had compiled for her, Chekhov and Gogol, de Maupassant, Anatole France and H. G. Wells. She joined a small Yiddish literary society that brought speakers in from Chicago to discuss the classics. And she spent hours babbling to her friends about Denver and savoring every word Morris wrote to her.

In her responses, Golda gushed out a torrent of insecurity. Am I smart enough? Pretty enough? Learned enough? Morris never seemed to tire of reassuring her. "I have repeatedly asked you not to contradict me on the question of your beauty," he wrote. "Every now and then, you pop up with these timid and self-deprecating remarks. I can't bear them."

Week after week, he limned his devotion. "If you were here now, oh how I would kiss you! But that can't be . . . so accept some kisses from me sailing through the ethereal blue," he wrote to Gogole, as he called her, in January 1915. By October, he was more ardent still. "If I don't get to see you soon I fear that I shall dwell in the bughouse forever!"

But neither Morris nor Golda had the time or money for visits. And as the months wore on, Golda began spending less time on Morris's literature and music and more on her old passion. Wars are notoriously bad for Jews, and World War I was a disaster for the masses in Eastern Europe. As the Russians battled the invading armies of Germany and Austria all across the Pale, both sides took out their frustrations on Jews along the way. Tens of thousands lost their lives; millions more were left homeless.

Together with her father, Golda threw herself into relief work, going door to door to beg nickels and dimes for the displaced Jews. As the news grew ever more grim, she made speeches on street corners and helped organize meetings in hired halls. Gradually, she grew angry. What's the point? she asked. How long will Jews continue to accept their plight and meekly be led to the slaughter?

When she began asking those questions in public, Milwaukee's leading Labor Zionist, Isadore Tuchman, suggested that it was time for her to do more than talk. "I noticed this striking girl on the platform surrounded by old women," he recalled. "I thought what on earth is this young girl doing with all these old women? She doesn't belong at all. . . . So I got in touch with her and asked if she'd join our movement."

That movement was Poale Zion, a small band of Zionist socialists more adept at arguing the future shape of Palestine's collective farms than at figuring out how to turn that concept into a national reality. Only twenty-three delegates, representing four chapters, showed up at their first national convention in 1905. At the second, Poale's nascent leadership found itself under assault for an alleged "collaboration" with non-Socialist Zionists, agents of the dreaded "bourgeois class."

Poale's goal was to prepare its members to move to Palestine, but few Americans were lining up to go, chastened by stories of pioneers starving for lack of work and of Turkish soldiers raping women. So in Milwaukee, Poale devoted itself to running a library and a *folkschule*, a weekend program of classes for immigrant children in Yiddish, literature, and Jewish history.

Golda, however, was not impressed with "parlor Zionists," as she called them, all talk and no action. While she worked in the library and taught classes for immigrant children, she didn't see any future in America. If she was going to commit herself to the cause, she needed to be serious about uprooting herself and contributing her own sweat to building Palestine.

In every place and generation, a handful of young people discover the exhilaration of being part of something larger themselves. They become religious, throw themselves into causes, or embrace new ideologies. While their friends follow the crowd, they march to the beat of that proverbially different percussionist. In Golda's day, Americans were less suspicious of those who dedicated themselves to principles, feeling no need to psychoanalyze the few impelled to go against the social grain. No one asked, then, whether Golda was attempting to live up to the image she'd built up of her older sister, trying to prove something to her overbearing mother, or exorcising the demons planted by her father's helplessness in the face of a threatened pogrom.

But her two causes—Zionism and Morris—collided. "I don't know whether to be happy or sorry that you are participating in the Zionist party and that you are, it seems, so enthusiastic a nationalist," Morris wrote in August 1915. "The idea of allotting Palestine or any other territory for the Jews is, in my eyes, ludicrous. Oppression does not exist because some nations have no territories but because nations exist at all."

Undaunted, Golda set out to convert Morris, penning him dozens of letters about the bravery of the Jewish pioneers in Palestine and pleading with him to at least agree to a trip there. "Yes, Gogole, I shall even consider a trip to Palestine (or the North Pole for that matter) with you," he replied indulgently. "But why we need go there now I don't know. We shall roll there, cost prepaid, after we have died of some Latin disease and been buried under four ft. of sod. As to that group of 'idealists'—why, you need not envy them. Six weeks of struggle with the virgin soil will cure them of their 'idealism' permanently."

Assuming that Golda would grow out of her Palestine fascination, Morris continued to write long letters about the world they had shared in Colorado, critiquing performances of Wagner's *Entrance of the Gods into Valhalla* and describing his research in zoology and paleontology. Golda replied excitedly with news about the revival of Hebrew in Palestine. Neither seemed to notice the growing disconnect.

Zionism consumed Golda's every waking moment. Asked to join Poale's Executive Committee, she threw herself into organizing their annual fundraiser for pioneers in Palestine. On weekends, she and her friends held picnics at Lincoln Park, auctioning off packed lunches to raise money for their *folkschule*.

In the United States, Zionism had attracted the energies of few young people, and even fewer who were both female and moderately American. Golda, then, who never powdered her nose or donned a flapper outfit, became a rising star in that tiny universe, news of her celebrity traveling well beyond Milwaukee. "There was a rumor afloat that there was a kookoo girl who was barnstorming for Labor Zionism in Milwaukee," recalled Judy Shapiro, whose father, Ben, was an activist in Chicago. "*Meshugenah* Myrtle, (crazy Myrtle), they called her."

By the time Morris finally found a way to free himself from his family and move to Milwaukee, Golda was basking in the attention, the plaudits, and the energy. Morris and Zionism, she was about to have it all. Shrewdly, at first she carved out plenty of time for them to fall back into their old routine. She and Morris read together, went to the opera and symphony. In the summer, they took the trolley to Bradford Beach or strolled through Lake Park with a basket lunch.

But as Morris settled in, working occasional jobs as a painter, Golda was constantly caught between him and her passion for Zionism, which he stubbornly refused to share. The two collided dramatically when the first important pioneers from Palestine arrived in Milwaukee as part of a thirty-city tour. Yitzhak Ben-Zvi and David Ben-Gurion were scheduled to address a public meeting on a Saturday night and have lunch at the Mabovitch house the next day. Golda was dying to hear their speeches, but Morris had already bought tickets for a concert by the Chicago Philharmonic. "I just didn't have the courage to say that I would not go to the concert," Golda explained. After all, she knew that she'd meet them the next day. But Ben-Gurion canceled the lunch because "somebody who could not come to listen to him speak is not deserving of having him as a guest."

Abashed, Golda vowed never to be unworthy again. She dropped out of teachers' college. What was the point of preparing for a career when she would soon join a collective farm in Palestine? The mounting reports of the hardships and dangers facing young Zionists there only made the prospect more tantalizing. Craving release from the anger and humiliation she'd felt facing Cossacks, Golda imagined herself fighting off the fellahin, the Arab peasants, and striking a blow for Jewish independence.

"The truth is that I didn't have exact information, but I knew very clearly what I wanted," Golda recalled years later. "My mind is not so complicated."

Unfortunately, Morris' was. He was searching for the meaning of life, not Jewish liberation. And he dreamed of a quiet life of children and family in America, not in some malarial swamp.

Golda tried reason, persuasion, and manipulation. When all else failed, she resorted to the tactic that would become a staple in her arsenal, an ultimatum: Move to Palestine with me or there will be no wedding. There was something hard in that ultimatum, a blindness or deep indifference to who Morris was, to anything but what she wanted.

Sheyna, with whom Golda had mended fences, was appalled. "I don't want to shatter your dreams," she wrote from Denver. "I know what it means. But, Goldie, don't you think there is a middle field for idealism right here on the spot?"

Morris refused to be blackmailed, and, in a snit, Golda accepted an invitation from the Chicago chapter of Poale to work in their office. But after a few months, like almost everyone in Golda's life, he gave in. In the fall of 1917, he offered to move to Palestine with her if she would marry him immediately.

On the day before Christmas 1917, a small group of friends and family gathered at the home of Golda's parents to witness the ceremony. Golda had no bridesmaids, no caterer, and no fancy dress. In keeping with her self-image as a pioneer, she wore a plain gray crepe de chine outfit—"the plainest of plain," as she described it—and Bluma served boiled potatoes, herring, and sponge cake. But for once, Golda bowed to her mother's wishes, forgoing a civil ceremony for a rabbi and a chuppah.

To Golda's dismay, and Morris' relief, the young couple couldn't leave immediately for Palestine because all transatlantic passenger service had been canceled for the duration of the war. They had time, then, time to save money for the trip, to develop as a couple, to establish a routine. Time, in Morris' fantasy, for Golda to change her mind.

To any objective observer, that was unlikely since the dream of a Jewish homeland seemed tantalizingly close to fruition. The British foreign secretary, Arthur James Balfour, had just issued a letter announcing a new foreign policy initiative. "His Majesty's Government view with favour the establishment in Palestine of a national home for the Jewish people," he wrote," and will use their best endeavors to facilitate the achievement of this objective."

The declaration was, in part, a gift to Chaim Weizmann, a British Jewish chemist who'd developed a new process to synthesize acetone, an essential component in the production of cordite essential for the ammunition Britain needed to win the war. Lord Balfour had asked what payment Weizmann wanted in return for use of his process. A national home for my people, he replied.

Balfour's letter was an extraordinary document in which "one nation solemnly promised to a second nation the country of a third," as author Arthur Koestler put it. But the British weren't quite as brazen as Koestler and most Zionists thought. The declaration contained a clear caveat: nothing should be done in pursuit of that homeland which might prejudice the *rights* of existing communities in Palestine.

Morris painted signs when he could find work, and Golda found a job at the Lapham Park Library for 20 cents an hour. But they never spent much time together. There was always a Zionist emergency. Faced with a weekly newspaper that was bleeding money, the Poale board debated whether to close it. Undaunted by either finances or lack of manpower, Golda proposed that they turn it into a daily—and volunteered to do more work.

When the American Jewish Congress was organized, she campaigned tirelessly for Zionist delegates. Shortly before the community election, Golda and her cohort showed up at a major synagogue and asked to speak. The synagogue leadership refused, so Golda waited for the service to end and then stood on a bench outside to harangue the congregants filtering out. "My fellow brethren, it is painful to hinder you as you leave a holy place. It is not us but rather your leaders who are guilty for this. The president and the members who closed the door in my face are to blame."

A fish out of the Zionist waters, Morris had married a movement. At least three nights a week, Golda was at Poale meetings and rarely got home before midnight. When she was home, her activist friends were constantly stopping by to continue a political argument, organize a rally, or chat.

And it only got worse after Golda was elected as a Milwaukee delegate to the American Jewish Congress convention called to develop a unified American Jewish position for the Paris Peace Conference. The Zionist demand that Britain be given a mandate over the old Turkish territory of Palestine and use it to fulfill the Balfour pledge had deeply divided the Jewish community. Fearing that support for a Jewish state would prompt governments to question their loyalty, most of America's most prominent Jews—German-Jewish bankers and lawyers, rabbis and politicians—vehemently opposed what they called "the segregation of the Jews as a nationalistic unit in any country."

As she always would, Golda listened silently for hours before she spoke. But when she finally rose to her feet, the young girl from Milwaukee brought the Congress to a hush. "I tell you her mouth was gold," said Tuchman. "Every time Goldie opened her mouth, it had an impact somewhere."

*   *   *

Just twenty years old, Golda had found her calling. "I tell you that some moments reached such heights that after them one could have died happy," she wrote Morris. "You should have been in the Hall when the resolution for Palestine was adopted. There were only two votes against it. . . . I didn't miss a single session." To Regina, she said simply, "This is the life for me!"

Impressed by Golda's fire, national Poale Zion asked Golda to become a traveling fund-raiser and organizer. Without a moment's hesitation, she quit her job at the library and signed on. Morris complained quietly, both to Golda and to their friends. Her father was more vocal: "A few minutes after the wedding, you are leaving your husband and going?" he yelled. "Who leaves a new husband and goes on the road?"

Golda's response was curt. "At Poale Zion, whatever I was asked to do I did," she said. "The party said I had to go, so I went."

The party, of course, did not order Golda to go. She wanted to go. So her relationship with Morris was reduced to a series of postcards, from Cleveland and Philadelphia, Indianapolis, Buffalo, and Youngstown. At times Golda nagged him about taking care of himself. But most of her communiqués were too brief for intimacy and suggested that she remained oblivious to Morris' disinterest in her cause.

"Dearest, I arrived in Buffalo last night," she wrote. "I got a very warm reception. Chaverim are waiting for me now to take me to Niagara Falls. . . . I expect to do quite a bit here. Will write more later. Your, Goldie."

While she crisscrossed the country, Morris decorated the house with framed magazine photographs and buried himself in the public library. When she came home, she always found fresh flowers waiting for her. And while she raced from meeting to meeting, he shopped, cooked, and cleaned—for Golda and the dozens of her *chaverim*, comrades, who were constantly in the house.

In the midst of her travels, her organizing work, and her frantic arrangements for their departure for Palestine, Golda got pregnant and had an abortion. No documents remain indicating whether she told Morris, either about the pregnancy or about the termination. But she confided, as always, in her big sister, explaining crisply that her Zionist obligations simply did not leave room for a child.

Sheyna was far from supportive:

> I was hurt and angered to the outmost depths to learn that you people did it. And taking all your "considerations" into consideration I still cannot see any good, strong reason for it. You are not sick, you have no defect. Your poverty I think is also a matter of mismanagement . . . and your social activities remind me [of] society ladies. They too cannot bear children, for it will take them away from their activities. . . . If you were really in earnest about your nationalism and mean the wellfair [sic] of your nation, you are the ones that ought to have children. By the time you'll be ready to go to Palestine you'll have a child with you to be brought up on that soil.

Sheyna repeatedly urged Golda to slow down, to stay home, cook the family meals, and take care of her and Morris' clothes. "It may not sound as independent as a young wife going to work, but you profit by it much more and you may spare an hour a day in preparing to be a good, wise mother and wife some day. And believe me, Goldie, it will pay not only for your personal well fare [*sic*], but for humanity in general."

But being a wise mother and wife in Milwaukee had no place in Golda's plan.

CHAPTER THREE

A Zionist doesn't make it conditional.

In Golda's Zionist fantasy, heavily influenced by a steady stream of re-
cruiting pamphlets, Palestine was gushing with milk and honey, or at
least they were poised to flow once she and the *halutzim*, the pioneers,
put their backs to the desolate soil. And Tel Aviv, the first new Jewish city
to be built in two millennia, was the Hill of Spring, conjuring up cool
breezes and tranquil gardens.

No one had mentioned the flies—swarming over the bread sold off old
blankets in open-air markets and laying eggs in the eyes of children. Or
the filth. Or the burning *hamsim* winds, blindingly relentless sand spouts.
Or the stench of donkeys, rotting meat, open latrines, and unwashed humanity.

When Golda, Morris, and their ragged band of seventeen pioneers
stepped off the train from Alexandria, Egypt, into Tel Aviv's ankle-deep
sand on July 14, 1921, Sheyna's daughter, Judy, started bawling. Morris
fell speechless. Regina's fiancé announced flatly, "I'm going back home!"

Golda looked past the ramshackle frame train station to the single tree in sight and pronounced it a symbol of the Jewish ability to flourish in adversity.

By then, no one was in the mood for inspiring Zionist platitudes.

Two months earlier, stirred by that sort of breathless rhetoric, they had gathered in New York, their departure fixed for May 22. The time was anything but ripe. Rioting had broken out in Jerusalem when Great Britain was given a mandate over Palestine with responsibility for implementing the Balfour Declaration. Six Jews had been killed and more than two hundred injured by small bands of fedayeen, Arab suicide warriors. The *yishuv*, the Jewish community of Palestine, was tensely preparing for the next wave of assaults.

The families of the would-be emigrants pleaded with them to delay their departure, but Golda wouldn't hear of it. Everything was ready, their possessions carefully packed into thirty-three trunks. Assuming they'd be sleeping on the ground, she had bought ten blankets as cushions. At the last minute, she'd sold her curtains since they seemed unnecessary to the tent she expected to call home. But she took her good suit, although only after she removed its little fur collar. Morris packed only two boxes, one for his windup gramophone and the other for his books.

Hundreds of passengers streamed aboard the SS *Pocahontas* on the morning of May 22, most bound for Italy on vacation. As Poale Zion activists made speeches about the bravery of the intrepid pioneers who would soon be reclaiming the Jewish homeland, Golda stood proudly on deck singing, "Come Home to Zion."

But their grand departure turned anticlimactic after the ship's mates went on strike. When they finally sailed, the crew mutinied and crippled the engines, leaving the rickety *Pocahontas* listing badly as it limped north to Boston. There they were forced to wait nine days while the engines were repaired and engineers replaced. But the voyage continued to be damned. In the middle of the Atlantic, pumps and boilers repeatedly broke down, followed by the condensers and refrigerators. Fires broke out in two holds and the engine room flooded. By the time the *Pocahontas* made its way into Ponta Delgada in the Azores for repairs, it was no longer seaworthy.

On the trip from the Azores to Italy, things grew even worse. The captain threw four crew members into a makeshift stockade after they threatened to sink the ship. An engineer was mysteriously thrown overboard. The captain's brother went crazy and had to be chained in his cabin. One passenger broke his leg. Another died.

After a week of eating nothing but rice and tea made out of brackish water, Golda's group finally shuffled off the boat in Naples, only to find that they'd been stranded by a strike of Arab boatmen in Jaffa against any vessel carrying Zionists. "Christians and Moslems can go to Palestine, but they can't sell tickets to Jews," Sheyna wrote Shamai. "So here you have a sad joke—no Jew can enter Eretz Israel."

Their only alternative was to sail to Alexandria, Egypt, from Brindisi. The voyage might have been a recuperative hiatus since they finally found a trustworthy vessel and bought comfortable third-class cabins. But the other Zionists on board, hard-muscled, Hebrew-speaking *halutzim* from Lithuania, were sleeping on the deck. When they snubbed Golda as the soft American, she insisted that her own group move out of their cabins and join them in pioneering solidarity.

By the time they arrived in Tel Aviv, fifty-three days after their departure, they were exhausted, filthy, and demoralized. Founded in 1909 by sixty Jewish families swept up in the romance of building a real Jewish city, Tel Aviv had sounded almost magical: Jewish policemen, Jewish businesses, a Hebrew high school, and Herzl Street, a Jewish avenue commemorating Zionism's founding father. But up close it was a dusty frontier town with few paved streets, no power station, little work, and a mayor who rode around on a white horse.

Golda, Morris, Sheyna, and Sheyna's children, ten-year-old Judy and three-year-old Chaim, rented a two-room apartment behind the Cinema Eden, the only movie house in town. They had no electricity or running water and shared a kitchen and outhouse with forty other people. But in a town where there was little work, Golda and Morris both were lucky. He found a job as a bookkeeper for a British company in Lydda, which meant that he commuted home on weekends. Golda was quickly offered a position as an English teacher at Herzliya High School, the Hebrew high school, but she refused to accept it. "I didn't come to Palestine to teach English," she declared flatly.

In the fantasy she'd been nurturing, Golda saw herself building Jewish socialism on a kibbutz, a collective farm, and she'd set her sights on Kibbutz Merhavia, up north in the Valley of Jezreel, where an old friend from Milwaukee was living. While she waited for Merhavia's thirty-two members to vote on her and Morris' applications for membership, she tutored children in English, worked on her broken Hebrew, and optimistically readied herself to reclaim the soil.

"We don't know what will be, but there is only one way," she wrote Shamai. "Whoever calls himself a Zionist and hasn't found comfort in his soul in exile must immigrate to the land of Israel. . . . Of course, this is not America, and one may have to suffer a lot economically. There may even be pogroms again, but if one wants one's own land, and if one wants it with one's whole heart, one must be ready for this."

Golda was not prepared for anything, however. In September, she and Morris were rejected by Merhavia, receiving only two votes, one from her old friend.

*　　　*　　　*

Golda's most enduring quality was her refusal to take no for an answer, so she got on a bus to Merhavia to plead her case. The reception was anything but sympathetic. The kibbutz was dominated by *halutzim* of the Second Aliyah, Palestine's most dogmatic and disdainful wave of immigrants. Seeing themselves as the Chosen, the most dedicated, the strongest in mind and body, they disdained those who did not live up to their standards—and Golda and Morris, they were convinced, couldn't begin to meet them. Everyone knew that Americans, especially American women, were soft individualists.

It didn't help that Golda and Morris were married, tarring them with the dreaded brush of hopeless middle-class conformity. For the young kibbutzniks, marriage reeked of conventional gender relationships, religion, and property rights. They had hoisted the banner of "a rebellion against the bourgeois norms of Eastern Europe," explained Esther Sternman, an early resident. Sex should be a free and loving act between two people with "a purity of heart," not a written contract.

The members' dogmatism only intensified Golda's determination. She argued, pontificated, and nagged. But when Merhavia voted a second time, she and Morris were again rejected. Still Golda didn't give up, wearing the kibbutzniks down with constant visits and her own brand of dogmatism. On the third vote, they succumbed, giving Golda and Morris one month to prove their worth.

The community Golda fought so hard to call home was in an area of the Jezreel Valley that local Arabs called the Death Swamp. Surrounded by a cement fence dotted with openings for guns, Merhavia was a collection of ramshackle frame barracks and a communal kitchen. The latrine, a quarter of a mile from the barracks, consisted of four holes in the ground with no partitions between them. No one ventured there alone because snipers from nearby Arab villages made a sport out of taking potshots at needy kibbutzniks.

Merhavia raised fruit and vegetables, and the kibbutz's cows produced enough milk for community needs with a surplus sold to neighboring villages and distributors in Haifa. The Jewish National Fund, founded at the Fifth Zionist Congress in 1901 to buy and develop land for Jewish settlement, gave members a regular income in exchange for the planting of trees. But the land was more marsh than soil, and where there was soil, it was filled with rocks. In the summer, work began at 4 A.M. because the *barhash*, swarms of tiny flies, turned afternoon work into a nightmare. Even with long sleeves and lavish coatings of Vaseline, field workers came in for lunch with insects plugging up their eyes, their ears, and their noses.

In late September, Golda and Morris settled into the spartan private room they were allotted. Their first morning, Morris was sent to dig rocks and boulders out of a field being cleared while Golda was assigned to pick almonds. "When I returned to my room in the evening, I couldn't so much as move a finger," Golda later wrote, "but I knew that if I didn't show up for supper everyone would jeer. 'What did we tell you? That's American girls for you!'"

Less needy of group approbation, Morris felt no such compunctions about avoiding the dining hall or the chickpea mush that was served at most meals, as soup, salad, or some sort of stew. Building a table and some cupboards out of orange crates, and hanging up strips of flowers on the walls to provide a bit of color, he fashioned a nest in their plain wooden room. His records were carefully organized, the windup gramophone set up and waiting.

All that was missing was a glass of tea, and for that he had to trek over to the communal dining room. Morris didn't mind the physical labor, but he hated the communal dining room, just as he hated the communal toilet, the communal shower, and the communal laundry, which doled out the communal clothes. It wasn't only the incessant togetherness that annoyed him. No one had the slightest interest in art or music or philosophy. After dinner, all they talked about was work, about Zionism, and socialism. He would have been content to stay in their room and listen to music. But if he refused to socialize, he was branded as aloof. If he tagged along with Golda, everyone frowned disparagingly at their togetherness. And he couldn't go alone because Golda was always there.

On Friday nights, she happily sat through long political discussions and then joined the other kibbutzniks as they sang pioneering songs and danced the hora. On weekday evenings, she stopped by for a glass of tea and lingered for hours. What was she supposed to do? Sit in their room and listen to endless classical music? she asked, seemingly puzzled at her husband's distaste for the groupthink and the absence of privacy.

"[Morris] was not able to tolerate . . . the sense of belonging to a community," she complained. "He was too individualistic, far too within himself."

Finally part of the world she had dreamed of for so long, Golda fed off her legendary stamina. Swinging a pickax was agony, but she never winced in front of her comrades. If the kibbutz ethic demanded gossiping in the shower or wearing communal underwear, Golda merrily gossiped and wore communal underwear, although she was, by nature, a hygiene fanatic. When she was asked to run the chicken coops, she never mentioned that she was deathly afraid of chickens. Nobody was ever going to say again that Golda Meyerson wasn't strong enough, that she didn't have enough spirit, or that she couldn't fit in with the cream of the Zionist elite.

Still, too domineering not to try to impose her will, Golda made waves, or at least a few ripples. The other women tried to shirk kitchen duty, deemed less important than the "real work" of tilling the soil. "Why do you regard this work as demeaning?" Golda asked haughtily. "Why is it so much better to work in the barn and feed the cows, rather than in the kitchen and feed your comrades?" Never very sympathetic to women's concerns, she seemed unaware of how sensitive the issue of kitchen duty still was for the veteran women pioneers, or indifferent to the long struggle they had waged.

In the first Jewish agricultural settlements, women caught up in the same "back to the land" fervor as Golda had been treated like maids. Kibbutzim limited their admission, arguing that women weren't sufficiently productive. "We young women did not encounter hardship in work but in the humiliating treatment and apathetic attitude toward our aspirations," wrote Sarah Malchin, a Russian immigrant who'd founded the first agricultural training school for women. For decades, women like Malchin had shunned the kitchen, where they were told they "belonged," and fought for the right to plow and dig ditches in the belief that only equal work would win them full equality.

Golda was as disdainful of the kibbutz's rustic ethos as she was of the politics of the kitchen. When she was on dining room duty, a chore rotated among the members, she put tablecloths and flowers out for Sabbath—for many, a clear sign that she was hopelessly middle class. When she peeled the canned herring served cold for breakfast, the other women mocked her.

"How would you serve herring at your family table?" she asked, her sarcasm resembling her mother's. "This is your home! They are your family!"

When the time came for the kibbutz to vote on the couple's permanent membership, ironically, Morris was admitted without debate. But while Golda was also admitted, the women members complained about Golda's behavior in the kitchen, about the fact that she wore stockings to dinner, that she ironed her kerchief and dress every day, not only on Sabbath, and that when she received her allowance, she used it to buy a hat. The sentiment that she didn't quite fit kibbutz life lingered. Years later, when Golda's friend Marie Syrkin asked another Merhavia member what Golda had been like, the woman recounted the story of a rainy day when a few residents were gathered in the kitchen peeling almonds. "Golda sat and appeared a little regal," she recalled.

*   *   *

The kibbutz ethos didn't put much stock in conferences and conventions. Sweat, toil, blisters, and calluses were the coin of that realm, not lofty rhetoric. No one objected, then, when the newcomer offered to represent Merhavia at the first convention of kibbutzim at Kibbutz Degania. In fact, everyone was relieved they'd found someone willing to attend.

Degania was the mother ship, the lodestar of Labor Zionism. There, the notion of a collective community—no money, no salaries, no private property, no hired workers—became reality. And the convention was filled with the leading figures in the Second Aliyah, Golda's Zionist icons: David Ben-Gurion, general secretary of the new Histadrut, the General Federation of Laborers; Levi Eshkol, leader of Degania's spinoff, Degania Bet; Berl Katznelson, the philosopher king of the movement and the strategic genius behind the new worker's party, Ahdut HaAvoda, Unity of Labor; David Remez, one of Katznelson's handpicked geniuses, who went on to help draft the declaration of independence.

Those were heady days in Palestine, as the *yishuv*, the Jewish community, laid the first building blocks of an independent Jewish nation. Although Jews might be decades from having their own country, men like Ben-Gurion and Katznelson were already locked in struggle for control over the *yishuv*'s current realities and the shape of the future nation.

Surrounded by Palestine's luminaries, and halting, at best, in Hebrew, the language of all respectable pioneers, Golda was awed into silence. Just once did she dare address the group—to defend the honor of kibbutz kitchen work. "Why is giving food to people no less an honor than giving food to cows?" she asked, in Yiddish, repeating the argument she made endlessly at Merhavia. "It's bad enough you speak Yiddish in Tel Aviv," one of Degania's founders scolded her. "In Degania, no." The female delegates, veterans of the struggle for gender equality, booed. Golda burst out into tears.

It was the second time she had been slapped down by her idols, the first time by Ben-Gurion in Milwaukee. She straightened her spine and vowed to show them what she was made of.

The chicken coop didn't feel very exciting after the gathering at Degania, and Morris had fallen into depression, worn down both by work and by Golda's indifference to his needs. To make matters worse, he was suffering from malaria despite the quinine pills served with the chickpea mush for breakfast.

"Ah, Palestine, Palestine, you beggarly little land, what will become of you?" he wrote his mother. "How ironic sound the lovely words at Poale Zion meetings about a free workers' Palestine."

But Golda had caught a glimpse of the Palestine she wanted to be part of and ignored Morris' complaints. Careful to phrase her offer as another selfless act rather than as ambition, she suggested to her comrades that she represent Merhavia at another meeting, of the Moetzet HaPoalot, the Women Workers' Council. Everyone but Morris seemed happy to see her go.

The Council had been created a decade earlier to develop new approaches to female employment, child care, and the training of female construction and agricultural workers. After World War I, it was clear they hadn't made much headway. Irate at the continuing resistance to women's equality, Ada Maimon, a Women's Council activist, decided to force the issue at the founding convention of the Histadrut, the new worker's cooperative organization launched by Ben-Gurion. She publicly demanded that the women's group be merged into the Histadrut and be guaranteed representation on its executive bodies. If Ben-Gurion refused, she threatened, the Council would run a separate list of candidates in the next Histadrut election.

An uneasy alliance of a dozen divergent ideological tendencies and competing power blocs, the Histadrut exercised only precarious control over Jewish workers. If Maimon carried out her threat, she might embolden a dozen other interest groups, from the Orthodox to the Yemenites, to follow suit, undermining the hegemony of the Labor Party leadership. So Ben-Gurion bowed to Maimon's demands, assuming the women could be more easily controlled from within than from without.

By the time Golda became involved in the Council, the women of the Second Aliyah, who'd envisioned a strong, independent body speaking boldly for workingwomen, were in open rebellion. Only the newer arrivals seemed content to be the Ladies' Auxiliary that Ben-Gurion hoped for.

Golda made her own sentiments crystal clear at the second Histadrut conference, held in Haifa in 1923. "It is a sad and shameful fact that we are forced to create a special organization to deal with matters of the woman worker," she submitted. The goal, she said, should be for the Council to become redundant.

Maimon and the other women were livid at the heresy, especially coming from a delegate allegedly representing the Council. Ben-Gurion, Remez, and Katznelson, on the other hand, were dazzled. Golda had made her first bid for membership in the boys' club.

Back home at Merhavia, she was greeted with grumbling, the other members growing impatient with her absences and her requests for new dresses or sandals for outside meetings. So when the leadership of Ahdut, Ben-Gurion's political party, asked her to accompany a prominent British suffragist married to a member of the British Parliament on a tour around Jewish Palestine, she demurred. Hobnobbing with the wife of a British grandee didn't seem important enough to risk the wrath of her comrades. But Berl Katznelson himself appeared at Merhavia to persuade her. Golda was beside herself; she'd been noticed by the sage, the father figure to whom Ben-Gurion himself deferred. No kibbutznik could deny him.

By the time Golda returned, Morris had reached his limit. When she broached the subject of having children, he offered *her* an ultimatum: If you want kids, we're leaving. Or at least that was one of several stories that Golda offered up over the years. The other was that she, the dutiful wife, left her beloved kibbutz because Morris' health was in serious jeopardy. "For him, I made the greatest sacrifice of my life," she told Oriana Fallaci in a 1972 interview. "You see, there was nothing more I loved than the kibbutz. I loved everything about it—working with my hands, the social life, lack of materialism . . . but he was not able to tolerate it—physically or psychologically."

*   *   *

Golda hadn't left Milwaukee to become a Tel Aviv housewife, but that's where she ended up, living with Sheyna and working as a part-time cashier for Solel Boneh, Histadrut's building cooperative. Morris assumed that he and Golda would soon return to America. No matter how miserable she was, though, Golda wasn't budging beyond the Mediterranean Sea. The two were barely speaking.

Their friends wondered how long the couple could keep going, or why they tried. Golda insisted theirs was an unbreakable emotional bond. "Ours was a great love," she said. "It lasted from the day we met till the day he died." Her friend Marie Syrkin didn't buy that romantic argument. "To her affection for Morris was added a sense of guilt," Syrkin wrote. "On account of her he had become deracinated and emotionally dependent. She was bound by his bondage."

Despite Golda's protestations of devotion to Morris, in Tel Aviv she began seeing other men, and soon the gossipy *yishuv* was alive with rumors that Golda was involved with an older man, who later worked as a physician in Tel Aviv, and that something had sparked between her and David Remez, one of the Histadrut leaders she'd met at Kibbutz Degania. Always discreet about her personal life, Golda never talked about the physician, Remez, or any of the other men in her life. "I was no nun," was all she would say. "There are people who generally don't talk about their intimate affairs, and I am one of them."

Remez had been Golda's sponsor into movement circles and he had arranged for her employment at Solel Boneh. Shortly after she began working there—perhaps tellingly—he offered Morris a job at Solel's office in Jerusalem. For several months, he saw Golda only if he commuted down to Tel Aviv and nagged Golda about finding a job closer to him. She resisted until she discovered that she was pregnant. Again, Remez saved her by giving her a transfer to Morris' office.

For most Jews, Jerusalem is the golden city, the emotional center of the Jewish universe. It was from Jerusalem that Saul, David, and Solomon ruled, and the old Temple wall had been a magnet for the Jewish imagination for centuries. But Golda couldn't stand it. Where others saw historical continuity, she saw religion, superstition, and defeat. Jerusalem was the Jewish past. Her eyes were on the Jewish future.

Nonetheless trapped, she left Tel Aviv, and she and Morris rented two rooms with an outside tin shack for a kitchen and a toilet in the backyard in a shabby neighborhood near the Orthodox quarter of Mea Shearim. For light, they had a smoky kerosene lamp. The cistern water was so foul that it had to be boiled.

Solel Boneh, their new employer, was collapsing, a detail Remez had neglected to share. Morris, Golda, and the rest of the office staff kept working, but they were usually paid in *mashbit*, credit slips, that few merchants accepted. On payday, Golda was forced to negotiate endlessly with grocers in the hope they would give her 80 piasters' worth of margarine and flour for a one-pound chit.

Golda's plight wasn't unique. The *yishuv* was reeling economically, the number of immigrants too great to absorb. A handful of viable farms had been created, but industry was primitive. And despite the pressure of the Histadrut, even Jewish businessmen preferred Arab workers to Jewish ones, who expected higher wages.

For Golda and Morris, things grew desperate once Menachem was born, on November 23, 1924, because Golda stopped working. "They were practically starving," recalled Regina. "The question wasn't ever what to cook but that there should be enough to cook!" Sheyna sent up boxes of fruit and vegetables from Tel Aviv. But more than once, the neighbors watched Golda weep because she had no money for oil.

They eased the financial crisis by taking in a boarder, but they were still left with the stark reality that their marriage was a disaster. Morris loved Jerusalem, which captured his poetic imagination. The city might lack cafés packed with socialists arguing whether Israel could avoid the class struggle, but it had a small opera company and plenty of dusty bookstores. Most important, he had a wife at home and a beautiful son.

Golda felt like her grandmother, living in poverty in a grimy slum, surrounded by Orthodox Jews who thought Zionism was an affront to God. Or her mother, stuck with a hapless husband who could never earn a living. "In Jerusalem I was a sort of prisoner," she wrote. "Instead of actively helping build the Jewish national home and working hard and productively for it, I found myself cooped up in a tiny apartment."

The Zionist action was in Tel Aviv, and Golda struggled to keep her contacts there. When her parents arrived in 1926 and began building a house in Herzliya, north of Tel Aviv, she made the long trek down the Judean Hills frequently. But weekend doses of politics weren't enough, and Golda finally decided to return to Merhavia, taking six-month-old Menachem with her. Morris didn't bother trying to change her mind.

But her old comrades did not greet Golda with open arms, chiding that their kibbutz wasn't a shelter for women running from broken marriages. Golda had dreamed of losing herself in toil. Instead, the work committee assigned her to work as the *mitapelet*, the kibbutz nanny, to care for five children, day and night. Within six months, she was back in Jerusalem, and the ever-patient Morris.

Golda was trapped, and the birth of a daughter in May 1926 cinched the noose tighter. Sarah was sickly and needed medicine, Morris' wages were unpredictable, and Golda could feel all of her dreams evaporating.

When she and Morris couldn't find the money for Menachem's nursery school tuition, Golda took in the school's laundry, spending hours boiling water by lamplight, soaking and scrubbing towels, aprons, and diapers in the bathtub in the living room. Finally, she swallowed her pride and accepted a job as an English teacher at a private school.

In Golda's account, just when she was about to hit rock bottom, she happened into Remez and, out of the thin air, he offered her a job as secretary of Moetzet HaPoalot, based in Tel Aviv. "They were interested in the services of someone like myself," she wrote. "I had already worked in Tel Aviv for Solel Boneh and had gone on working for it—though only briefly—in Jerusalem."

It was an odd explanation since Golda had been nothing more than a cashier in the Tel Aviv office of Solel Boneh, so insignificant that years later, her boss could barely remember her. The truth was that Golda was the Histadrut's solution to their recurrent problems with the obstreperous women of the Council. Despite Ben-Gurion's promises, not a single woman was serving on any of Histadrut's major policy bodies and wage discrimination remained rampant. The men of the Histadrut were trying to turn the Women's Council into a female enclave while remaining the gatekeepers of who could leave it to enter the establishment, and its leader, Maimon, had resigned in disgust.

When Remez ran into Golda, the Histadrut leaders were frantically searching for her replacement, for an obedient woman who'd allow the Histadrut and the Ahdut Party that controlled it to set the Council's priorities and keep it loyal to the emerging establishment. The young American was the answer to their problem.

CHAPTER FOUR

Not being beautiful forced me to develop my inner resources.
The pretty girl has a handicap to overcome.

If Jerusalem represented the resignation of the shtetl, for Golda, Tel
Aviv sung with the vibrancy of the new Jew. Rothschild Boulevard was
turning into the city's own Hyde Park, where heated discussions about
Russian politics competed with pronouncements about the perils of capitalism and a dozen competing ploys for forcing the British to fulfill the
promise of the Balfour Declaration. The entrepreneurial energy of the
34,000 immigrants of the Fourth Aliyah exploded into rows of bookstores, haberdasheries, cafés, and cinemas. In the shimmering light of the
desert, Golda left behind Morris and the drudgery of a babushka to throw
herself into that future.

Except for the waking minutes she had to spend on Menachem and
Sarah. That conundrum Golda wrote about two years later in a collection
of stories by pioneer women:

> Taken as a whole, the inner struggles and the despairs of the mother
> who goes to work are without parallel in human experience. But within

that whole there are many shades and variations. There are some mothers who work only when they are forced to. . . . In such cases, the mother feels her course of action justified by compulsion. . . . But there is a type of woman who cannot remain home for other reasons. . . . Her nature and being demand something more. . . . She cannot let her children narrow her horizon. And for such a woman, there is no rest. . . .

She of course has the great advantage of being able to develop. . . . Therefore, she can bring more to her children than if she were to remain at home. . . . But one look of reproach from the little one when the mother goes away and leaves it with a stranger is enough to throw down the whole structure of vindication. . . .

Having admitted all this, we ask: Can the mother of today remain at home all day with her children? Can she compel herself to be other than she is because she had become a mother? . . . Can we today measure our devotion to husband and children by our indifference to everything else? . . . If a woman does remain exclusively with her children and gives herself to nothing else, does that really prove that she is more devoted than the conventional mother?

Golda's plaint was more than a tad disingenuous, or perhaps self-deluding, for she was no ordinary working mother. She didn't go off to work for eight or ten hours to keep her mind lively or make a minor contribution to her society. She made her play to become somebody by throwing herself into an unending series of meetings and caucuses, conventions, strategy sessions, and tactical debates.

Menachem, five, and Sarah, three, were enrolled in a kindergarten run by the labor movement and picked up after school by their babysitter, who fed them and put them to bed. If Golda didn't make it home by midnight, which she usually didn't, the nanny slept over. Weekends provided no respite because Golda didn't have weekends. Her schedule was so tight that she made appointments to take her children to concerts and the cinema.

Although Golda knew that her departure from Jerusalem meant the end of her marriage, Morris kept up the pretense, or perhaps the hope. He painted the furniture in the children's new room, created whimsical decorations, and faithfully showed up every Saturday morning to cook them breakfast. All day he read to them, filled the apartment with music, took them for walks, and made their favorite fudge.

"Morris was father and mother to their children," said Judy Shapiro, Golda's old friend from Chicago.

Clara, Golda's younger sister, who had stayed in the United States, praised Golda for being a warm and loving mother "when she was around. But she wasn't always around."

In fact, she was almost never around.

Golda's time wasn't sapped by the Women's Council, for hers was a minor job in which she had little interest. "I never had sympathy for the women's organizations as such," she acknowledged later in life. But the Histadrut was the heart of the revolution, what Katznelson called one of the world's greatest revolutionary movements, the "plot on which contemporary Jewish history hinges," as he put it. And her Council position gave Golda entrée to that crusade.

When Ben-Gurion created the Histadrut, he went the Industrial Workers of the World, the Wobblies, one better. Israel would not only have *one big union* that would guarantee workers the right to strike, decent wages, vacation, and sick leave. It would have a labor movement that would also be the economic engine fueling the national economy and controlling the means of production. Eight years after its founding, Histadrut was already Palestine's biggest employer, producing one-third of the gross *yishuv* product.

The rest of the world might have been skeptical about the prospect of an independent Jewish country in Palestine. But inside Histadrut headquarters, the new nation was already taking shape. From its small offices in a low red building downtown, Ben-Gurion limned the outlines of the future state, David Remez struggled to revive a dead language that had no words for "bulldozer" or "road sign," and Levi Eshkol searched for water for new settlements.

Golda was the most junior member of that coterie, but she quickly proved herself to be more than the expected lackey. Every year, the Histadrut sent representatives on fund-raising tours across America, an experience that most Zionist revolutionaries found so painful that volunteers had become impossible to recruit. To Palestine's Eastern European socialists, the Jewish residents of the fat capitalist land, comfortable and complacent, were anathema, and they hadn't proven all that generous.

Pontificating either in Yiddish or in heavily accented English with flowery Eastern European prose out of step with the American vernacular, few of the visitors from Palestine managed to fire American passions. They were too out of touch with the local zeitgeist and too contemptuous of "dollar land," as Manya Shohat, one of the first Histadrut emissaries, called the United States.

Golda, however, was, by upbringing, an American Zionist. Unschooled in lofty Eastern European rhetoric, she knew the American audience, where their fears and dreams lay buried, what made their emotions vanquish their intellects. Nonetheless, the first time she was sent back to the United States as a Histadrut emissary, when she rose to the podium to address the closing session of the launch of the annual 1928 tour, her knees trembled as she faced an audience of 585 delegates. She was thirty years old and hadn't been in the United States in seven years.

In a plain dress with her hair pulled back in a low bun, she spoke in a monotone with little cadence. But instead of high-minded platitudes, the delegates heard a young woman paint a picture of the struggles of Jewish workers in Palestine bravely fighting off Arabs, facing down the British, to keep themselves—and the Jews of the world—safe for another generation. She didn't bully the delegates—that would come later. She didn't berate them for staying comfortable on the western shore of the Statue of Liberty. Unlike her predecessors, she simply reached out her hand.

"The Palestinian workers have begun to dig a tunnel to reach their American brothers," she said, "and . . . the American Jewish workers have heard the chop of our hammers. . . . The wall of separation is being broken down and the union is coming."

The applause was long and fervid, the reviews fantastic. Golda quickly became the most popular Histadrut liaison to the United States.

But the situation back home in Palestine distracted Golda and the rest of the leadership from tasks like fund-raising and nation building. While she was gone, well-organized Arab groups had seized control over the Western Wall and attacked Jewish settlements near Jerusalem. In Hebron, they went house to house, hacking at women and children with hatchets. The violence spread to Safed in the north and the Gaza Strip. For six days, the British had allowed the mob frenzy, the first serious jihad in the region, to continue. By the time it was quelled, 135 Jews were dead, more than 300 wounded.

Anxious to be more than the Histadrut's token female, Golda didn't go home to unpack or check on the children but went directly to a gathering point for the Haganah, the Jewish self-defense force created clandestinely after the 1921 riot, to help resist the rioting she called a pogrom.

Although the Haganah didn't need her, a woman with no military training, the Histadrut did, and the leadership knew it. Her popularity in the United States became her entry ticket to the upper echelon. She might have lacked the education or intellectual heft of the men around her—Ben-Gurion and Remez, after all, had studied law in Istanbul, and Moshe Sharett, the emerging *yishuv* diplomat, had studied at the London School of Economics. But her perfect Yiddish and American English, and her power over an audience, made her their most formidable translator and emissary. In 1930, when Ben-Gurion's dream of a united labor political party came true with the merger of his Ahdut HaAvoda party and Hapoel Hatzair into Eretz Yisrael Workers' Party, to be known as Mapai, Golda proudly sat in the hall as a delegate.

Within months, however, she was off again, to the Conference of Socialist Women in England, looking for friends for the Jewish community in Palestine. The British government was dragging its feet on the creation of a Jewish state in Palestine, and the *yishuv* sent emissaries to every possible forum to drum up political support for their cause.

Later in the year, she returned to London as a Histadrut delegate to the annual Imperial Labour Conference, this time with Ben-Gurion. BG, as everyone called him, was a difficult person, diffident and unyielding, and Golda was still nervous around him. At the meeting, Ben-Gurion spoke first and was promptly shouted down by scores of Arab delegates. "Don't waste your breath," he told Golda when it came time for her to address the gathering. But she refused to be cowed.

"I trembled at her daring words," Ben-Gurion reported in *Hapoel Hatzair* newspaper. "Her speech shook the convention. She spoke with genius, assertively, bitterly, with hurt, and sensibly. Although I had heard of her success in the women's convention and other gatherings arranged for her by the labor movement in different places, her speech was a great surprise to me."

Within a year of her departure from Jerusalem, Golda was a traveling celebrity. She spoke in cities all across the United States, hammering away at her message and raising sorely needed funds. In Great Britain, she won the backing of members of British cooperative societies. In Brussels, she wooed the big shots of the Socialist International, including Leon Blum, who was to become France's first Socialist and Jewish prime minister.

Back home, Golda became firmly established as the sole woman in the old boys' club, a fixture among the senior movers and shakers. Always with her trademark unfiltered Chesterfield cigarette in her hand, she never missed a meeting or discussion, anxiously catching up after a monthlong absence lest she be sidelined.

Menachem and Sarah paid a high price for her success. Late one night, Golda was chairing a meeting at Histadrut headquarters that ran on and on, as labor meetings did, often not breaking up until the early morning. People wandered in and out for coffee, for caucuses, for breaks from the dense cloud of cigarette smoke. At home, the babysitter was long gone, having been told that Golda would return early. For hours, Sarah and Menachem entertained themselves by singing songs. Finally, restless and worried, they wandered over to the Histadrut building and silently took seats on a bench. Wrapped up in argument, Golda didn't notice them until she called a vote and realized that two of the hands she was counting as FOR belonged to her own children.

"She did these things because she felt it was her duty," said Menachem in later years, stoically defending his mother. "She would always say, 'If you want things, you have to go and do them.' But if we really needed her, she was always there." Sarah, who'd been born with serious kidney problems and was regularly confined to bed, never became that sanguine, admitting that she was "orphaned of my mother."

Perhaps the most telling glimpse into what they both felt was how vividly they remember the excitement they experienced when Golda was felled by one of her regular migraines. "Even though she was sick, we were happy just to have her home," Sarah recalled.

Morris, Sheyna, and Bluma picked up the slack when Golda was gone, often for months at a time. But furious at Golda's neglect of her offspring, Sheyna finally fired off a particularly pointed letter to Golda in Brussels about the importance of motherhood and the relative insignificance of her outside work.

"Believe me, I know I will not bring the Messiah, but I think that we must miss no opportunity to explain what we want and what we are to influential people," Golda shot back. "I ask only one thing, that I be understood and believed. My social activities are not an accidental thing: they are an absolute necessity for me. I am hurt when Morris and others say that this is all superficial, that I am trying to be modern. It is silly. Do I have to justify myself?"

While Sheyna berated Golda for the family aspect of her private life, Golda was simultaneously dogged by gossip about its intimate side. The *yishuv* was like a small town, with scuttlebutt the currency of social cohesion. The neighbors were too nosy, the culture too ingrown for anyone to have much privacy. Ben-Gurion's affair with Rivka Katznelson, the editor of the monthly magazine of the Histadrut, for example, was an open secret, as his relationships with a string of other women would become over the years. Outside the religious community, such affairs rarely caused much scandal unless they led to the breakup of a marriage.

But as the only woman in the political hierarchy, Golda was subject to a different standard. Gossip spread that she was "easy to get," recalled Aviva Passow, the daughter of one of Golda's oldest friends. People sniggered at her nickname, the Mattress. "Men have always been good to me," Golda commented. "I've always moved within a circle of intellectual giants. I've always been appreciated and loved." Those in the know put it differently: she was sleeping her way to power.

Some of the gossip was fed by envy at the rapid rise of such a young woman. But Golda was anything but celibate. Although he had both a wife and a longtime mistress, David Remez still maintained his relationship with her, on and off. And her affair with Zalman Shazar, the poet, essayist, and intellectual who would become Israel's third president, wasn't exactly a secret; he was always at her house.

Golda had caught her first glimpse of Shazar when he spoke at a May Day rally in Tel Aviv, and the description of him she gave an interviewer almost half a century later suggests how captivated she had been. "He did not speak with his mouth but with both arms and legs, with his elbows and knees, with every rise and fall of his back, rising and quivering and leaping and filling the entire ambience of the stage," she recalled. "His every bone spoke. Spoke? Shouted, raged, threatened, exhorted, adjured."

While Remez served as Golda's confidant and political guide, Shazar became her inspiration. But the triangle was hopelessly complicated because Remez was Shazar's best friend. "We always avoid talking about serious things," Remez berated Golda in one letter. "I don't know why, but first there was no opportunity and then I didn't want to make a scene, and Zalman had the same feeling. Talking to you would make a scene. . . . The trouble with you is you were raised on praises. No question you are successful. I don't question you in your social earnestness. You are not a mensch. . . . Zalman's wife loves him and he's everything to me. Gershon [Shazar's code name for Morris] has a wife that ignores him. . . . True, Gershon is a very difficult person, but you made him what he is today."

Listening to the increasingly ugly rumors, Regina asked Golda why she didn't simply divorce Morris. "Morris won't give me a divorce," Golda told her. After a pause, she added, "What do I need it for? I'm not going to marry anyone."

*   *   *

Golda was in Europe, yet again, when Morris cabled her grim news: SARAELE DESPERATELY ILL. COME HOME AT ONCE.

After years of consultations, treatments, low- and no-carbohydrate diets, Sarah's kidneys seemed to be giving up. Her face was swollen. She could no longer climb stairs. And she was subsisting on little more than sweetened tea. The doctors predicted that the six-year-old wouldn't live beyond the age of twelve.

In Golda mythology, she, the frantic mother, raced home, took one look at Sarah, and implored the Histadrut to send her to America as an emissary so Sarah could receive proper medical care. Sarah's medical crisis, however, was a well-timed coincidence. The Histadrut leadership had already asked Golda to go to the United States for two years to revitalize the Labor Zionist movement there. Caught between her fierce loyalty to her comrades and her fear of being isolated from the action for such a long period, Golda had hesitated—until Sarah's illness pushed her out of indecision.

Less than a month after Morris sent his frantic telegram, Sarah was admitted to the new Beth Israel Hospital in Manhattan, and Golda and Menachem were unpacking their meager belongings in the Brooklyn apartment of Fanny and Jacob Goodman, old friends from Poale Zion. Six weeks later, Sarah, who'd long been misdiagnosed in Tel Aviv, was discharged and settled in with Menachem at the Goodmans'. Golda was free to begin her travels.

In the wake of the Holocaust, support for a Jewish state became such an integral facet of American Jewish life that it's all too easy to forget how tepid and ambivalent American Jewry was toward Zionism earlier in the century. The messianic fervor of European Zionism was driven by the conviction that only an independent state could ensure Jewish survival and liberate Jews from centuries of victimhood. But that belief struck little chord in Boston or Kansas City. Despite job discrimination and anti-Jewish housing covenants, the United States had proven to be the safest of harbors. So while few American Jews opposed some sort of Jewish nation in Palestine, and many were willing to throw a few dollars into a blue Jewish National Fund box to purchase land there, only a handful saw Palestine as their homeland.

Even mainstream Zionist groups like Hadassah and the Zionist Organization of America practiced a brand of support for Palestine that was anathema to Golda. For them, Zionism was nothing more than a charity on behalf of Jews who had fled Eastern Europe in the wrong direction. The sort of Socialist Zionism preached by Golda was a pie-in-the-sky fantasy few took seriously. Who was going to give a country to a group of crazy Jewish Socialists? And who was going to live there?

Golda's mission to fire a Labor Zionist movement, a Socialist movement, then, sent her to the fringe of the American Jewish community, into a tiny immigrant ghetto. While most American Jews were studying at night to move up and out of manual labor, there they made a cult of it.

Broadening the appeal of the movement, especially of Pioneer Women, the American subsidiary of the Women's Council, was a tough sell. For two years, Golda rode buses and trains across the country, meeting with chapters of six or eight women in places like Canton, Ohio, Sioux City, Omaha, and Rock Island, to visit Zionist summer camps, and attend conventions, while publishing a monthly magazine and trying desperately to scare up funds for the Histadrut.

Over time, she refined the *yishuv*'s story until it was near-biblical: The members of the First Aliyah, penniless schoolteachers and shopkeepers, trudged over the rugged Caucasus Mountains in the deep snow, making their way through hostile Turkey and Syria only to find a land eroded and seared by the centuries; the Second Aliyah pioneers, young and fired by ideology, cleared the Emek, built kibbutzim throbbing with Socialist spirit despite Arab snipers, malaria, and British betrayal. Golda spoke softly about Jewish women struggling to protect their children during riots, about new hospitals built where no medical care had ever existed, about training schools, desert settlements, the first Hebrew University.

"Palestine today is a place where there are no gangsters, policemen, or flappers, and men and women work side by side in the fields," she reported breathlessly.

She so stirred hearts and awakened such pride that Goldie Meyerson Clubs sprung up all across the country. "Goldie brought us a waft of fragrant orange blossoms, sprouting veggies, budding trees, well-cared-for cows and chickens, stubborn territory conquered, dangerous natural elements vanquished, all the result of work, work, . . . just for the ecstasy of creation," wrote Anne Mellman from St. Louis to the Pioneer Women's newsletter. Her eloquence and sincerity "have instilled in her hearers a reverence for our cause."

Despite her popularity, Golda seethed with frustration at how little concrete interest she managed to inspire. At thirty-four, she was blazing with newfound self-confidence, but she was devoting her days to a bevy of parlor Zionists, for whom she had nothing but contempt, and not raising very much money. Never one to suffer in silence, Golda took out her annoyance on her members. In her Circular No. 4, February 1933, she berated Pioneer Women chapters for sponsoring "ten-dollar luncheons *instead* of other tasks, when the intent of the luncheons was that this should be an *addition* to all those diverse drives and undertakings that our members have carried out."

Holding back none of her irritation, she went on to lambaste their programming. "I . . . noticed that there are yet certain branches that carry out the cultural work in a decidedly incidental manner. They support themselves solely by means of lectures by marginal individuals, devoid of a calculated method, such that the topics are constantly incidental, and not each lecture importing significant meaning for our branch."

In one midwestern city, Golda attacked members for sponsoring card games instead of organizing the usual dances, picnics, or raffles. "For

Palestine, you play cards?" she asked indignantly, revealing her streak of pioneer Puritanism. "This is the kind of money we need? If you want to play cards, you can play as long as you like, but not in our name!"

Simultaneously, Golda was caught up in a running feud with mainstream Zionists worried that the *yishuv* was being taken over by socialists who would leave no room for individual entrepreneurs. Where were those rugged individualists when we were clearing the swamps in the Emek? she shot back in *The Pioneer Woman.* "It is our firm conviction that through our methods it has become possible today to speak of statehood."

Nonetheless, Golda refused to give up on the mainstream. Unlike Ben-Gurion and other *yishuv* leaders, she understood that Palestine's pioneers needed the deep pockets and political support of American Jewry. For them, Golda went so far as to dress up. "When I first met Golda in Toronto in 1932, she looked like a femme fatale with a big fluffy hat and a long cigarette-holder," recalled Meyer Steinglass of the Zionist Organization of America. "She was very striking, very good looking and she had this certain air of mystery about her, always."

Golda's only respite occurred during periodic trips to New York. At 1133 Broadway in New York's garment district, where a group of Zionist organizations shared offices, someone was always willing to listen to her vent. And she often snuck out to spend time with Shazar, who was studying at Columbia University.

But Golda was away so much that the children, left with a family they barely knew, were on her the moment she walked through the door. "Sometimes weeks would pass and we didn't see each other," recalled Sarah. "My brother suffered a great deal from this. He quarreled with mother and tried to stop her from leaving the house. I also felt lonely without her." One night Sarah asked Golda what she did on her long trips. I go to meetings and talk to people, Golda answered. "So why can't you stay home and talk to me?" Sarah responded wistfully.

Occasionally, she took the children to the zoo or the planetarium, to concerts, museums, movies, and her favorite Chinese restaurants. For the first time, they met their aunt Clara, Golda's youngest sister, who lived in

Bridgeport, Connecticut, and Morris' mother and sisters, who lived in Philadelphia.

But it was never enough. "It was a terrible position for both of them," said Judy Goodman, who often cared for the children when her parents were busy. "They knew no one here, no one."

Golda's old friend Regina put it bluntly. "She certainly never should have had children."

CHAPTER FIVE

*I never did anything alone. Whatever was accomplished
in this country was accomplished collectively.*

When Golda left Milwaukee in 1921, she never mentioned how long she thought it would take to build a Jewish Socialist paradise in Palestine. But when she returned from the United States in 1934, that chimera seemed even more elusive than it had thirteen years earlier. The Jewish population had increased 400 percent, but almost three-quarters of the residents of Palestine were still Arabs. *Yishuv* leaders churned out reports about a 200 percent increase in exports, a tenfold growth in the citrus industry, and a 1,000 percent rise in electricity. But since they'd started from zero, the figures weren't all that impressive. Tel Aviv might have been burgeoning, with a population of 150,000, but the *yishuv* was still welfare dependent, begging for handouts from a skeptical international Jewish community.

Increasingly well-organized Arabs were lashing out at the Jewish presence not only with semi-organized and random attacks, but with strikes, assaults on travelers, and the burning of thousands of the trees the pioneers had lovingly planted. Desperate to bolster the economy, *yishuv* leaders inflamed their anger further with campaigns to pressure Jewish-owned businesses to replace cheap Arab workers with "Jewish labor" and consumers to "buy Jewish." The Jewish press railed against employers who deprived Jewish immigrants of jobs by employing "alien" workers and housewives who purchased Arab-grown tomatoes.

Hostile Arabs were the tip of an iceberg of problems. Despite Britain's pledge to the League of Nations that it would use its mandate to create a Jewish homeland, London had been inching away from its own promises almost as soon as the Balfour Declaration was issued for fear of damaging its position in the Middle East. After each outbreak of Arab rioting, the Foreign Office dispatched a commission of inquiry into its cause. The verdict was always the same: the Arabs resented the growing Jewish presence and the only way to calm the situation was to avoid "a repetition of the excessive immigration."

Emboldened by such conclusions, the Arabs took to regular rioting in the hope that the next commission would persuade Britain to end Jewish immigration entirely. And a long series of reports and White Papers about the alleged scarcity of open land and the royal obligation to the Arabs led *yishuv* leaders to believe that the government was laying the groundwork to move in that direction.

"You know, the trouble with you is you want a national home but all you're getting is a rented flat," Victor Adler, head of the Socialist International, cautioned Golda.

To make matters worse, Ben-Gurion and the Labor Zionists were fending off constant power plays by other Zionist groups. It wasn't just the Jewish capitalists tying all of Ben-Gurion's plans in a knot. Kibbutzniks accused their old comrades of selling out their pioneering spirit for urban decadence. Religious Jews demonstrated against the desecration of the Sabbath. And Zionist revisionists, the *yishuv*'s right, declared war on labor's hegemony.

The problems were messy and contentious, and Golda couldn't have been happier debating them from inside the Histadrut's Va'ad HaPoel, its executive committee, a sort of shadow legislature in the government-in-the-making, to which she'd been appointed after her return from America. By 1934, Histadrut members traveled to work each morning in buses run by a Histadrut company along roads built by Solel Boneh; toiled for Histadrut-owned stores, offices, or factories; and bought food produced by Histadrut members at Histadrut cooperative stores. Their children studied at Histadrut-run schools, where they learned the fundamentals of Labor Zionism. When they fell ill, they looked for care at a Kupat Holim clinic, also run by Histadrut. If they had money to save or wanted loans, they made their deposits or applications to the Histadrut bank.

What the Histadrut couldn't provide, another arm of a Zionist structure dominated by Labor Zionists did. Immigrants arrived on visas arranged by the Jewish Agency, the administrative body of the *yishuv*, which Ben-Gurion also ran. They looked to Mapai, the Labor Zionist political party, for both political direction and patronage.

Decked out in shorts or austere dresses with no hint of makeup, Labor Zionist leaders wore and lived their power lightly. But with little tolerance for disloyalty or public shows of disagreement, they knew how to wield it. Their closed circle controlled virtually every significant institution in the country.

Golda had earned her seat at the table with her perfect English, her way with foreign Jews and non-Jews alike, and her natural gift for oratory. But she was the junior-most member, given the most insignificant of jobs, the establishment of a tourism department to guide Protestant missionaries, quizzical overseas Zionists, and a rotating cast of British Labour Party activists around kibbutzim and cooperative marketing ventures.

Golda hated every minute of it. But membership in the Va'ad allowed her to involve herself in every corner of *yishuv* life. She helped set up a program to send recruiters to Eastern Europe to train future immigrants in Hebrew and farming. When Ben-Gurion tried to abolish the tradition of Histadrut employees receiving salaries based on need, she stood him down, horrified at the prospect of replacing a Socialist pay scale with Ben-Gurion's more "flexible" one. And as immigrants began streaming out of a Germany where Jewish civil rights were being eroded by Nazi militants, she threw herself into the problem of figuring out how to care for the children who'd arrived from Berlin without their parents, or what to do with the famous research chemist who didn't want to lay bricks.

A font of energy, Golda happily raced off to Haifa for the morning to calm the charged air between employers and workers, inspected new Histadrut housing in the afternoon, and then spent the evening with her comrades figuring out how to finance their latest brainstorm. Within a year, she was elected to the secretariat, the inner cabinet.

In 1936, the Arab Higher Committee called a general strike, urging Arabs to cease work until the British prohibited Jewish immigration and land purchases. With construction, transportation, and food delivery at a standstill, the Histadrut vowed to replace every Arab worker with a Jewish one.

Faced with the resulting paralysis of the port of Jaffa, Remez asked, "Why not build a Jewish port and a Jewish shipping company?" It was another starry-eyed Histadrut idea, more the romance of seeing a Jewish star on a steamship than a practical necessity. Nachshon, Remez called his imaginary shipping line, borrowing the name of the first Israelite to throw himself into the Red Sea when others hesitated to obey Moses' command.

Golda picked up the Nachshon banner and went on an American tour to raise the money. "The sea is an organic, economic, and political part of Palestine, and it is yet almost unpossessed," she told audiences in the usual two dozen cities. "The force which drew us from the city to the farm is now driving us from the land to the sea.

"This is one more step toward the independence of a nation."

On her way home, she rendezvoused with Remez in London to shop for vessels, although theirs was as much an intimate interlude as a buying trip. "We stayed quite a while in London," she wrote of that trip. "We didn't have very much to do, and we would sit up all night at the Lyons Corner House in Oxford Circus. . . . We used to walk for hours in the night."

Golda thrived on the nitty-gritty of translating her vision of a Jewish

Socialist state into the fabric of daily life. She began running Histadrut's complex of mutual aid programs and chaired the board of Kupat Holim, a medical system that covered 40 percent of the Jewish population. When the British began building army camps across Palestine, she negotiated the wages and working conditions of Jewish laborers—and made sure that the British lived up to their agreements.

And, still, she traveled, her Mandate passports filling up with stamps from Switzerland and England, the United States and Czechoslovakia, as the representative of the Histadrut at international meetings and the Actions Committee of the World Zionist Organization.

It was the ideal life for a woman who couldn't abide voids, who needed constant motion and constant company. Golda could never stand to be alone. If she was in town, her house was always filled with people. If it wasn't, she called someone to come over to talk or wandered by a friend's house.

Still, Golda's schedule was brutal, frequently sending her to bed with a migraine or sheer exhaustion. "What I need is a wife," she told Menachem one day. In fact, she had a series of proxy wives, children of friends who needed a place to stay in Tel Aviv, Americans who'd recently arrived in Palestine, all of whom were sucked into the vortex that was Golda.

*   *   *

Casting off her role as overseas emissary and fund-raiser, Golda took on a new life as an organizer and a serious political player with her trademark style, an inimitable fusion of idealism, moralizing, and arm twisting that would define her for decades. Golda pursued what she wanted or believed was right—and for Golda there was usually little distinction—with molten single-mindedness. Shrewd and dogged, she always denied that she was motivated by any trace of personal ambition, as did every other Labor Zionist leader. The movement ethic demanded that they pretend to be called to service rather than aspire to power. So after each call, Golda ritually protested that she wasn't worthy. But she never flinched from saying yes.

The trappings of power held no interest for her. She lived simply in a two-bedroom apartment in a workers' cooperative building and owned just two dresses, one drying while the other was worn. Nor was she driven by conventional political egomania. Golda was a Zealot, in the original sense of the word, a spirit kindred to the Jewish underground of the first century AD that opposed the Roman occupation. She sought power for its uses, not its perquisites, absolutely confident that she knew precisely how it should be applied.

Curiously, in light of the near adoration she evoked abroad, Golda neither sought nor received similar popularity at home. "She did not court her public," observed her friend Marie Syrkin, dramatically understating Golda's indifference to the approval of the masses. Given the *yishuv* political system—and the Israeli governmental political system that grew out of it—Golda had little incentive to buff off her hard edges, play the populist, or pretend patience with a recalcitrant public. If Golda wanted to serve on the Zionist executive, on the Histadrut secretariat, or a party ruling body, she didn't need to woo voters. What counted was the backing of a party hierarchy since they chose candidates for office. In a system carefully designed to centralize power in the hands of party bosses, she needed only to make sure that she was one of them.

Her lack of concern for her own popularity served her well since Golda was always willing to risk public anger by delivering bad news and pounding the populace into submission, and Ben-Gurion regularly sent her into that fray.

Her most brutal fight in those years was for a new unemployment tax for all Histadrut members, who were already paying their regular union dues as well as contributing to pension and sick funds. The proposed Mifdeh B, designed to raise money to ease the burden of the unemployed, met a storm of opposition from workers already barely feeding their own families, and Golda went on the stump to stave off rebellion. Plodding from factory to factory, she idealistically appealed to a people's pride in sharing and caring for its own. "We have among us not only grown-ups who go hungry, but also children who are hungry, hungry for bread," she said, mixing high-minded rhetoric with a hefty dose of guilt. "It is absolutely necessary that one of the first and main things we do will be to wipe out this shame, this blot on the community and chiefly on the Labor Federation."

But when she invoked the workers' "debt of honor" with their comrades or the redemptive power of social responsibility, rank-and-file members jeered and shouted that she should go back to America "where she belonged." Why can't overseas Jews pitch in the money, workers asked. Golda wouldn't hear of it. "We shall not go to the Jews of the Diaspora to seek relief. Our topic of discussion with them cannot be the matter of unemployment, the matter of hunger. . . . We go to the Jews of the Diaspora to ask them to do their utmost to enable us to build, to create new things."

When the revival meeting tone failed, Golda talked tough, conjuring up ominous visions of a curtailment in medical services. Sharp and caustic in debate, she was a battering ram and pushed the new tax through.

Next she faced down the nurses, who opposed the prevailing pay scale, based on family size rather than on experience or skill. "We're professionals and expect a professional wage," they insisted. Golda accused them of greed and moral irresponsibility. After all, although she earned less than the charwoman in her office at the Histadrut, she wasn't complaining.

As she did battle on behalf of the Histadrut, Golda was denounced from the right and the left, often brutally. "She was always against the workers," one critic said. "There is a sadistic streak to her."

Golda never seemed to care.

She clung to peculiar blind spots, particularly when harsh realities collided with her ideological illusions, and the harshest of those collisions occurred when she tried to juggle her Socialist ideals against her commitment to Zionism. Her support for the Hebrew labor campaign and Histadrut's exclusion of Arab workers from its ranks, for example, left little room for expressions of Socialist solidarity with Arab workers. Nonetheless, Golda loudly proclaimed such camaraderie, bragging about Histadrut efforts to organize Arab workers, ignoring the fact that they were excluded from Histadrut benefits.

Gradually, Golda won deep respect from the *yishuv*'s movers and shakers for her drive and dedication. Given that reputation, Golda's "feminine" side, especially her proclivity for crying, caught both her admirers and her adversaries off guard. "In social questions she has a genius for belligerent opposition," said one of her colleagues. "In regard to a political or ideological issue, she is strong as iron. But if, in the course of the debate, someone offends her personally, she can begin to cry like a high school girl or a spoiled child." Over time, however, except at the movies, those tears rarely flowed without design, just as her girlish flirting was rarely uncalculated.

Despite the respect she garnered, Golda did not elicit a similar level of affection. She could be snappish and dismissive. One afternoon during a Va'ad secretariat meeting, for example, she complained that Shamai, who'd joined Sheyna and the children in Palestine, had been fired from his job because of her position. "I can't accept the fact that my sister's family has to suffer because I am a member of the secretariat," she barked. Until she received proof that she was wrong, she said, "it isn't possible to continue on my job."

Impatient with those who refused to agree with her, she lashed out bitterly at anyone who expressed an opposing point of view. Her cold streak frightened even her friends. "God forbid if she didn't like you," said Nomi Zuckerman, whose family had been close to Golda for more than a decade. "She could be intensely vindictive. And she remembered. She never missed anything anybody ever said or did."

The expert in Golda's ability to turn to ice was, no doubt, Morris, whom she'd continued to string along despite their physical separation. When she and the children returned from New York, both Menachem and Sarah hoped that Golda and Morris would reunite, but Golda asked for an "unofficial separation." Morris tried everything to change her mind. But when he arrived on Friday nights to spend weekends with his family, she barely acknowledged his presence, ignoring him for the friends who always crowded her house.

By then, Morris was working in Jerusalem, boarding with Nomi's family. "I can still see his room," Nomi recalls, "a metal bed, a keyhole desk, and Golda's photograph, which he stared at for hours."

In 1938, Morris reached his limit and demanded that they either reunite or separate officially. "I came to Palestine for one reason only—to be with Goldie," he told Nomi. "But she was never there." Unaccustomed to Morris' calling the shots, Golda fled to the house of a friend, where she spent two weeks of crying. Then she went home to her children's anger and Morris took a job with Solel Boneh in Persia. "What has Tel Aviv to offer except my bare, cheerless room on Maaze Street and the meetings with you?" he wrote the children.

While Morris was the most dramatic example of Golda's penchant for freezing out people, he wasn't unique. Golda had a sort of personal blacklist, "and once you got on that list, you never got off," said David Passow, the grandson of one of Golda's old friends. "But, on the other hand, if you were on her white list, she welcomed you like a real Jewish mother and she'd go to the mat for you every time. If you were on her white list, you could do no wrong. There was no gray list. She was incapable of gray. It was always either black or white."

Her dismissiveness colored every relationship Golda ever had, whether personal or political. Knowing that there was no going back once you crossed her invisible line, people hesitated to disagree with her. Is it worth risking her wrath over *this* issue, they'd ask. Nomi Zuckerman still recalls the day she dared disagree with Golda about something trivial, the quality of an opera singer. "She looked at me as if I were a piece of paper flying around on the floor," says Zuckerman. "She could give you a look when she wanted to. It froze you. It almost got colder next to me."

All of Golda's personal relations were problematic. The men she was involved with were inevitably married, and she had issues with her children as well: much to her chagrin, her son, Menachem, dropped out of both the Labor Zionist youth group and high school. Sarah, too, left school without finishing. And her mother, Bluma, on whom she depended for child care, criticized her constantly, about her smoking, about the children, about her clothes.

But it was Sheyna who remained Golda's greatest challenge. Proud enough of her sister to keep a scrapbook of her press clippings, Sheyna was too envious of the woman for whose success she claimed credit to offer any of the approval Golda had craved since childhood.

"Golda's mother was impossible, and Sheyna took after the mother," says Zuckerman.

CHAPTER SIX

To be or not to be is not a question for compromise.
Either you be or you don't be.

In retrospect, the implications should have sent a more tremulous shudder through the *yishuv*, the boycotts of Jewish businesses, the prohibitions against land ownership, and the Nuremberg Laws that stripped Jews of their German citizenship. But products of Eastern, not Western Europe, *yishuv* leaders had the ingrained anti-Semitism of Polish peasants and the blood libel of the Ukrainian petty bourgeoisie as their points of reference. Adolf Hitler's rantings about the Jewish peril sounded too familiar to panic men and women born in Minsk or Plonsk or Bobrinsk during the era when the Russian secret police began circulating *The Protocols of the Elders of Zion*. To them, even *Kristallnacht*, that three-day spate of coordinated attacks on Germany's Jews, looked like a German version of Kishinev, or a dozen other pogroms.

Unlike the Jews of Germany, Golda, Ben-Gurion, and the rest of the *yishuv* political elite were inured to anti-Semitism's most violent manifestations. They expected Diaspora Jews to be murdered, to lose their rights, to be humiliated simply because they were Jews. That certainty, after all, was the Zionist raison d'être.

Coming to grips with the reality that this outbreak of anti-Semitic horror was of a more lethal order was a gradual process, made all the more painful by the realization that it would mean an end to all their steady progress, to cautious diplomacy, tree planting, and road building. The shift from decades of single-minded focus on the gradual rebuilding of Palestine to turning it into an immediate safe haven was wrenching. But too many Jews were in peril. They needed a Jewish homeland. Immediately.

*   *   *

Hitler had been chancellor of Germany for three years when the British sent Lord Earl Peel to the Middle East to investigate what had become its intractable Palestine dilemma. No matter how high London set the annual Jewish immigration quota, Zionist partisans denounced them as anti-Semites; no matter how far they lowered the bar, the Grand Mufti of Jerusalem vilified them as Jew-loving Arab haters. Although no prior investigatory commission had found any way out of the impasse, the Foreign Office seemed to believe that Peel, an experienced colonial administrator, and the five members of the team might pull a rabbit from his diplomatic headgear.

During the commissioners' sixty-six meetings, held in late 1936 and early 1937, the Peel Commission heard nothing that hadn't been said a dozen times before. Speaking for the Zionists, Ben-Gurion and Weizmann, the British scientist who'd become the leader of the World Zionist Organization, pleaded for an increase in Jewish immigration quotas, assuring the investigators that a modus vivendi with the Arabs would develop if neither Arab nor Jew felt dominated. And the Grand Mufti demanded an immediate end to the "experiment of a Jewish national home" in favor of an independent Arab state.

Professor Reginald Coupland, an Oxford don, listened closely to the Grand Mufti's response when he was asked what would become of the

400,000 Jews already in the country if Palestine became an Arab state. The Arab new nation couldn't possibly absorb them, the Mufti responded. They would have to be expelled.

Coexistence wasn't an option, Coupland concluded. "Surgery" was imperative. What he meant became clear when the commission issued its final report in July 1937. "The problem cannot be solved by giving either the Arabs or the Jews all they want. The answer to the question which of them in the end will govern Palestine must be neither. . . . Partition offers a chance of ultimate peace. No other plan does."

*   *   *

Ben-Gurion could hardly contain his excitement at the Peel proposal for a Jewish homeland in a piece of Palestine. "We are given an opportunity which we never dared to dream of in our wildest imaginings," he enthused at a Mapai Central Committee meeting. Golda, who rarely disagreed with him, was horrified. The proposed Jewish homeland was less than a quarter of historical Palestine, just the Galilee, the Jezreel Valley, and the central coastal plain. And that small piece of territory would be cut in half by a strip, including Jerusalem and Bethlehem, that the British proposed assigning to themselves. The rest would be attached to Transjordan, which the British had severed from Palestine in 1921.

"Some day my son will ask me by what right I gave up most of the country and I won't know how to answer him," Golda complained.

Three weeks later, she and Ben-Gurion battled it out over partition at the Twentieth Zionist Congress in Zurich. By then, the British Parliament had accepted the principle of partition as laid out in the Peel report, and many of the Zionist delegates assumed that the division of Palestine was imminent. Ben-Gurion painted the choice as stark and clear: a Jewish majority in a small Jewish state or a Jewish minority in an Arab Palestine.

Golda dismissed that formulation out of hand, calling the partition plan a "grotesque proposal." It wasn't just that she could not let go of her vision of an *Eretz Yisrael* large enough to welcome millions of Jews. More attuned to the nuances of the British, she suspected that if the Arabs rejected the plan, the British would forget about partition but use the Zionist agreement to it as a wedge against them in future negotiations.

To stop Ben-Gurion's pro-Peel steamroller, Golda made common cause with an odd alliance of religious Zionists appalled by the notion of giving up what God had set aside for the Jews, left-wing Zionists dreaming of Arabs and Jews living harmoniously in a binational state, and realists frightened that a tiny state could neither assimilate all the Jews who needed a new home nor resist the inevitable hostility of the Arab state to be created next door. Let's accept the concept of partition, she argued, but we must reject the Peel specifics.

Ben-Gurion, however, was in the thrall of his vision of a homeland, no matter now minuscule. For him, the only question remaining was how the Jewish homeland should deal with the 250,000 Arabs who would be left inside its borders. The Peel Commission had suggested transferring them to the Arab state, and Ben-Gurion concurred.

Golda had no problem with transfer in *concept*. "I would agree that the Arabs leave *Eretz Yisrael* and my conscience would be clear," she said. After all, she continued, the Arabs don't need Palestine; there were plenty of other countries the Arabs could call home. But she knew that the Arabs would never accept transfer. Wishful thinking, she called Ben-Gurion's plan. Dangerously wishful thinking that would inflame the Arabs.

After two weeks of stormy debate, nearly two-thirds of the Congress delegates accepted Golda's formulation, rejecting the specifics of the Peel proposal without repudiating the principle of partition. But as Golda had predicted, all the angst was for naught. The Grand Mufti had already flown to Berlin and the leaders of the surrounding Arab countries were declaring the preservation of Palestine as an Arab country as "sacred duty." Prime Minister Neville Chamberlain formed another commission, which concluded that partition was "impracticable."

*   *   *

Night after night, from the small balcony of her apartment overlooking the sea, Golda watched and worried. Had the boat been stopped by the British before its passengers could glimpse the shore of their Promised Land? Had the rickety vessel been caught in a storm and torn apart by an angry sea? So much could go wrong. Too much already had.

In the year since the fruitless partition debate, the Nazis had entered Austria, home to 200,000 Jews. Hundreds of shops had been looted, Jewish leaders arrested, old women impelled to clean streets with raw acid. All over the expanding Grand Reich, the official noose had tightened; poised to add Czechoslovakia, with 180,000 Jews, to the Nazi empire, Hitler stood virtually unopposed.

Day after day, Golda sat with her colleagues from the Histadrut and Mapai trying to figure out how to reach out to the Jews of central Europe. But what could they tell them? Come to Palestine? The Nazis were more than willing to get rid of Jews, but the British had virtually slammed the door on their entry into Palestine. Every year, the immigration quotas dropped, from 61,800 in 1935 to 29,700 in 1936 and a meager 10,500 in 1937.

Every afternoon, she donned a proper dress and played the perfect lady in her negotiations with the British as the intermediary between the *yishuv* and the Mandate authorities, beginning with the colonial officials to help save European Jewry. "Who else did we have?" explained Gershon Avner, who later worked for Golda in the Foreign Ministry. "Golda was one of the few who could speak fluent English, who could cope with British officials, who had the confidence of our leadership. . . . And she had the added capacity to explain Jews and Jewish Palestine to non-Jews, most of whom knew nothing about us."

But at night, knowing how lightly the British took her treaties, Golda watched from her balcony, not as a party or Histadrut official, but as a member of a secret network, the Mossad, set up to smuggle Jews out of Europe. From an office on Marc Aurel Street in Vienna across the street from Gestapo headquarters, organizers searched for seaworthy ships, captains and crews and ports from which they could embark without raising

British suspicions. They became masters of bribery, of forging consular stamps, of locating what one of them called "honest" crooks.

On Golda's end, members of the network traveled up and down the coast scouting out the least obvious places to land. They forged identity cards and arranged housing at kibbutzim from which the new arrivals could disappear into the population. Then Golda watched, for dilapidated cargo ships, ramshackle riverboats, and barely seaworthy fishing vessels that had snuck out of inlets on the Greek coast, or Yugoslavian ports crammed with sixty or one hundred, or as many as four hundred Jews from Germany or Poland, Austria, Romania, and Czechoslovakia.

On good nights, the nights when the ships slipped in past the British, the heaviness in her heart lifted as she radioed the central Mossad office, and then raced down to the beach, forged papers in hand, to welcome the desperate immigrants.

Ben-Gurion was staunchly opposed to the Mossad's activities. The numbers they could rescue were too small to offset the harm to the Zionist relationship with the British, and he believed that without the British, the Arabs would wipe out all they had built. "We shall never be able to fight both the Arabs and the British," he declared.

Ben-Gurion was thinking strategically. Golda followed her gut. For her, the numbers weren't important. It was the doing, her only relief from the grim realization that the world was willing to let Europe's Jews die, her only respite from the image of her father's helplessness in the face of a pending pogrom.

"We know, we mothers, that there are Jewish children scattered everywhere in the world, and that Jewish mothers in many different countries are asking for only one thing," she wrote in the newspaper of the Women's Labour Council. " 'Take our children away, take them to any place you choose, only save them from this hell!' "

*   *   *

The Hotel Royal at Evian-les-Bains, the gayest resort on Lake Geneva, was a bizarre setting for Golda to broadcast that plea to the world. But in the arcane world of international affairs, diplomatic gatherings devoted to massacres and other human disasters rarely occur on the mean streets or near the killing fields under discussion. The elegant old spa on the French south shore of the lake across from Lausanne, then, was the site chosen for delegates from thirty-two countries—the Western European powers, Canada, New Zealand, Australia, South Africa, and twenty Latin American nations—to gather in the summer of 1938 in response to President Franklin Delano Roosevelt's call for an international solution to the growing catastrophe of Hitler's expulsion and oppression of the Jews of the Reich.

Neither the refugees in question nor their envoys had been given a seat at the horseshoe conference table. Weizmann had fought for a delegate from the Jewish Agency, the *yishuv*'s official representatives under the League of Nations Mandate. But the British had objected. Golda, then, was relegated to the status of observer.

She had not yet given in to despair. When Menachem and Sarah complained that she was leaving again she'd promised them, "This will not be just another conference. It may turn out to be one of the most important conferences in all of Jewish history."

So she listened intently to the opening speeches, momentarily heartened by all the sympathetic words about the "millions of people . . . actually or potentially without a country." Then the sessions deteriorated into two days of polite arguments between the United States and France over who would chair the sessions, dashing any hope she'd had that the congress would offer her people any relief. Never taking a break during the nine days of sessions, she sat "disciplined and polite," in her own words, caught between rage, frustration and horror.

The Swiss delegate bragged about his country's tradition of helping political refugees, never mentioning that Switzerland had already decided to block the immigration of Austrian Jews. With 200,000 refugees already on French soil, France's spokesman lamented that his country had "reached, if not already passed, the extreme point of saturation." The Netherlands advised that the climate in its colonies was "unsuitable," and the Canadians, Peruvians, Paraguayans, and Brazilians offered to welcome refugees who were farmers, a suspect invitation since virtually all the refugees in question were urban. Colonel T. W. White, the Australian minister for trade, remarked with striking candor, "It will no doubt be appreciated that, as we have no real racial problem, we are not desirous of importing one."

The United States announced that it had raised its immigration quota from greater Germany to 27,370 people a year, but the quota was already filled until 1940. Lord Winterton, head of the British delegation, noted that England was already "fully populated and is still faced with the problem of unemployment." The government would, he promised, consider admitting some refugees to British colonies and territories. Those territories, however, did not include Palestine. Winterton was speaking of Kenya and Northern Rhodesia.

The gathering, reported the *New York Times*, had the air of "a poker game" and the stakes could not have been higher. More than half a million Jews were already living under Hitler's command, and experts estimated that their evacuation would take four years. No one dared mention Poland, with four million Jews. Excluding Russia, a total of six million Jews lived east, north, and west of Switzerland. Where would they go?

Golda haunted the grounds of the hotel asking that question of every delegate she could find, grabbing them as they entered the theater or waited to ride on the funicular. "I wanted to yell, 'Don't you know that these numbers are human beings,'" she said. But, by then, she knew no one was listening.

When the delegates finished their work, Golda appeared before a press conference in an ornate private dining room at the hotel. It was her first contact with the world of international diplomacy, and she was angry and scared. "We have very little bread," she told the reporters. "However, we will share the crumbs of that bread with [the refugees]." The journalists peppered her with questions, but it was clear that they felt sorry for Golda, one woman standing against the tide of the history and the indifference of the world, one Jew safe from the pending Holocaust.

Their sympathy grated at her. "There is one thing I want to see before I die," she said, "that my people should not need expressions of pity any more."

* * *

Arm in arm, they filed through the streets of Tel Aviv, Golda and Remez, Ben-Gurion and Katznelson, behind them hardened kibbutzniks, the denizens of the city's café scene, and Orthodox Jews who normally flinched at contact with the nonreligious. In Jerusalem, Haifa, and dozens of small settlements, they marched as well, 175,000 members of the *yishuv* denouncing the British in one voice.

It was May 1939, and the government of Prime Minister Neville Chamberlain was about to issue its latest White Paper, its final solution to two decades of strife in Palestine. Jewish immigration would be capped at 75,000, spaced out over five years, and Jews would be barred from purchasing property outside a narrow strip. After that period, no Jew would be permitted to immigrate without Arab permission. Within ten years, Britain would allow the creation an independent state of Palestine. "His Majesty's Government therefore now declare unequivocally that it is not part of their policy that Palestine should become a Jewish State," it concluded.

The new policy openly violated the conditions of the mandate given Britain by the League of Nations, but the League was dying, so Britain traded in the mantle of trustee for that of colonial power. "Another victory for Hitler and Mussolini," the British Labour Party dubbed the White Paper. What is the watchword now? opposition members asked, providing their own answer. "The watchword is, 'Appease the Arabs, appease the Mufti.'"

The strike was a first step, a steam vent for public anger and an attempt to hold the revisionists, the radical right, at bay. Their retort to London had been the storming of the district commissioner's office and the smashing of the windows of a British-owned shop. By the time they were done, a British constable lay dead and more than 100 Jews were wounded.

For the nascent *yishuv* government, the strike was only the first step. They were already organizing the community for a bold display of passive resistance. Rather than shrinking into the Jewish enclave the British had carved out for them, they planned to build settlements all across the country and flood the shores with illegal refugees to claim all of Palestine as their own.

For nine months, Golda had been buffeted from depression to terror. But the dejection that had haunted her since the Evian meeting had given way to anger. "It is in the darkest hour of Jewish history that the British propose to deprive the Jews of their last hope and close the road back to their homeland," read the official *yishuv* response that Golda helped draft. They would not allow it.

"It is inconceivable that we shall not succeed in our work here, in our toiling to defend every single settlement, even the smallest, if we have before us the picture of thousands of Jews in the various concentration camps," she wrote in *D'var HaPoalot*, the magazine of the Women Workers' Council. "Therein lies our strength."

Inflamed with anger and steely purpose, Ben-Gurion began building a secret arms industry, reorganizing the Haganah, and mobilizing the entire adult *yishuv*. Don't sneak refugees in at remote landing points, he urged. Escort them off their boats in the middle of Tel Aviv, defended by armed guards.

He's going too far, some whispered. Jews are a peaceful people. How can so few fight the mighty power of the British Empire and still defend themselves from the wrath of the Arabs?

Not Golda. *"Ein brera,"* she declaimed, a phrase that became the community watchword. "We have no alternative."

*   *   *

The final plan for the *Ma'avak*, the struggle the *yishuv* planned to wage against both the British and the Arabs, was scheduled to be hammered out at the Zionist Congress in Geneva in August. Normally, when Golda was preparing to go overseas, Sarah and Menachem begged her not to leave. "This time, we asked nothing," said Menachem. "We knew for ourselves that if she was needed in Geneva, that was where she had to be."

Ben-Gurion was at his finest, the tenacious prophet boldly proclaiming, "The Jews should act as though we were the state in Palestine and should so act until there will be a legal Jewish state." The overseas Zionists who footed the bills applauded wildly although they thought it lunacy to believe that the tiny *yishuv* could challenge the combined might of the British Empire and the Arabs. Yes, we've been betrayed, they counseled. But we must be prudent.

"Prudence" was a word Golda could no longer abide. Millions of Jews were in danger. Consequences be damned.

In the midst of the arguments, conducted during hours of general debate, in the hallways and over tea at Lenin's Café, Stalin and Hitler agreed to a nonaggression pact, and the murmur of war swelled to a rumble. "If war breaks out, don't worry, I'll come home at once," she wrote Menachem and Sarah. One week later, Germany invaded Poland.

Still Golda lingered in Europe while world Zionists devised ways to stay in touch during the pending war and Labor Zionists debated the gut-wrenching choices confronting them: Should they cease their struggle against the White Paper and throw their support to England in the hope that the British military could save the Jews that diplomats in London were refusing to shelter from harm? Or should they continue battling the British to bring European Jews to safety in Palestine?

It was an impossible dilemma, so they settled on an impossible strategy: "We shall fight the war as if there were no White Paper, and the White Paper as if there were no war," Ben-Gurion vowed.

"It was a very nice slogan, but not so simple to implement," Golda later quipped. They had to rouse the Jews of Palestine to enlist in the British army to help defeat Hitler and simultaneously organize them to defy the British army locally by sneaking in illegal immigrants.

The first priority was to flood the British army with Jews. Who better to free Jews in danger than their brethren from Palestine? Golda submitted to the British, suggesting the formation of a Jewish army, equivalent to the Free Poles. Golda was not without guile: every Jew who fought in the world war would return more experienced, more combat-hardened to wage the battle for a Jewish state.

Community leaders established a draft, or as much of a draft as a non-government could institute. During the first five days of their recruitment drive, more than 135,000 men and women signed up to save the European Diaspora from annihilation and to defend Palestine from a feared German assault on the Suez Canal.

Distrustful of the Jews and fearful of the Arabs, the British refused most of them and established a system of parity, one Jewish volunteer for every Arab. Theirs was a naive pipe dream, a colonial master expecting the colonized to remain loyal. The Grand Mufti had already cemented ties to the Nazis, who were sending him both money and supplies. And even without that alliance, few Arabs would have stepped up to help the British, who they hated as much as the Jews did. The White Paper, which they called the Black Paper, hadn't mollified them. It offered them too little, too late.

* * *

Less than a week after the publication of the White Paper, the steamer *Atrato* sailed out of Constanta, Romania, hanging low in the water, 410 men, women, and children penned up on the narrow deck in the hope of reaching Palestine. Near Cyprus, they ran into a storm and were forced to drop anchor in a bay. While waiting for the winds to calm, they received a warning by wireless: APPROACH TEL AVIV. STOP AT A DISTANCE OF 2 MILES AND LOWER THEM IN BOATS. WARN THEM TO BE READY FOR ANYTHING.

The tension was thick the next night when the *Atrato* made its way toward Tel Aviv. Just as the coast seemed to creep over the horizon, the glaring searchlights of a British cruiser blazed through the darkness. "Proceed to Haifa and your passengers will be disembarked unharmed," a voice from the loudspeaker of the HMS *Sutton* announced, breaking the stillness. "Then they shall be detained."

That was the final voyage of the *Atrato*, which had already successfully smuggled 2,414 Jews into Palestine under the noses of the British.

The *yishuv* was already in an uproar about the detention of refugees. In April, the mandatory authorities had seized the Greek cattle boat *Assimi*, carrying 263 passengers. Twelve days later, when they tried to force the ship back out to sea, without food or medicine, the passengers had vowed to die rather than return to Europe. As hundreds of Jews on board prayed, the residents of Haifa called a sympathy strike, marching with signs reading, "Open the gates to Jewish illegals." Unfazed, the British pushed the rickety vessel out of Haifa harbor, forcing the passengers back to Europe.

Publicly, the British wouldn't admit that they were trying to mollify the Arabs and flailed around looking for more noble motives: Maybe Nazi spies will sneak in among the Jewish refugees, they told Golda. The ships you're using aren't safe and all the seaworthy ones are needed for the war. The *yishuv* scoffed and redoubled its efforts.

Every week, the leaders of the smuggling network plotted new strategies for evading the British, telegraphing coded instructions across a far-flung radio network. One center directing the illegal traffic was set up in the apartment of a Haganah fighter they knew was immune to British searches. Code-named Pazit, it was Golda.

"There is no Zionism save the rescue of Jews," she proclaimed at a meeting of the Executive Committee of the Mapai.

The illegal immigration occupied a tiny part of Golda's growing portfolio. As one of the five or six most important *yishuv* leaders, she had a hand in almost everything. Appointed to membership in the British War Economic Advisory Council, she sold new austerity and rationing measures to the Jewish population. As the new head of the Political Department of the Histadrut, she dealt with routine labor disputes, kept Kupat Holim going, and trekked through factories, railways, and ports to rally support for a new tax to fund the Haganah.

Maintaining a working relationship with the British was her greatest challenge. But charming and cajoling the authorities into flexibility was an impossible task, especially when Golda was simultaneously stumping the country to inflame the population against them. "For twenty years we were led to trust the British government," she remarked in a typical speech. "But we have been betrayed."

It was an odd sort of schizophrenia, inherent in the *yishuv*'s need to fight the British on domestic issues while maintaining an international alliance with them, but Golda proved adept at judiciously balancing wheedling, pleading, and threatening to preserve it. When the British followed the White Paper immigration quotas to the letter, factoring in their estimates of the number of illegals who'd slipped by them, Golda first wielded guilt. Then, playing on their traditional chivalry, she pleaded in the most womanly possible fashion—and no one in the *yishuv* underestimated Golda's ability to use femininity to advantage.

When they were recalcitrant, she turned to public humiliation, deriding them at marches and protests well supervised by British troops: "They . . . fight like lions against the Germans and the Italians, but they won't stand up to the Arabs," she said mockingly. In an offhanded manner, she referred to the camps where they kept illegal refugees as concentration camps. But the next morning, she'd cordially show up for tea with British officers and colonial officials, careful to preserve a facade of civility.

The conflicts went well beyond immigration. On the eve of Shavuot, the Va'ad sent seven teams of young people into isolated areas of the country well outside the proposed "Jewish" zone in the dead of night to erect watchtowers and stockades. By the time the British awoke, it was too late for them to destroy the settlements since British law defined an existing community that could not be destroyed as one with a tower and a wall. Almost fifty settlements were erected along that model to strengthen the Jewish presence, often so far off the road that building materials had to be hauled in by donkey.

"Our way in this country, our Zionist war, must be different from the war of every other nation," Golda preached. "Any other nation entering a war doesn't dream . . . that in the war period itself it will have to build, to

construct, to develop. We cannot afford to wage war without construct-
ing. Such a war has no meaning for us, no reason.

"Britain is trying to prevent the growth and expansion of the Jewish
community in Palestine, but it should remember that the Jews were here
two thousand years before the British came."

*　　*　　*

The British couldn't take the insurrection lying down. It wasn't just that
the Jews were flouting the law; they had begun to act as if they were no
longer subjects of the Crown. As in so many anticolonial rebellions, the
blend of arrogance and defiance that drove the rebels was almost unbearable for men who thought of themselves as masters of the realm. A crack-
down was as essential to their sense of authority as it was to their control
over Palestine.

When a Jew named Eliahu Sacharoff was caught with thirteen bullets,
one more than the law permitted, he was sentenced to seven years in jail,
although an Arab found with a rifle and eighty-six bullets two days later
received six months' imprisonment. Suddenly, kibbutzim were searched
for illegal weapons and travelers stopped and deprived of the arms they
needed to defend themselves. The *yishuv* was put on notice that Jews
could no longer flout the law.

As the head of the political department of the Histadrut, in October
1943, Golda took the stand at the trial of two men rounded up during this
period, charged with the illegal possession of 300 stolen rifles and 105,000
rounds of ammunition.

"You are a nice, peaceful, law-abiding lady, are you not?" Major Baxters, the prosecutor, asked her.

"I think I am," Golda responded flatly.

Then Baxters read from a speech Golda had delivered three years earlier, quoting her as saying, "We never taught our youth the use of firearms
for offense, but for defensive purpose only. And if they are criminals, then
all the Jews in Palestine are criminals."

"What about that?" he pressed.

"If a Jew who is armed in self-defense is a criminal, then all the Jews in Palestine are criminals."

"Were you yourself trained in the use of arms?"

"I do not know whether I am required to answer to that question. In any case, I have never used firearms."

"Have you trained the Jewish youth in the use of firearms?"

"Jewish youth will defend Jewish life and property in the event of riots and the necessity to defend life and property. I, as well as other Jews, would defend myself."

Again, Baxters quoted the witness. "There are not enough prisons and concentration camps in Palestine to hold all the Jews who are ready to defend their lives and property." Did she say that? he asked.

"If a Jew or Jewess who uses firearms to defend himself against firearms is a criminal, then many new prisons will be needed," Golda answered.

"And are they ready to do all that you said in your speech, which I read before?"

"They are prepared to defend when attacked," she answered. "Everybody in Palestine knows, as do the authorities, that not only would there have been nothing left, but that Jewish honor would have been blemished . . . if brave Jewish youths had not defended the Jewish settlements."

When Golda was turned over to the defense counsel, he led her back into the history of Jewish self-defense, beginning with the massacre of Jews in Hebron a decade earlier.

"That was in 1929, and the same thing happened the same year in Safed," she said. "In 1936, there was a night of terrible slaughter in the Jewish quarter of Tiberias, and all this could only happen because there was no Haganah in those places."

The president of the court cut her off. "I ask you to limit yourself only to what concerns this case and not go backwards, or otherwise we'll soon be back to a period of two thousand years ago."

Golda interrupted him. "If the Jewish question had been solved two thousand years ago . . ."

"Keep quiet!" the president ordered her.

"I object to being addressed in that manner."

"You should know how to conduct yourself in court."

"I beg your pardon if I interrupted you, but you should not address me in that manner."

*   *   *

"Zionism and pessimism are not compatible" was one of Golda's favorite phrases, but by 1943, even die-hard Zionist optimists were rattled. Reacting to British pressure, European countries had cracked down on the smuggling of Jews from their ports, virtually halting illegal immigration. And after the surrender of France and the German alliance with Italy, Palestine itself seemed in imminent danger of attack by the German and Italian armies. The Haganah drew up its own defense plan and launched a mobilized fighting force, the Palmach. If the Germans come, Ben-Gurion decreed, all Jews in the British army should desert with their weapons, the *yishuv* should gather around Haifa, fortify Mount Carmel, and fight to the death.

Every piece of news from Europe left them to wonder if anyone in the European Diaspora would survive to join them. In a forest near Riga, 35,000 Jews murdered. In Odessa, 19,000 Jews burned alive by Romanian and German troops. Then, in August 1942, Jewish leaders in Switzerland sent word about rumors of the extermination of Jews deported from Warsaw and Lvov and gas chambers built in concentration camps.

When that information was brought before a Histadrut executive meeting, most of its members refused to believe it. Even for Jews who'd lived through pogroms, such systematic slaughter seemed fantastic. Golda, however, was certain that those first suggestions of the nature of the Holocaust were true.

"What can we do, first of all, to let them know that we know and that we're doing something about it?" she asked. "And second, we need to do something."

Clinging to denial, her comrades couldn't face planning for such a drastic scenario, so they dismissed Golda's pleas. That night, she returned home to her new apartment in a workers' cooperative by the sea, put on her housecoat, and said to the children, "It's bad enough that the rest of the world doesn't and won't help us, but our own people, some of them, just don't understand what is at stake."

Fixated on making contact with the Jews in Europe, Golda didn't distract herself trying to convince her colleagues that what they were hearing was true. Instead, she put out feelers everywhere, both within Palestine and abroad: How can we open a line of communication to Jews inside Nazi Europe? What can we do to support whatever underground exists?

Finally, a Histadrut emissary reported to her that he'd found an "unsavory contact" who had offered to become the *yishuv*'s liaison with Polish Jewish freedom fighters for 75,000 British pounds. Golda talked the Histadrut and the Jewish Agency into giving her part of that money. Then she went on a mission to raise the rest. "How do you know the money will ever reach the Jews?" asked a wealthy businessman she approached. Golda didn't dance around the issue. "They'll be lucky to get 10 percent," she said. "But who knows what weapons or influence that small sum might buy?"

Simultaneously, Golda barraged the British with new pleas for relaxed immigration quotas, for help rescuing Europe's remaining Jews, besieging every soldier, official, or visitor from London. When she described the horrors of the Warsaw Ghetto and the mounting number of mass graves to one officer, a man she liked, he cut her off. "I could see in his worried, rather kind blue eyes that he thought I had gone quite mad," she said. "He told me not to believe everything I heard."

There was, of course, little that the *yishuv* could do, and the sense of helplessness was excruciating. In April 1943, the Histadrut Executive Council received a message from Warsaw, through one of the convoluted networks they'd established with Poland. Should we take a last stand? Ghetto resistance fighters asked. By then, more than 300,000 ghetto residents had been packed into cattle cars and shipped to Treblinka. Only 55,000 or so remained, and they knew that their deportation was imminent.

Surrender or die was the oldest Jewish dilemma, and *yishuv* leaders were alternately exhilarated and chilled by the image of Masada, where a band of first-century Jews chose the latter course when Romans stormed the fortress where they had taken refuge. Others debated how to reply, but Golda refused to consider sending ghetto organizers advice. "How can we, in Tel Aviv, tell them to die?" she screamed in agonized frustration. Before anyone could respond to her outburst, an underling interrupted the meeting. The news had come on the radio that the ghetto uprising had begun. With a handful of pistols, seventeen rifles, and homemade Molotov cocktails, 750 young people had fired on 2,000 well-armed German troops, forcing them to retreat.

In their need to take action, Weizmann and Ben-Gurion still tried to induce the British to launch a Jewish brigade in Palestine itself. But Golda couldn't take her mind off Europe. The ghetto uprising persuaded her that cells of resistance must exist, not only in Poland, but also in Russia, Germany, and Hungary. Each act of sabotage, each refusal to be carted off to death would spread hope and, perhaps, create some sense of insecurity among the Nazis. The *yishuv* needed to find a way to send its own fighters to assist.

The Haganah was training a small band of men and women born in Europe to parachute in behind enemy lines, and they offered to help Allied prisoners of war to escape, to make contact with partisans and sabotage British-chosen targets. But knowing that the Haganah had its own agenda, the British refused to send them in. With her singular tenacity, Golda pushed and prodded until the British relented, although they selected only a token group of thirty-two. The parachutists were given strict orders to confine themselves to Allied war orders. But none of them was about to ignore their other brief, given to them by the Haganah, to find the Jewish resistance.

But no matter what Golda did or how hard she worked, a crippling sense of helplessness ate away at her. Her migraines kept coming, and she was exhausted. But she couldn't stop. She had to figure out how to handle complaints about the rationing of everything from sugar and oil to shoes, how to get to meetings during blackouts in Tel Aviv. Rommel was moving toward Palestine from North Africa, Jews were being gassed in Poland, and she had to do something, no matter how small.

"I have sometimes wondered how we got through those years without going to pieces," she later wrote. "But perhaps physical and emotional stamina is mostly a matter of habit, and whatever else we lacked, we did not lack opportunities for testing ourselves in times of crisis. . . . One can always push oneself a little bit beyond what only yesterday was thought to be the absolute limit of one's endurance. . . . I don't recall ever having felt 'tired' then, so I must have gotten used to fatigue. Like everyone else, I was so driven by anxiety and anguish that no day (or night, for that matter) was long enough for everything that had to be done."

CHAPTER SEVEN

*Those who don't know how to weep with their whole
heart don't know how to laugh either.*

The colored lights danced off the fountain in Tel Aviv's Dizengoff Circle for the first time since 1940, as a mile-long procession wended its way into the center of the crush outside the Habima building. A few hours earlier, the chief of the operations staff of the Wehrmacht had signed an unconditional surrender on behalf of Germany, and the Jews of Palestine, like the citizens of London, New York, Paris, and Moscow, celebrated May 7, 1945, as the end of the long war.

But the gaiety was undermined by the massive black column erected at the center of the gathering, the burning torch atop it a stark reminder that world Jewry could not yet begin to count its dead.

More than 600,000 Jews remained in the newly liberated concentration camps, and the *yishuv* was intent on bringing them "home."

"At this solemn hour we fervently trust that in the shaping of the new international order, justice may be done to the Jewish people and that its age-long yearning for national rehabilitation in the land of its origin may be fulfilled," wrote Moshe Shertok—who later changed his surname to Sharett—to Lord Gort, commander in chief of the British Expeditionary Forces in Western Europe on behalf of the Jewish Agency.

Although the cynics and skeptics—and after the Nazi obliteration of the Jewish communities of Europe, they were legion—warned that the 600,000 Jews of Palestine could not count on the British to rethink the White Paper of 1939, few really believed that London would remain unmoved by the Holocaust. And when the Labour Party, led by Clement Attlee, ousted Winston Churchill in elections held two months after the end of the war, even the skeptics began believing that the doors of Palestine were poised to swing open. After all, during ten consecutive party conferences, Labourites had passed resolutions in support of a Jewish homeland.

Six months later, when Golda and eleven other Jewish leaders were called to the Jerusalem office of John Shaw, chief secretary for the Mandatory Government, for a preview of the new government position on Palestine that Ernest Bevin, the recently installed foreign minister, would announce in the House of Commons the following day, they anticipated that a new era was about to begin. But reflecting none of Labour's long solidarity with Zionism, the new policy bore a striking resemblance to Churchill's.

> *Civilization has been appalled by the sufferings which have been inflicted in recent years on the persecuted Jews of Europe. . . . On the other side of the picture the cause of the Palestinian Arabs has been espoused by the whole Arab world and more lately has become a matter of keen interest to their 90,000,000 co-religionists in India.*
>
> *We have inherited, in Palestine, a most difficult legacy and our task is greatly complicated by undertakings, given at various times to various parties, which we feel ourselves bound to honor. Any violent departure without adequate consultation would not only afford ground for a charge of breach of faith against His Majesty's Government but would probably cause serious reactions throughout the Middle East and would arouse widespread anxiety in India.*

When he was finished reading, Shaw grimly informed the twelve Jews before him that any resistance to the government's policy would be met with "force of arms."

When news of the British betrayal got out, 50,000 people gathered at the parade grounds in Tel Aviv, jeering at every mention of the Labour Party from the podium. A gang of three hundred young people stormed British government offices, hauled out papers and furniture, and created a pyre. The Histadrut council in Haifa plastered the city with posters reading, "The walls will be destroyed, and there will be immigration, even over our dead bodies."

Golda stood before the Workers' Committee of the Histadrut and declared war:

> Jewish life is precious but we do not want to be slaves to another nation. . . . not only because every man wants to be free in his own country, but because we have learned that for Jews, living as a minority in someone else's nation is not real life and, in the end, in that position, we would be killed. . . . We need freedom, to be independent. . . .
>
> We understand that in the Land of Israel today we don't have power equal to that of our enemies. . . . We don't have a naval fleet. We don't have an army or airplanes. We don't have the secret for the atomic bomb. We only have sacrifices, the people who have died, the people who are hungry and sick facing death in DP camps. . . . But we also have millions of Jews who are ready to live and not willing to accept this. We have a yishuv of 600,000 Jews. Yes, we are a little weak. But we will fight, and we will succeed. . . .
>
> We will not accept their document, and we will not remain silent.

It's hard to imagine how Bevin convinced himself and the rest of Prime Minister Clement Attlee's government that Britain could stay the White Paper course. Hitler had infused world Jewry with a new militancy forged on the anvil of "Never Again." And even without a twenty-four-hour news cycle and crisp video footage of skeletal survivors in striped camp uniforms, the grainy black-and-white images of concentration camps and mass graves printed in newspapers across the globe had shifted international sentiment toward the Jews.

But Bevin was clinging to the last gasps of the British Empire. And in his determination to keep it alive in Palestine, the Zionists would have been nothing more than a pesky inconvenience dismissed with dispatch but for the support they received from President Harry S. Truman.

"The American people, as a whole, firmly believe that immigration into Palestine should not be closed," Truman wrote to Prime Minister Attlee.

Bevin assumed that the Arabists in the U.S. State Department would bring Truman around, and they tried, churning out report after report arguing that giving in to Zionist demands would undermine Arab confidence in the United States. Truman's concession was agreeing to Bevin's request that an Anglo-American commission be formed and sent to Palestine and Europe to investigate the situation.

The Jews of Palestine understood that the commission was Bevin's stalling tactic. So while politely promising to meet with its members, they didn't wait for their arrival before launching a full-scale rebellion. Stage one was Aliyah Bet, a coordinated effort to sneak thousands of Jews out of displaced persons camps in Europe and into Palestine. Let the British try to stop us, *yishuv* leaders said, unable to imagine the Royal Navy shooting men and women still wearing the ragged remnants of concentration camp garb. Stage two was an uprising of civil disobedience, noncompliance, and the sabotage of Britain's colonial infrastructure by the newly formed Jewish Resistance Movement, a coalition of the Haganah, Irgun Zvi Leumi (Etzel), and Lehi (the Stern Gang), the latter revisionist guerrilla groups, Etzel led by Menachem Begin.

The Haganah's alliance with right-wing terrorist groups was a bitter pill for Golda to swallow because she'd been urging their obliteration for almost a decade. Although hardly a pacifist, she had long supported the Labor Party policy of *havlaga*, self-restraint, worried that shedding British blood would backfire.

Etzel, which had seceded from the Haganah in 1931, had long advo-cated offensive actions to shatter British power, and in 1944, after they defied the *yishuv* ruling bodies by moving from advocacy to action, the Mapai political committee had convened to discuss "liquidating the pos-sibility of another such attack in the future," in Golda's words. "I have no moral restraint as to that group," she said.

Nonetheless, the Stern Gang stepped up the violence by assassinating Lord Moyne, the British resident minister in the Middle East. Again, the Histadrut Executive considered what measures they should take, worried both about the *yishuv*'s relationship with the British and about the growing popularity of the Stern Gang and Etzel. As Labor's leaders wrestled with their dilemma, Golda was the hardest liner in the room. "We have to liquidate by killing a few boys," she insisted. Seemingly taken aback, Ben-Gurion asked, "They won't retaliate?" Golda was un-moved. "They bring catastrophe on us, not only on the British officers," she argued. "Therefore, everything is permissible when it comes to them."

Rejecting Golda's drastic recommendation, the Histadrut and Mapai had settled instead on a milder two-pronged approach. Using the full power of their institutions, they launched a propaganda campaign against the Jewish terrorists. And they sent the Haganah to capture as many Etzel and Stern Gang commanders as they could locate, detaining some, turning others over to the British. During that period, dubbed the *Saison*, the hunting season, 20 members of those groups were abducted and held, some for months; 91 were kidnapped and released after interroga-tion; and the names of 700 other individuals and groups were given to the police.

Bevin's new policy, however, changed everything, especially for Golda. While the alliance with the terrorist groups provoked a storm of controversy, she vigorously defended the marriage of convenience:

> *We have no alternative but to follow a new path. . . . I believe that in*
> *a war for Jewish existence, every road must be an option. No road is*

*unethical because there is nothing more ethical than helping those
who survived the Holocaust to remain alive.*

*I have always opposed the policies of Stern and Etzel because I
thought their way would damage and destroy Zionism. But now I am
saying that with common sense and by thinking cautiously before we
act, we should go their way—we have no choice.*

So it began, a cycle of attacks and reprisals that edged Palestine toward
civil war. On November 23, 1945, the British seized the refugee ship *Berl
Katznelson* upon its arrival at Sharon beach. Two days later, the Haganah
blew up coast guard radar stations used to detect illegal vessels. During
December, the united resistance attacked airfields, police stations, and
armories. On Christmas, they fired mortars at the Jaffa headquarters of
the Central Intelligence Division and killed ten people. Two days later,
they hit the police headquarters in Jerusalem, police stations in Tel Aviv,
and a military depot in Tel Aviv, leaving nine British soldiers dead

The British responded by turning Palestine into an armed camp,
sending Bren gun carriers, armored cars with manned machine guns,
and mobile artillery to patrol mountain roads. Bevingrad, as the Jews
called the neighborhood around colonial headquarters, was ringed with
barbed wire. The 60,000 British troops already in Palestine were reinforced by members of the Sixth Airborne unit, paratroopers in distinctive
red berets.

But they could not stop the onslaught of Jewish fury. When the British
became adept at capturing illegal immigrants on the beaches, Aliyah Bet
began unloading them offshore—until the British moved their patrols
out to sea. The government brought in radar to spot the vessels and un-
load their passengers at refugee camps. But the Jewish resistance didn't
give up. One night, the Palmach infiltrated the refugee center in Athlit,
bound and gagged the guards, and released 208 prisoners. "Jews will no
longer tolerate the deportation of their brothers from this country, whatever measures of force are used by Government," announced Kol Yisrael,
the *yishuv* radio station.

Meanwhile, veterans of the Jewish Brigade Group, the Jewish unit of the British army finally formed only months before the end of the war, remained in Europe to organize the *Briha*, the escape, and the Histadrut and Jewish Agency sent dozens more young Jews from Tel Aviv and Jerusalem into detention centers and DP camps to stir up a clamor for immigration and help arrange escape.

Despite their successes, the *yishuv* leadership knew that at any minute they could fall off the tightrope they were walking. Even as they blew up British military bases and faced down the Royal Navy, in London, Weizmann and Sharett, the head of the political department of the Jewish Agency, were trying to negotiate a new policy with the British and, at home, Golda still spent her days haggling with colonial authorities.

But Golda no longer had enough faith in the British to waste time cajoling them. At one point, she heard a rumor that the mandatory authorities were going to use German prisoners of war to work in a military camp and approached John Shaw, the colonial chief secretary, to complain. We don't have enough British workers for the job, he explained. "What's wrong with workers from the Histadrut?" she asked, not hiding the venom she felt. "We have workers." Shaw laughed bitterly. "We can't count on Jewish workers. Whatever they build, they turn around and blow up."

Golda stormed out. "So go depend on your Nazi workers," she said, wondering for the thousandth time how the British could be so insensitive to Jewish sentiment.

*   *   *

On Monday morning, March 25, 1946, Golda donned her usual simple, dark dress and took a seat before the mahogany table in the lecture hall at the YMCA in Jerusalem to testify before the Anglo-American Commission. Ben-Gurion had opposed cooperation with the investigation Bevin had forced on Truman, but Golda had prevailed, arguing that the *yishuv* could ill afford to offend the Americans.

The committee's charge was to investigate how conditions in Palestine might impact on Jewish settlement and to gauge whether the Jews remaining in Europe really wanted to immigrate to Palestine. Before they left London, bluff as ever, Bevin had pledged, "If you come back with a unanimous report, I'll grant a hundred thousand visas."

The Zionists planned their lineup carefully, offering both diplomacy and defiance. Weizmann testified with succinct diplomacy that the only guarantee of Jewish survival was a homeland. Ben-Gurion angrily asserted that the time had come for World Jewry to create an international commission to investigate why anti-Semitism ran so rampant among gentiles. But after three more presentations, about the economic potential of Palestine, the commission members were clearly growing bored.

Then Golda spoke on behalf of the Histadrut, without a single note in front of her:

> *The people . . . who . . . laid the foundation for what we call the labor commonwealth in this country . . . grew up in an environment of persecution, massacres, lack of security, helplessness. . . . It was this generation that decided that there must be an end to senseless living and senseless death among Jews. . . . They came to Palestine because they believed then, as we believe now, as many millions of Jews believe now, that the only solution for the senselessness of Jewish life and Jewish death lay in creating an independent Jewish life in a Jewish homeland.*
>
> *The question has been brought up at these hearings several times about . . . whether the Jewish labor movement was prepared to sacrifice something in order to have large Jewish immigration. Gentlemen, I am authorized on behalf of the close to 160,000 members of our federation to say here in the clearest terms, there is nothing that Jewish labor is not prepared to do in this country in order to meet and accept large masses of Jewish immigration, with no limitation and with no condition whatsoever. . . . Otherwise, our life here, too, becomes senseless.*
>
> *I don't know, gentlemen, whether you who are fortunate enough to belong to the two great democratic nations . . . can . . . realize what it*

means to be a people that is forever being questioned . . . in regard to its very existence. . . . To be forever questioned whether we have a right to be Jews as we are, not better, but not worse than others. . . . We only want that which is given naturally to all peoples of the world, to be masters of our own fate, only of our fate, not of others . . . to live as of right and not on sufferance, to have a possibility to bring up Jewish children—of whom not so many are left now in the world—as our youngsters that were born here, free of fear.

The commissioners were openly moved, but they were just as clearly worried. In a Jewish state, wouldn't Jews dominate or exploit their Arab neighbors? several delegates asked. "It would be foolish to expect that we could live here in comfort and peace with the Arabs in our midst neglected," Golda responded confidently.

But doesn't your use of Hebrew undercut your ability to forge the type of harmony you say you are seeking? asked Sir Frederick Leggett, former deputy secretary of the Ministry of Labour and National Service. "But Hebrew is our language, as English is yours," Golda answered, patiently. "This is what I've been trying to make you understand. Would anyone question the use of French, or Chinese?"

As if to underline the seriousness of what Golda said about *yishuv* commitment to immigration, fifteen hours later the Jewish resistance landed a refugee ship, boldly sailing a vessel into Tel Aviv under a screen of armed attacks on police stations and the waterfront. Gangs of young people blocked the streets of the city to allow the illegal immigrants to escape detection.

While the commissioners debated the future of Palestine, the British made a serious miscalculation that shifted international sentiment more drastically against them. In La Spezia, a small port on Italy's Ligurian coast, the British occupying forces blocked the harbor to prevent the sailing of the ship *Fede*, carrying 1,014 Jewish refugees. The plight of the concentration camp survivors quickly became an international cause célèbre.

"2000 Jews in a Ship that Cannot Leave," trumpeted *Corriere della Sera*, one of Italy's largest newspapers. "The Daughters of Israel marked by the fire of Teutonic barbarians invoke the Promised Land."

The passengers began a hunger strike, and the Aliya Bet organizer on board sent a cable to Prime Minister Attlee holding him personally accountable for any loss of life.

Sharett was caught in a quandary. The Jewish Agency, the diplomats of Zionism, never associated itself openly with "illegal" activity. And Ben-Gurion thought the entire affair was nonsense, a distraction from the important work of moving Jewish refugees out of the camps. But sensing that something important was happening, not only to the passengers on the boat, but in world opinion, Golda convinced her colleagues that the leaders of the Jewish community in Palestine should mount a hunger strike in solidarity with the La Spezia refugees to further publicize their plight.

In the Jewish National Council Building, cots were set up for thirteen strikers, who vowed to eat no solid food, nothing more than tea without sugar twice a day. ("Fortunately," said Golda, "cigarettes were permitted.") Crowds gathered outside to cheer on their leaders by singing Hasidic songs, and the international press flocked to the spectacle.

Still, the authorities in London dithered, and the refugees in Italy announced that if their ship were not permitted to sail, ten refugees would commit suicide on deck in public every day until the British were left with a coffin ship. With the international spotlight broadcasting British obduracy, Prime Minister Attlee finally bowed to international pressure, and after 104 hours without food, the strikers, both in Italy and in Palestine, agreed to call off their action. On May 20, the passengers from the *Fede* were welcomed into Palestine.

By then, however, Golda was too distracted to celebrate. On May 1, the findings of the Anglo-American Commission were published and the commissioners unanimously called for a binational state under a continued British mandate, although they also issued a unanimous recommendation that Britain immediately grant one hundred thousand visas for

Jewish immigration. No one, on any side, paid much attention to the former recommendation since it amounted to maintenance of a failed status quo. But latching onto the visa proposal, *yishuv* leaders demanded that Bevin live up to his earlier pledge. When he refused, they carefully began preparing their rejoinder, maximizing British tension over what the Jewish resistance would do.

Six weeks later, on the night of June 17, the Palmach blew every road and rail bridge connecting Palestine to its neighbors. "In place of every hundred or thousand who will be arrested or who will fall, there will rise other hundreds and thousands," the Jewish leadership vowed in a declaration read on Kol Yisrael. "We are the military Jewish people. We will confront the British government with a choice: accept our vital demands or destroy us. We will not surrender."

Neither would Bevin. Shortly before dawn on June 29, 17,000 British troops backed by tanks and armor sealed off Palestine's borders, blocked internal roads, cut telephone service, and imposed a curfew. Then, more than 100,000 soldiers and policemen fanned out across the country. Troops stormed the Jewish Agency building, ransacked the offices, loaded all the documents onto three trucks, and drove them to the King David Hotel, which housed the government secretariat and military command.

Emergency regulations suspended virtually all civil rights. Those found harboring illegal immigrants could be jailed for up to eight years. The penalty for carrying arms without a license or membership in a group whose other members had committed crimes was set as death. The high commissioner was given the authority to order the detention of anyone for an indefinite period. Organizations could be banned, civilians tried before military courts, and property confiscated if the owner was suspected of having broken the regulations.

Military units swept into more than two dozen settlements to search for weapons. At Kibbutz Yagur, members were dispersed with tear gas and incarcerated inside barbed-wire enclosures set up by the soldiers. Those who resisted were sprayed with hot oil. Day after day, the British tore the kibbutz apart. When they discovered 300 rifles, 100 two-inch mortars, 5,000 grenades, and 78 revolvers, they arrested all the male members.

Simultaneously British troops broke into homes, stopped cars, and grabbed men at random off the streets in their search for community leaders. At 4:15 A.M., a policeman accompanied by an army officer and two military vehicles filled with troops surrounded the home of Rabbi Judah Fishman-Maimon, chairman of the Jewish Agency Executive, and arrested him. Explaining that religious law prohibited him from riding in a car on Sabbath, Rabbi Fishman offered to walk to prison. Rather, they picked him up and shoved him into a vehicle.

By the end of the day of what came to be called Black Sabbath, Fishman, Sharett, Remez, and most of the rest of the *yishuv* leaders were in custody at a special detention center—the VIJ camp, the Very Important Jews camp—at Latrun, a fort near Jerusalem. Ben-Gurion escaped capture because he happened to be in Paris. Moshe Sneh, the head of the Haganah, eluded the dragnet, as did the commanders of the Irgun and Lehi.

Golda was the only leading figure spared the round-up, much to her chagrin. The first morning, she had waited expectantly for the roar of a car or military jeep coming to get her. After a while, every time the telephone rang, she flinched, as if her freedom were a sign that the British didn't consider her sufficiently dangerous or significant. When she finally realized that she was being ignored, she showed up at Bevingrad and confronted Sir Alan Cunningham, the high commissioner. "If Shertok [Sharett] is guilty, then I am guilty too," she screamed. Cunningham looked at Golda with amusement. "Perhaps you should be arrested too," he responded. But he did not call in his guards.

Golda's freedom was clearly intentional, but the British never revealed their reasoning. Other women were taken, so her gender was an unlikely motivation. Had they misjudged her, assuming that she was a moderating force, an appellation more appropriate to Sharett, the diplomat, than to Golda?

"Maybe I wasn't important enough," she later admitted. "I was very annoyed."

*　　*　　*

Even as a young woman, Golda exuded self-confidence, an air of utter certainty about her own ability, about her right to lead and the correctness of her positions. But after Black Sabbath, she fell apart. Running a section of the Histadrut, negotiating with the British, and manipulating workers were the tasks of a lieutenant. Golda had been a superb aide-de-camp, but accustomed to collective political decision making among a close-knit group of comrades that occurred, as often as not, over endless cups of coffee at her house. She'd always been able to turn to Remez for counsel, to rely on Ben-Gurion to adjust her thinking, and on Sharett for guidance in diplomacy.

But by the evening of June 29, the silence was deafening, as was the isolation. With Ben-Gurion in Paris, Weizmann old and sick, and the other key leaders locked away in Latrun, however, it fell to Golda to lead the *yishuv* through the crisis by taking over the political department of the Jewish Agency, essentially to become president of Jewish Palestine.

"As long as Golda's outside, the only man in the Jewish Agency is still free," people joked.

But the religious Zionists, a key part of the coalition that enabled the Labor Party to govern, balked. "Kudos to a smart and energetic woman," read an editorial in *Hatzofeh*, an Orthodox newspaper. "But it is impossible to put Golda at the head of the most important thing of the Jewish people. This is not a position for a woman."

The Orthodox weren't alone in their concern about Golda's ability to lead the community. When the question of who would stand up to the British arose at a meeting of American Jewish leaders, Golda's name came up. "Golda?" one man sniggered. "A lovely lady, a good speaker, but are you kidding?"

While men like Sharett and Ben-Gurion were educated, scholarly, experienced in diplomacy, and accustomed to navigating subtleties,

Golda was like a blunt force—emotional, caustic, and almost allergic to any shades but black and white. Within hours of her appointment by the Jewish Agency Executive, she was bickering with everyone left unjailed. At the first meeting of the Va'ad, her imperious demeanor so infuriated leaders of the Palmach that one of them, Miriam the Red, accused her of setting up a *Judenrat*. The Palmachniks decided to ignore her and open a direct channel of communication with Latrun.

From behind the walls of the fort, Shertok tried to calm the victims of Golda's sharp tongue and to nudge her into a modicum of tact. But the challenges were overwhelming. The year's crop was about to rot in the fields because hundreds of kibbutz members had been locked up. The prisoners in the detention camp were threatening to go on hunger strike, Weizmann was appealing for an immediate end to all resistance, and the resistance movement was strategizing an escalation in the violence.

Before any decisions could be made, the *yishuv* needed reassurance that someone was in charge and, back on the familiar turf of a podium, Golda offered it. "What we want is complete independence," she proclaimed. "British policy has put us in an impossible position. With the detention of our leaders and thousands of other Jews accused of no crime, we have reached the limit."

Golda's oratory stirred the populace, but she lacked the clout to impose her view of what that limit was or how far the *yishuv* should go once it was reached. Early in the struggle against the British, the lines had been easy to draw. If British soldiers searched a kibbutz or neighborhood for illegal immigrants, of course people would block their way or refuse to help them; no one had any moral qualms about passive resistance. Even blowing up the patrol boats that hounded refugee ships or sabotaging rail lines to interfere with troops movement found easy ethical backing in a community willing to risk virtually anything to help Europe's surviving Jews. But with every action, they'd crept closer to the terrorism of Etzel.

Weizmann and many other overseas Zionist leaders were horrified. We need to pull back, they clamored. Stop the resistance and return to the bargaining table. Golda wouldn't hear of it. I never lived through a pogrom in Russia, she said, but "what happened on June 29 was what I imagined a pogrom to be. . . . Something happened. The *yishuv* just can't pass this over as if nothing did."

At the funeral of a settler killed during the searches, she reiterated that position. "We have hated death," she said. "We are for life. But we have no choice."

Every day, the dispute was replayed at Histadrut meetings, Mapai caucuses, over coffee, and by telephone with Ben-Gurion, who couldn't return from Paris without risking arrest, Golda battling for an activist stance. Isn't it wiser to preserve what we've built than to court destruction? others asked her. Are we getting drunk on our own rhetoric?

Golda mounted all of her powers of persuasion, pointedly reminding her colleagues of the dangers of losing touch with the public mood. "If we don't do something active, Lehi and the Irgun will," she insisted.

But gradually, Golda was worn down, above all by notes from Remez, who was reading Mahatma Gandhi. Under his influence, widespread civil disobedience became her new mantra, a middle ground between the violence of the *Ma'avak* and the inaction advocated by Weizmann.

Reflecting the mood of the populace, however, the Haganah ignored her directive to pull back from violence, and Golda had neither the political savvy nor the moral authority to sway them. Then, on July 22, 1946, Etzel crossed the *yishuv*'s ethical line with the bombing of the King David Hotel, which housed the British secretariat, the British military command, and a branch of the Criminal Investigation Division of the mandatory police. The warning, which Etzel said they delivered by phone minutes before the explosion, was either not received or never reached the right people. And the explosion, seemingly much more powerful than intended, killed ninety-one people and left hundreds more wounded.

Unnerved by the casualties, *yishuv* leaders knew they had to pull back from the precipice. On August 5, convening in Paris, the Jewish Agency Executive formally voted to end armed opposition. Begin announced that he would ignore that decree.

It was a grim time for Zionists. Despite all of the cautious diplomacy and incautious resistance, the building and organizing, the Jews of Palestine were still helpless to control their own fate. Europe's surviving Jews were on the verge of another harsh winter in bleak DP camps and the few who had been smuggled out were crowded into a new British internment camp in Cyprus. Their leaders remained in prison, Ben-Gurion holed up in Paris, and the country under virtual martial law.

Before they could begin to regroup, the British put forth yet another solution to the Palestine problem, the creation of a federalized state under British control with a small Jewish district. The entire Zionist community rejected the plan out of hand. But when Bevin called an Anglo-Jewish-Arab conference in London to discuss his proposal, that unity unraveled, every fissure in the political fabric of Zionism threatening to tear the movement to shreds.

From Latrun, Sharett repeated the advice he had offered for years: we should never miss an opportunity to present our position. Having absorbed that lesson, Ben-Gurion agreed that they should send a delegation to London. Golda, however, was adamantly, almost maniacally, opposed. She'd already retreated from her position on resistance. Unless Bevin agreed to release the *yishuv*'s leadership and allow the discussion to move beyond federalization, she said, she would not countenance Jewish attendance.

"Golda is inflexible and is making the party look bad," wrote Sharett from Latrun. "She won't admit it, but she's acting as if Ben-Gurion is a traitor."

*   *   *

By the time the train pulled into Basle in early December 1946, Golda was teetering. For months, she'd been caught in the endless spate of debates over restraint versus resistance, over negotiation versus defiance, everyone yelling at her for not enforcing their will. Whenever Etzel of the Stern Gang dared British power by bombing railroad tracks, robbing a bank, or attacking a police station, she was called in for a lecture or an open threat, and she couldn't exactly admit that she was powerless. When the British weren't yelling at her to control her people, her people were fussing at her to control the British. After troops fired on a crowd trying to free refugees about to be shipped off to Cyprus or seized weapons settlers had stockpiled for their self-defense, everyone expected Golda to *do* something.

Yet no matter what Golda did or said, she was heaped with an unending wash of complaints—from Latrun, from Weizmann, from Ben-Gurion. And she had no one to turn to for comfort. Her son, Menachem, was studying music in New York, and her daughter, Sarah, was on Kibbutz Revivim in the middle of the Negev. (The sight of the barren settlement left Golda aghast. "I thought I would die," she said. "For miles around, there was nothing, not a tree, not a blade of grass, not a bird, nothing but sand and glaring sun.")

Remez was available only by letter. And when he was released in November, he offered little support. "You and Ben-Gurion are destroying the last hope of the Jewish people," he berated her for supporting resistance.

"Look, you are afraid and I am afraid," she pleaded. "But which way is more dangerous, yours or mine? Yours won't amount to anything, which will really bring us to destruction. My way means people's lives, but one thing I am sure of—we can't forfeit things necessary to our existence for the promise that if we'll be in trouble we can get help from the outside."

The only people who seemed to approve of her were foreign journalists, who hung on her every word. "She had this gift of propaganda, and I use this word unashamedly," said Gershon Avner, her aide. "In those days, there was a crisis every day and the journalists pulled no punches. . . . But she could handle them, all of them."

Even on days when curfews emptied the streets, her phone rang incessantly and she was forced to sneak out to meetings. Between recurrent attacks of gallstones and migraines, she was surviving on semiregular morphine injections. One afternoon, she fainted dead away and was taken to Beilenson Hospital, where doctors ordered her to rest. "A lot of us die at around fifty," she replied flippantly. They secured her promise to stop smoking, but she started again within a week.

Golda had traveled to Basle for the Twenty-second Zionist Congress, dreading the anticipated brawl over the London conference. Weizmann and his supporters thundered both against the resistance and against those urging a boycott of the conference. With the support of a large swath of the American and Palestinian delegations, the revisionists opposed any negotiations with the British until they ceded all of Palestine as a Jewish homeland.

Ben-Gurion planted the Labor caucus firmly in the middle. "One thing is prohibited under any circumstances—murder," he announced in a closed meeting of the political committee, clearly leaving other methods of resistance on the table.

But the resistance-negotiation dilemma didn't turn out to be the major question that wracked the congress. For years, international Zionism had shied away from spelling out the precise size and shape of a Jewish homeland they sought. Faced with a plea from Truman that they clarify their positions, many delegates believed that the time had come to tell the world exactly what the Jews wanted.

Ben-Gurion asked the congress to endorse a demand for all of Palestine as a political position but be willing to accept partition if Britain offered it. After the Holocaust, Golda had resigned herself to partition, but she worried—vocally and repeatedly—that accepting it could lead them down a slippery slope to federalization. "The question is, what is our starting point?" she said in a caucus meeting. "[Eliezer] Kaplan [the Jewish Agency treasurer] really wants a state, but he wants to begin by requesting a province. This is dangerous. . . . Truman might say, if you agree to a province, then make a start with it, and do you really believe that the British government, in a couple of years' time, will change its views in our favor and agree to a state?

"If Truman really is prepared to offer us a State of Israel, then let's request it *now!*"

Gradually, the notion picked up steam although almost half of the delegates, including Ben-Gurion, balked at making such a bold declaration. While they thought of themselves as the "new Jews" creating their own future, they instinctively shirked directness, as if fearing that they might be seen as insolent.

Golda had no such inhibition. "Why are we NOW pressing our demand for a Jewish state?" she asked in eloquent Yiddish. "When did it become clear that we must have total control over our lives and over immigration, that these matters must be in the hands of Jews, not as a distant aim, but as a desperate, immediate need? . . . When we, the 600,000 Jews of Palestine, despite everything we had created in the country and endured during the long years of war, found ourselves powerless to rescue millions of Jews from certain death. . . . The White Paper was like an iron wall erected between us and Hitler's victims. It was then, when our helplessness was so tragically revealed, that the argument among us as to the goals of Zionism came to an end. Zionism, redemption, and rescue became a single concept . . . and we knew that there was only one way of fulfilling Zionism now—and that was by creating a Jewish state."

Ironically, all the angst at Basle over the London conference and the shape of Palestine's future turned out to be irrelevant. With Labour backbenchers revolting against a policy that seemed to go nowhere, unrelenting pressure from Truman, and a public wearying of the daily media diet of floggings and executions in Palestine, Prime Minister Attlee abruptly reversed British policy.

In February 1947, his cabinet announced that it would submit the question of Palestine to the newly formed United Nations.

CHAPTER EIGHT

*We refuse to disappear, no matter how strong and brutal
and ruthless the forces against us may be.*

They made an odd pair, the sixty-five-year-old Hashemite king born
in Saudi Arabia and the forty-nine-year-old Jewish woman from
Pinsk meeting secretly to negotiate the future of a country to which neither held a birthright. But the labyrinthine diplomatic wrangling sparked
by the British decision to refer the Palestine question to the United Nations had generated dozens of bizarre alliances and odd bedfellows as it
degenerated into a series of petty public squabbles and backroom deals
over the United Nations Special Committee on Palestine, UNSCOP,
charged with issuing recommendations for the fifty-five-nation General
Assembly to consider in the fall of 1947.

By the time Golda snuck across the border into Naharayim, Transjordan, to meet King Abdullah, the General Assembly was ready to consider
the committee's proposal for political partition, with Jerusalem as an international city. At a press conference, Golda had been asked if she
thought blood would be spilled if that recommendation were adopted.
"We don't anticipate spilled blood, but we are ready if it happens," she

said, knowing full well that the tiny *yishuv* could not hold off a concerted attack by the armies of all the Arab countries threatening invasion. The only hope was that Abdullah would not send his Arab Legion—15,000 men trained and led by British officers—across his borders and that he would deny other Arab armies easy access to them.

The pressure on Abdullah to join the Arab front against a Jewish state was fierce, but the king was the most unpredictable of the Arab leaders. Granted his kingdom by British fiat in 1921, he had set his sights on ruling more than that desolate stretch of territory with no outlet to the sea, no oil, and no historical grandeur. He dreamed of presiding over an empire, Greater Syria, the union of Transjordan, Palestine, Syria, and Lebanon.

Such brazen designs didn't win him much affection or trust among Arab leaders, but Abdullah understood patience. For the moment, he was ignoring Syria and Lebanon while he figured out how to swallow up the rest of Palestine, which he could see from the window of his palace by the Dead Sea. His greatest threat was not his fellow Arab leaders but Haj Amin al-Husseini, the former mufti of Jerusalem, who envisioned Palestine as the base for spreading his own influence across the Arab world. Since the Jews hated al-Husseini even more than Abdullah did, according to the idiosyncratic king's impeccable logic, the Zionists were his natural allies.

"There are already enough Jews in Palestine," Abdullah trumpeted in public. But he had been negotiating with Zionists since 1922, and had privately agreed to accept the establishment of a Jewish state in exchange for Jewish support of his annexation of the Arab part of Palestine.

Golda had almost no experience dealing with Arabs and none whatsoever negotiating with monarchs. But Sharett, Zionism's leading diplomat, was stuck in New York leading the lobbying effort at the United Nations, and one of the *yishuv*'s leaders needed to meet with the king. So at the end of November 1947, she slipped across the border and waited for the king to arrive, as planned, at the compound of the Palestine Electric Corporation.

King Abdullah had been told that he would be meeting with the second most important Zionist diplomat, but the last person he expected was Golda, or any other woman. "Although he respected the opposite sex, King Abdullah was a conservative," wrote Sir Alex Kirkbride, the British high commissioner in Transjordan, who'd caught wind of the meeting. "In his eyes, women could not be the equal of men, particularly in politics."

But Abdullah was a natural charmer, a raconteur of desert lore who dabbled in poetry. Recovering quickly, he shared thick Arab coffee with his guest and invited her to visit him at his palace in Amman—at some unspecified time in the future, of course. Then he launched into a soliloquy about partition and his recent discussions with the Arab League Council, reporting that he had told them that he would not collaborate in the destruction of a Jewish state.

"Over the past thirty years you have grown and strengthened yourselves and your achievements are many," he told Golda. "It is impossible to ignore you, and it is a duty to compromise with you. . . . Any clash between us will be to our own disadvantage. . . . I agree to partition that will not shame me before the Arab world."

That was sweet music to Golda's ears, although Abdullah then returned to a suggestion he'd been making for decades, that he annex all of Palestine and allow the Jews their own republic within it. Brushing off that notion, Golda tried to lead the conversation around to the partition resolution about to go to the United Nations.

But Abdullah wouldn't let her do much talking. How would the Jewish community in Palestine feel if I captured the land set aside for the Arabs? he asked, radiating confidence.

Favorably, Golda responded, if his moves didn't interfere with the establishment of a Jewish state. "But you should declare that your sole purpose is to maintain law and order until the United Nations can establish a government," she advised, an odd recommendation, at once naive and patronizing.

"But I want this area for myself," he huffed. "I want to ride, not to be ridden."

Dismissive of the intrigues of both the mufti and the other Arab rulers, Abdullah announced that he could make no concrete promises until after the United Nations decision. But with all the rhetoric stripped away, Golda's translators believed that Abdullah had pledged that he would not allow his army to "collide with" the Jews, as they put it.

Golda was taken with the king, "a small, very poised man with great charm," in her words. But it was impossible to know how far to trust him. Not far, Ezra Danin, one of Golda's two translators, suggested, cautioning her that Abdullah was notoriously unreliable and that Bedouins had their own concept of the truth.

*   *   *

Despite their request that the United Nations consider its future, Britain was not done with Palestine. Bevin refused to commit to turning over a mandate that had effectively expired with the demise of the League of Nations. And he told the House of Lords that Britain retained the right—by what legal authority was not specified—to reject any UN recommendations.

And while the UN debated, the British colonial authorities set out to prove that they were still the masters of the moment. In February, shortly after Etzel kidnapped a judge, they'd evacuated all nonessential personnel and launched another all-out effort to suppress political violence. Declaring martial law, they put more than a third of the Jewish population and 90 percent of the economy under emergency regulations.

In June, Britain even rejected the UN's call for a truce during the proceedings of UNSCOP. And just when that committee was beginning its work in Palestine, the Mandatory authorities demonstrated how much respect they had for the new world body by announcing a death sentence—death by hanging—for three Irgun members who'd raided a police station and emptied its armory several months earlier.

Britain's last days as overlords in the Middle East were bitter times, and as the leader of the *yishuv* and its intermediary with the high commissioner, Golda was caught between the British rock and a Jewish hard place. Every seizure of an immigrant ship, every deportation to Cyprus, every image of the internment camp there, so reminiscent of Nazi concentration camps, further embittered the *yishuv*. When Etzel blew up the British Officers' Club in Jerusalem, thousands cheered. And when guerrillas dressed as British soldiers and blasted a breach into the walls of the old Acre fortress, where 163 Jews were imprisoned, they became folk heroes.

"The struggle against terrorism cannot be divorced from the political circumstances which have given rise to this terrible aberration," Golda wrote the British in one seven-point letter. "Terrorism has its roots in the despair and bitterness engendered by the White Paper policy."

Rejecting that argument, the government demanded that the Jewish Agency turn Etzel's leaders over to them. "We will not become a nation of informers . . . with each one watching a neighbor or friend and reporting on what appears to be wrong with the person under suspicion," Golda decreed.

All she would promise the British was that the Jewish Agency would do everything possible to stem the violence. But the organization cannot "be called upon to place itself at the disposal of the Government for fighting the evil consequences of a policy which is that of the Government's own making."

Golda did try. In political discussions and at public meetings, she railed openly against the Etzel, especially when they robbed Jewish jewelry stores to raise money for their exploits, hijacked Jewish trucks or taxis for transportation, or sent children to put up illegal posters, a crime punished by flogging. "Even if it means armed struggle, we'll stop them," she told the leadership, knowing full well that no one had the heart to plunge the *yishuv* into a civil war with Etzel when they were already fighting the British.

"Terrorism is assisting Palestine's British administration," she pleaded at a mass rally in Tel Aviv. "It has put Palestine Jewry on the defensive, whereas but for terrorism the Zionists could have pursued a more vigorous line in their political efforts."

Those exhortations drew the wrath of some terrorists, who lashed out at the Jewish Agency, planting explosives in its press office and tourist information center in the heart of Jerusalem. When they openly declared Golda a traitor, Golda's colleagues tried to dissuade her from walking alone, especially at night. One evening, she and Sheyna went to the theater only to have their evening's relaxation interrupted by a shower of leaflets denouncing Golda thrown from the balcony.

But Golda had won the admiration of most of the *yishuv*. "On the whole she had the right measure of independence and the ability to listen to people," said a man who worked closely with her. "And the community admired her ability to speak simply and directly to their hearts."

That respect soared when she launched an initiative to bring all the Jewish children living in British detention centers in Cyprus into Palestine immediately. By the middle of 1947, refugees were languishing there, in steamy tents behind barbed wire, for more than a year while they waited their turns for one of the 750 visas granted Cyprus detainees each month.

Since the British refused to increase that quota, Golda went to Cyprus to convince refugees high on the waiting list to step aside to allow families with children to leave first. It was an agonizing request to make of men and women who'd spent years in German death camps and British DP camps before being confined on the bleak island. But for three days, Golda trekked from tent to tent, asking elderly couples to give up their places on the departure list for families with babies and berating unmarried men for their indifference to the plight of children growing up without much hope.

The refugees bowed to Golda's request, and the British commandant knew better than to intervene. Before Golda arrived, he'd received a cable from Jerusalem, warning, "Mrs. Meyerson is a very formidable person. Watch out!"

By the time the United Nations was ready to vote on the partition of Palestine, Golda was the mistress of the *yishuv*, often dominating the contentious community single-handedly while her colleagues were in New York.

Those were days of boundless apprehension. The partition plan had engendered a tangle of mixed feelings, excitement, almost disbelief, that the Jews might finally be permitted to create their own country, and despair at how much of that country was being assigned to the Arabs. After Golda saw the map proposed by UNSCOP, which excluded the western Galilee from the Jewish state, internationalized Jerusalem, and was rife with enclaves and passageways between the two countries, she could barely speak to the press, muttering, "We can hardly imagine a Jewish state without Jerusalem." But, on balance, she fell back on her stock phrase, "We have no alternative."

That alternative was anything but solid. With the winds shifting daily in Lake Success, where the General Assembly was meeting, no one was confident that the Jews could garner the necessary two-thirds majority, and no one was convinced that the British would actually leave if they did. Even if they did, the Arabs weren't just threatening war should the UN vote for partition; they were massing troops on the borders.

"The Messiah hasn't come," Golda warned at a Histadrut meeting in Tel Aviv. "The Day of Redemption isn't here yet. . . . Let us . . . prepare for the worst. Then there is hope that it won't come to the worst."

It was well past midnight in Jerusalem by the time the General Assembly voted on November 29, 1947. Golda had spent the evening at a reception for Yugoslavia's Independence Day. But she needed to listen to the news from New York alone. Smoking endless cigarettes, she planted herself on her couch at home and listened to the session on the radio, checking off each nation on a piece of paper. Afghanistan was against, no surprise. But when Australia, Belgium, Bolivia, Brazil, Byelorussia, Canada, and Costa Rica voted yes, she began to smile. The final tally was thirty-three nations for and thirteen against with ten abstentions, including Great Britain.

The next evening, she stood next to Ben-Gurion on the balcony of the Jewish Agency building. All day, Jerusalem had rejoiced. Flags with the Star of David fluttered, trucks filled with young people from settlements honked their horns, and thousands of people danced, quite literally, in the streets. As night fell, they gathered to cheer their leaders.

"For two thousand years we have waited," Golda told the crowd. "We always believed it would come. . . . Now we shall have a free Jewish state."

The following morning, eight Jews were murdered on the Haifa-Jerusalem road and along the nebulous boundary between Jaffa and Tel Aviv. In Jerusalem, following a plan designed by the Arab Higher Committee, Arabs trashed Jewish shops and murdered random citizens. In Haifa, armed mobs attacked workers at the oil refinery. And in the countryside, funeral processions became a target. Hundreds of Arab irregulars were sneaking into the country from Egypt and Syria, Iraq and Lebanon to join the Army of Salvation led by Fawzi El Kaukji, who'd spent World War II at the mufti's side in Berlin. And Abdul Khader, the leader of the Palestinian resistance movement who'd been fighting the British presence in the Middle East for more than a decade, snuck back home from Iraq to take charge of the local resistance. "We shall keep our honor and our country with our swords," he pledged. And his first order: "We will strangle Jerusalem."

By the two-week anniversary of the partition vote, eighty-four Jews had been buried.

Hoping either to minimize their own casualties in the last days of their Mandate or to send the Jews a chilling "we told you so," the British would neither recognize the Haganah's right to defend the *yishuv* nor provide that defense themselves. Producing photographs of Arabs looting Jewish shops and houses while British police watched with folded arms, Golda repeatedly complained to the chief secretary. He promised an inquiry but did nothing. Authorizing Jews to patrol the roads or towns would be taking sides, he informed her curtly.

Taking advantage of that "malevolent neutrality," as Golda called it, the Arabs launched a major battle to cut off Jerusalem. Perched in the Judean Hills, an island surrounded by heavily Arab towns, Jerusalem was linked to Tel Aviv by a single approach, the cliffs above it dotted with Arab villages. Arab snipers hid in the pine forests at Bab el Wad and Kastel and took potshots at Jewish vehicles carrying supplies to the 100,000 Jews of the ancient city. Within days, the skeletons of overturned buses littered the side of the highway.

In charge of Jerusalem at Ben-Gurion's command, Golda was suddenly overseeing the city's defense, the distribution of food, the acquisition of arms, and the protection of the 2,000 ultra-Orthodox Jews living inside the walls of the Old City. When pharmacies ran out of medicine, when scientists needed money to build weapons, when British troops did nothing to protect unarmed civilians, Golda's office was besieged. She was so exhausted that she met visitors while soaking her feet. Sleeping four hours a night at best, she was in her element.

Golda needed to be in Tel Aviv twice a week for meetings and the only alternative to the Jerusalem Road was to hop into a phosphate truck loaded with illegal weapons heading to the northern shore of the Dead Sea and catch a tiny Primus from there. Most weeks, she ran the gauntlet instead.

One afternoon in December, she sat calmly in her seat while Arabs rained bullets down on her bus, but she covered her eyes. What are you doing? a companion asked. "I'm not really afraid to die, you know. Everyone dies. But how will I live if I'm blinded? How will I work?"

On December 27, her bus was ambushed on a curve six miles from Jerusalem. The man next to her was shot and died in her lap.

Several days later, Golda found herself in a convoy of 170 vehicles stopped by British armored cars. Her fellow passengers scurried to hide their weapons, Golda concealing grenades and pistols under her coat and inside her purse. But a soldier glimpsed the barrel of a Sten gun peeking out from underneath the jacket of a young woman and arrested her. Where are you taking her, Golda challenged him. "Fallujah," he responded. An Arab village, Golda thought. Impossible. "If you take her, you must take me too," Golda insisted. The captain didn't know who the insolent woman was, but her presence was menacingly commanding, so he hauled Golda and his captive off to the police station in Jewish Hedera.

"Mrs. Meyerson," the chief of police there exclaimed when they arrived. Apologizing, he offered her a drink in celebration of the New Year and escorted her to Tel Aviv. The death of Golda, he knew, could provoke serious trouble.

*　　*　　*

In mid-January, Eliezer Kaplan, the treasurer of the Jewish Agency, returned from the United States with bad news: there's no way the Jews of Palestine can raise more than seven or eight million dollars for weapons. After years of donating enormous sums for Palestine, for Jewish refugees in Europe and their own communities at home, American Jews were tapped out at the very moment when Arab armies were poised for attack.

"I suggest that Kaplan and I leave at once for the United States," Ben-Gurion announced. Golda flinched. Ben-Gurion was a genius with America's committed Zionists, but he was too impatient and too old-fashioned for more mainstream Jews.

"What you can do here, I cannot do," she told him. "But what you can do in the United States, I can also do."

Ben-Gurion wasn't convinced, but he was outvoted. Without returning to Jerusalem to pack, the next day Golda flew to New York with a shopping list from the Haganah quartermaster—weapons, ammunition, blankets, tents, and sweaters.

This time, Golda needed the support of Jews she'd long disdained, German Jews who owned department stores and factories, non-Zionists who wore fancy clothes, English-speaking Jews who had never had much truck with Yiddish-speaking socialists. At home, she had become "someone." But to those Americans, she was hardly a brand name. On earlier trips, she'd been "an impecunious, unimportant representative, a 'schnorer' for various little funds," said Henry Montor, then the director of the United Jewish Appeal. "When she came here she stayed in houses, not in hotels."

Arriving on a Friday afternoon in the midst of the worst blizzard in half a century without enough money to pay for a taxi, Golda called Montor, Ben-Gurion's best connection. "I've got to have money for arms," she told him. "I've got to get it from the UJA [United Jewish Appeal] because . . . the Arabs are not playing card games." Montor required no convincing, but worried what purchasing weapons would mean for their tax exemptions; the leadership of the UJA, as well as that of the other major Jewish charities, had already refused.

Undaunted by the opposition of the other leaders of the major American Jewish organizations, Montor decided that Golda's only chance was to go to Chicago for the annual meeting of the Council of Federations and Welfare Funds, one of the largest Jewish charities in the country. She wasn't particularly welcome; in fact, the executive director of the council wouldn't give her a speaking slot. But Montor, a notorious battering ram, mounted pressure through members in Minneapolis and New York until the council agreed to allow Golda to speak at lunch on Sunday, when he knew the *balabatim*, the rich guys, would be in the room.

The 800 wealthy men gathered at the Sheraton Hotel that day controlled the American Jewish fund-raising machine. Rich, cynical, and hard-nosed, they were most definitely not Golda's crowd, and Golda, in a simple blue dress, her hair pulled back tightly in a bun, was most definitely not one of them.

As usual, she spoke off the cuff:

> *The nations of the world have given us their decision—the establishment of a Jewish state in a part of Palestine. Now in Palestine we are fighting to make this resolution of the United Nations a reality, not because we wanted to fight. . . . The Mufti and his men have declared war upon us. We have to fight for our lives, for our safety, and for what we have accomplished in Palestine, and perhaps above all, we must fight for Jewish honor and Jewish independence.*
>
> *All we ask of Jews the world over, and mainly of the Jews in the United States, is to give us the possibility of going on with the struggle. . . . I want to say to you, friends, that the Jewish community in Palestine is going to fight to the very end. If we have arms to fight with, we will fight with those, and if not, we will fight with stones in our hands.*

*I want you to believe me when I say that I came on this special mis-
sion to the United States today not to save 700,000 Jews. During the
last few years the Jewish people lost 6,000,000 Jews, and it would be
audacity on our part to worry the Jewish people throughout the world
because a few hundred thousand more Jews were in danger. That is
not the issue. The issue is that if these 700,000 Jews in Palestine can
remain alive, then the Jewish people as such is alive and Jewish independence is assured. If these 700,000 people are killed off, then for
many centuries, we are through with this dream of a Jewish people
and a Jewish homeland. . . .*

*Merely with our ten fingers and merely with spirit and sacrifice, we
cannot carry on this battle, and the only hinterland that we have is
you. The Mufti has the Arab states—not all so enthusiastic about
helping him but states with government budgets. . . . We have no government. But we have millions of Jews in the Diaspora, and exactly as
we have faith in our youngsters in Palestine I have faith in Jews in the
United States. . . .*

*I want to close with paraphrasing one of the greatest speeches that
was made during the Second World War—the words of Churchill. I
am not exaggerating when I say that the Yishuv in Palestine will fight
in the Negev and will fight in Galilee and will fight on the outskirts of
Jerusalem until the very end.*

*You cannot decide whether we should fight or not. We will. . . . You
can only decide one thing: whether we shall be victorious in this fight
or whether the Mufti will be victorious. That decision American Jews
can make. It has to be made quickly within hours, within days. And I
beg of you—don't be too late. Don't be bitterly sorry three months
from now for what you failed to do today. The time is now.*

Montor listened from the back of the room, surveying the crowd carefully. "Sometimes things occur, for reasons you don't know why, you don't
know what combination of words has done it, but an electric atmosphere
generates, people are ready to kill somebody or to embrace each other,"

he later said. "That is still vivid in my mind, that particular afternoon. The delegates of Dallas were all . . . strongly non-Zionist, but they said they were going back to Texas to get so much money they wouldn't know what to do with it. . . . She swept the whole conference."

With the momentum from Chicago, Montor sent Golda on the road, to luncheons for women's groups and dinners with leading fund-raisers, for breakfasts with the wealthiest and private meetings with the skeptical in Omaha and Tulsa, Houston, Dallas, Cleveland, and New York. Honing her message, Golda shared accounts of the conversations she was having with Haganah agents in Europe searching for weapons. One called her to report that he could buy planes in Czechoslovakia and captured German equipment in France but had no money. "Money is not any of your business," she recalled telling him. "You stay there." Another cabled that he could buy tanks if he received $10 million immediately. "OK, I'll get it for you," she promised.

Stopping mid-anecdote, Golda then turned to her audience and shook her head coyly. "I had a lot of chutzpah. Where was I going to get ten million dollars?" The answer was left hanging.

Golda was not all sweetness and charm. When a crowd in Texas proved recalcitrant, she walked out. At a women's luncheon in Atlanta, a woman upset by the recent marriage of Paul Robeson's son to a Jewish girl asked Golda, "How do you feel about a Jewish girl marrying a Negro?" Flashing to Sarah's new beau, a dark-skinned Jew from Yemen, Golda lashed out at the woman about racism and the Jewish concept of justice.

Out of place among the American Jewish glitterati, Golda was always nervous, and no place was worse for her than Miami. "I remember coming down to the patio which was so beautiful and seeing the people there dressed with all that beauty," she later recalled. "I was sure that they couldn't care less. I was sure that when I began talking, they'd walk out of the room." She ended the day with $5 million in pledges.

On February 3, she reported to Ben-Gurion that she had already raised $15 million.

Back in New York, Golda again approached the leaders of the major fund-raising groups, the United Jewish Appeal, the Joint Distribution Committee, and the National Refugee Service. When they again raised concerns about their tax status, Golda began weeping. This time around, Henry Morgenthau, former secretary of the Treasury under Franklin Roosevelt and chairman of the UJA, intervened. "The UJA is here for the purpose of saving the Jewish people and we can't save the Jewish people unless the Jews in Palestine . . . are able to defend themselves. . . . If Golda Meyerson says that they have to have arms and we are the only place where they can get the money to buy the arms, I'm afraid . . . you're going to have to accept my decision. We're going to include Golda Meyerson and her request in this year's campaign."

By February 25, she had $25 million. A week later, Sharett recorded in his diary that she'd brought in $30 million and hoped for as much as $40 million before her departure.

By the end of her trip, Golda had become starkly blunt, and just as bluntly stark. "You have a choice," she told a crowd of New Yorkers. "Either to meet in Madison Square Garden to rejoice in the establishment of a Jewish state or to meet in Madison Square Garden at another memorial meeting for the Jews in Palestine who are gone."

In March, Golda flew home to Jerusalem, having raised $50 million that Haganah agents in Europe were using to purchase rifles and machine guns, ammunition, and airplanes.

"Someday, when history will be written, it will be said that there was a Jewish woman who got the money which made the state possible," Ben-Gurion wrote.

*   *   *

By the time Golda returned home, El Kaukji's troops were regularly attacking roads and isolated settlements. Volunteers from the Muslim Brotherhood of Egypt, including a young Egyptian whose parents had migrated from Palestine named Yasir Arafat, were training at bases in Hebron and Bethlehem. Forces led by the former mufti had laid siege to

Jerusalem, cutting water pipelines and keeping the road impassable. And the Arab League was readying twenty thousand volunteers to cross the borders the moment the British Mandate expired.

"You will hang on to Jerusalem with your teeth," Ben-Gurion instructed, despite the British suggestion for a wholesale Jewish evacuation. It was no easy task. The population was living under a hail of bullets but the city's Jewish defenders had just 500 rifles, 400 Sten guns, 28 machine guns, and a few mortars. Ben-Gurion couldn't send them more; the entire *yishuv* had only 10,000 rifles.

Golda imposed a brutal system of rationing, three ounces each of dried fish, lentils, macaroni, and beans each day, supplemented with 1.5 ounces of margarine and a few drops of water. Still, the city's stocks were almost depleted.

"DOES WORLD INTEND REMAINING SILENT?" Golda cabled the UN. She received no reply.

The *yishuv* hung on without surrendering a single settlement, but news of the attacks chilled what little enthusiasm the international community had for partition. For months, the U.S. State Department, led by Loy Henderson, chief of the Near and Middle Eastern Affairs Division, had tried to convince Truman that partition was unworkable and that American support for it would open the region to Soviet influence. Finally, Truman wavered, first succumbing to that pressure and then reversing his stance after visits from his oldest Jewish friends and from Chaim Weizmann himself. Nonetheless, on March 19, without the president's knowledge, Warren Austin, the U.S. delegate to the United Nations, asked the Security Council to impose a temporary trusteeship to allow the General Assembly to reconsider its decision.

Terrified that the chaos would lead the UN to do so, Ben-Gurion ordered the Haganah on the offensive. During April, they broke the siege of Jerusalem, capturing Tiberias, Jaffa, and Haifa. But those victories proved bittersweet since they intensified Arab panic and provoked mass flight. Wealthy landlords and merchants had departed well before the shooting started in earnest, some 20,000 waiting out the turmoil in Damascus or

Cairo. Then the calls for the departure of the workers began, issued by the mufti and the leaders of the surrounding countries: Don't work. Destroy the economy. Finally, as the guns of war echoed ever louder, so too did the message from El Kaukji, the Muslim Brotherhood, and the mufti: get out so that we can get in.

When the Haganah seized Tiberias, the entire Arab population of 6,000 was evacuated voluntarily under British supervision. Two weeks later, 25,000 Arabs bolted from Haifa during an offensive led by El Kaukji. Tens of thousands more began fleeing when Jewish troops took the city.

Ben-Gurion sent Golda to try to persuade them to remain. Again, she was a strange choice, a Jew who spoke no Arabic, a woman emissary to a severely patriarchal society, a Russian who called every riot a pogrom. But Golda was also a woman of common sense. A month before the partition vote, the Jewish Agency Executive Committee had met at her apartment in Tel Aviv to develop an "Arab policy" for their new state. "It would be more than foolish to expect that we can live here in comfort and in peace and not do everything for the Arab minority," she remarked. "We have no desire to be a master race and have people of a much lower standard among us. Look, we shall have to show the world how we are making up for our 2,000 years of suffering as a minority not by emulating what was done to us but by isolating every single method of making people suffer, and doing away with each of these methods, one after the other."

Golda asked the city's popular Jewish mayor to go down to the port to plead with the Arabs to remain. They are too afraid of bombs, he reported. She went down to the beach herself to beg them to return to their homes. They would not listen. She enlisted the help of the British army commander. But it was useless.

Golda's failure in Haifa was not the only demoralizing factor in her life during those hectic days. On March 2, the *yishuv* leadership had decided to form an interim government, and Golda assumed she'd be awarded a top post, the equivalent of a cabinet position. But while she was appointed to membership in the thirty-seven-member National Council, a sort of legislature, she was passed over for inclusion in the

Minhelet Ha'am, the thirteen-person national administration that would run the government until the new state could organize its first election.

It was a bitter blow, but that first cabinet, like every Israeli cabinet thereafter, was carefully stitched together to solidify a ruling coalition. After shoring up support for Mapai on the left, the right, and among the religious, Ben-Gurion had only six spots he could hand out to his own comrades and was forced to choose between Golda and Remez. When Golda complained about her exclusion, Ben-Gurion offered to give her Remez's spot, but she didn't have the heart to demand that her companion be pushed aside. Still, her isolation rankled.

"I have been home for several weeks and am not doing anything," she complained in a letter to Ben-Gurion signed "The Woman Who Saved the State."

*   *   *

In mid-April, after months of nonstop work, Golda suffered a mild heart attack and was ordered to bed. But two weeks later, Ben-Gurion had another mission for her, and Golda still could not resist Ben-Gurion. The signals from Transjordan had become contradictory, and the Arab world was rife with rumors that British foreign minister Bevin had agreed to allow King Abdullah to swallow up all of Palestine after the Jews were defeated and that Saudi Arabia, Syria, and Egypt had decided to let him do so, and then assassinate him and give the former mufti his throne.

Still hoping to keep the king out of the coming war, in late March the National Council had voted to send Golda for a second meeting with him. Abdullah delayed, but on May 10 he passed word through a British intelligence officer that he would see her, but in Amman, not at the border.

*   *   *

Golda and Abdullah had been passing messages back and forth for months, and in her last note, she'd admitted that she was worried. I'm hurt, Abdullah had responded. Remember three things: I am a Bedouin and therefore a man of honor. I am a king and therefore doubly bound to behave honorably. And I would never break a promise to a woman."

Golda remained skeptical, but the *yishuv*'s military situation was so desperate that she was willing to grab at any straw. "If there is any chance, even a very minor one, that I could save the life of even one Jew, I'm going to do that," she said, setting off to cross a frontier lined with Jordanian and Iraqi troops.

Abdullah devised a complicated stratagem for sneaking Golda into the country to ensure that no hint of the meeting would leak out. So Golda and her interpreter, Ezra Danin, changed cars repeatedly on the road to Naharayim. Once there, Golda donned an embroidered black dress and veil, was picked up by the king's driver, and crossed the Arab Legion checkpoints to Amman disguised as an Arab woman.

She was impatient by the time she was ushered into a stone house in the hills above the road to the Amman airport. "Have you broken your promise to me after all?" she asked King Abdullah abruptly, eschewing traditional Arab politesse.

The king sighed. "Then I was one," he said. "Now I am one of five. Why are the Jews in such a hurry to have a state?"

Already exasperated Golda snapped. "We've waited two thousand years. That's not my definition of in a hurry."

Again Abdullah broached the idea of the Jews joining Transjordan, promising them full representation in his parliament. Dismissing that notion out of hand, Golda warned, "You must know that if war is forced upon us, we will fight and we will win."

Abdullah seemed resigned. "It is your duty to fight. But why don't you wait a few years?"

Golda knew there was little she could say, but she reminded the king of their common enemy, the Mufti, and of how much the *yishuv* had already done to weaken him militarily, clearing Abdullah's path through Arab Palestine. "We're your only friends," she added.

When Abdullah didn't budge, Golda prepared to leave. But Danin and Abdullah lingered, chatting. You're embarking on a dangerous course, Danin warned the king. "You worship at the mosque and permit your subjects to kiss the hem of your garments."

Abdullah seemed unconcerned about the possibility of being assassinated as he escorted his Jewish visitors to the door.

When Golda reached Tel Aviv, Ben-Gurion interrupted a security briefing to hear about the encounter. "We met in friendship," she reported. "He is very worried and looks terrible. He didn't deny that we had spoken and had reached an understanding on a favorable settlement in which he will rule the Arab section. . . . This is the plan he proposed, a united country with local autonomy in Jewish areas, followed a year later by the country's unification under his rule."

Never known for her fluency in subtlety, Golda interpreted the meeting as a failure, a negation of all the optimistic signals Abdullah had been sending. And over the years, Moshe Dayan, who met with the king six months later, alleged that Golda herself had destroyed any opportunity for peace with her curt manner.

Most others read the encounter differently. "His Majesty has not entirely betrayed the agreement, nor is he entirely loyal to it, but something in the middle," Yaacov Shimoni told a meeting of the Arab Section of the Political Department of the Jewish Agency. After all, while Abdullah indicated that he would send the Arab Legion across the River Jordan, he said nothing about attacking Jewish forces or crossing into Jewish territory.

*   *   *

The British Mandate was due to expire at midnight on May 14, but the *yishuv* leadership was still feuding about whether they should delay their declaration of independence. U.S. secretary of state George Marshall had warned that if they proceeded, the Americans wouldn't rescue them from an Arab invasion. When that prospect failed to chill Zionist ardor, he threatened to stop the transfer of funds from U.S. Jews and ask the United Nations to withdraw the partition resolution.

It's suicide not to agree to Marshall's terms, cautioned Remez, backed by a full third of the Mapai bloc in the new cabinet. Yigal Yadin, the chief of operations of the Haganah, estimated their chances of surviving an all-out Arab assault at 50–50.

"We need to go all the way," insisted Golda, who had no vote but never allowed herself to be deprived of a voice. "We can't do a zigzag. . . . The world is waiting for our announcement. If we don't make it now, we never will."

With sixty hours until the scheduled declaration, six of the ten cabinet members present swung in Golda's direction.

Golda was equally unwavering on the other major issue confronting the government-in-the-making, Jerusalem. The UN had dragged its feet on internationalizing the city, and the Arabs had begun their attempt to seize it militarily. When the British high commissioner tried to arrange a cease-fire, Golda laid down firm conditions—free access to the Jewish Quarter of the Old City and the holy places, unimpeded transit on the road from Tel Aviv, and Jewish control over property abandoned by Arabs in West Jerusalem. Ignoring her, the British announced an unconditional truce and, for the moment, the Haganah had respected it.

Despite the British heavy-handedness, Golda supported the cease-fire, fearing that the Christian world would blame the Jews if fresh blood stained the Stations of the Cross. But she clung to the conditions she'd set down.

"Golda is too inflexible," Remez argued. "If she wants an armistice, she should forgo the condition which states that we will not restore the Arab quarters to the Arabs. . . . How can we refuse to allow them to return there? Will those houses become Jewish property?"

No final decision was made, nor were final decisions made on dozens of urgent issues. The leaders of the *yishuv* had been preparing for statehood for decades, but they still hadn't agreed what the new state should be called—Zion, Judea, or Yehuda. Ben-Gurion wanted the capital to be built in the desert; others were intent on Jerusalem; and Golda was pressing hard for Mount Carmel in Haifa. Scores of details—about stamps and passports, stationary and state symbols—were yet to be worked out, and the British Mandate expired on May 15.

Without a vote, without a portfolio, without any title beyond member of the National Council, Golda sat in on every meeting, assuming by action the status she hadn't been granted by Ben-Gurion.

The declaration of statehood ceremony was organized as a secret event lest the British try to prevent the creation of a Jewish government before their Mandate expired the following day. Only on the morning of the fourteenth, the day scheduled for the declaration, were the invitations delivered, by hand, urging the guests to arrive promptly at 3:30 and dress in "dark festive attire."

Just after noon, Golda carefully washed her hair, put on her best black dress, and waited alone for a driver to pick her up. Menachem was still in New York, her parents and sister in Holon, and Sarah digging trenches at her kibbutz, directly in the path the Egyptian army would undoubtedly take in a drive toward the heartland of the state.

Despite the attempt at secrecy, the *yishuv* was too small, too inbred for word not to leak out. As Golda made her way up Rothschild Boulevard into the Tel Aviv Art Museum, hundreds of people were already gathered outside, thousands more tuned in to Kol Yisrael for the station's first live broadcast. The hall was packed with men and women from the Jewish Agency and the World Zionist Organization, the leaders of political parties, cultural and religious institutions, and the international press.

Golda took her seat with the other members of the National Council beneath a massive portrait of Theodor Herzl. At 4 P.M., Ben-Gurion, dressed for once in a suit and tie, rapped his gavel. The Philharmonic Orchestra, hidden on an upper floor since there was no room for them below, poised to play "Hatikvah," the new national anthem. But the crowd beat them to the punch, spontaneously bursting into song.

"The Land of Israel was the birthplace of the Jewish people," Ben-Gurion read from the preamble to the declaration. "By virtue of the natural and historic right of the Jewish People and the resolution of the General Assembly of the United Nations, we hereby proclaim the establishment of the Jewish State in Palestine, to be called *Medinat Yisrael*, the State of Israel."

The members of the audience surged to their feet in a single, unbridled wave, disbelief and excitement erupting as tears and applause.

Ben-Gurion laid down the principles of freedom, justice, peace, and equal social and political rights that were to guide the new state, concluding, "With trust in the Rock of Israel, we set our hand to this declaration, at this session of the Provisional State Council, on the soil of the homeland, in the city of Tel Aviv, on this Sabbath eve, the fifth of Iyar, 5708, the 14th of May, 1948."

One by one, the signers of the declaration solemnly walked to the desk where Sharett held out the temporary parchment. Ben-Gurion had pleaded with the thirty-seven signers to adopt Hebrew names before the ceremony, and many had complied. Golda had not. Her hands shaking, tears streaming down her face, Golda signed Golda Meyerson.

"From my childhood in America, I learned about the Declaration of Independence and the geniuses who signed it," she said. "I couldn't imagine these were real people doing something real. And here I am signing it, actually signing a Declaration of Independence. I didn't think it was due me, that I, Goldie Mabovitch Meyerson, deserved it, that I had lived to see the day. My hands shook. We had done it. We had brought the Jewish people into existence."

CHAPTER NINE

*If only political leaders would allow themselves to feel
as well as to think, the world might be a happier place.*

The order for Golda to leave was issued by Ben-Gurion, who knew that the infant state's only hope of survival depended on the acquisition of serious weapons. The armies of five nations had invaded, and they were equipped with artillery, tanks, armored cars, and personnel carriers, and supported by the air forces of Egypt, Syria, and Iraq. Israel had fewer than 50,000 fighters, some who'd been trained in displaced persons' camps with wooden rifles and dummy bullets. Since the *yishuv* could not openly buy arms until the state was declared, Israeli troops were going into battle without cannons or tanks.

"We now need much larger sums to finance the armaments, so it is necessary to send Golda to America to raise the funds we need," he wrote in his diary. "Now we need planes and tanks."

The last place Golda wanted to be was in New York. The Egyptian advance toward Tel Aviv was bringing its troops close to Kibbutz Revivim and she knew that her daughter, Sarah, and Revivim's twenty-nine other poorly armed pioneers couldn't possibly defend it. Jerusalemites needed water, food, and medical supplies. Overwhelmed by refugees suddenly free to flood the country, the Histadrut had to organize housing and food. And almost 30,000 Jews were penned up in Cyprus waiting to be brought to Israel.

But on May 16, two days after independence was declared, Henry Montor cabled to say that American Jewry had been moved by the dramatic ceremony in Tel Aviv. If Golda came personally, he thought she'd be able to coax another $50 million from the community. So she threw a hairbrush, toothbrush, and clean blouse into her bag, received a laissez-passer, the first travel document issued to a citizen of the State of Israel, and caught a plane to America.

This time, Golda didn't arrive in New York as a beggar for a band of starry-eyed Jewish optimists but as the official emissary of a Jewish government. Based out of a suite at the Sulgrave Hotel, her first luxury accommodation, she plotted a one-month tour to hammer home the stakes to America's Jews.

"We cannot go on without your help," she begged until she was hoarse, in St. Louis and Dallas, in Tulsa, Denver, Kansas City, and Philadelphia, at United Jewish Appeal dinners and at hastily arranged meetings with bankers and industrialists whose Jewish pride had been stirred for the first time. "What we ask of you is that you share in our responsibility, with everything that this implies—difficulties, problems, hardships and joys. Make up your minds and give me your answers."

They responded with $75 million.

Her mission complete, Golda returned to New York to find a way back to Israel. The day before she was due to fly home, she made a quick ride out to Brooklyn to say good-bye to old friends. En route, she stretched out her tired legs in the backseat of a taxi, worrying about what would come next. Sharett, the new foreign minister, had asked her to go to Moscow as Israel's first ambassador and she still hadn't figured out how to change his mind.

Her trip to the United States had been agony enough, especially with the news that the Jordanians had pushed the Jews out of the Old City of

Jerusalem after house-to-house fighting. Every attempt by the new Israel Defense Force to dislodge Arab troops from Latrun, which dominated the Jerusalem–Tel Aviv road, was failing. And the Syrians controlled much of the Galilee.

Even if the fledgling state had been triumphing on all fronts, Golda would have balked at her new assignment. For three decades, she'd dreamed of throwing herself into the nitty-gritty of building a Jewish homeland. Negotiating with the Russians on its behalf felt like a poor substitute, no matter how critical Ben-Gurion deemed such diplomatic negotiations to be.

As she was considering how to convince Sharett to send someone else, a car struck her cab and Golda was rammed forward, twisting her right leg, which had not healed correctly after it was hit by a cart during her time at Merhavia. She wound up at New York Hospital for Joint Diseases.

The accident could not have been more perfectly timed, or at least that's what Golda initially thought. Since Sharett and Ben-Gurion were anxious to get their new ambassador to Moscow quickly, she had an ideal excuse for suggesting that someone else be sent. But her hopes were quickly dashed. Forget it, Sharett cabled from Israel. Her name had already been submitted to the Russians and the press notified. No matter what, Golda was going to Moscow. Soon.

But Golda's recovery was anything but swift. Although her leg began to heal, she developed phlebitis and blood clots. And as she lingered in her hospital bed, her illness sparked a minor political scandal. Mapam, the Marxist Zionist party, decided that her leg problem was a "diplomatic illness" staged by Ben-Gurion to delay an exchange of ministers with Moscow to allow the U.S. ambassador to arrive in Israel before the Soviet ambassador and become the dean of the diplomatic corps.

Under pressure from both Mapam and the Soviets, who had decided to feel snubbed, Sharett barraged Golda with cables only partially disguised as polite inquiries about the state of her health. HOW DO YOU FEEL? WHEN CAN YOU LEAVE NEW YORK?

He also contrived a way to deal with the rumor that Golda was still hoping to wiggle out of the Moscow assignment. "DO YOU HAVE ANY OBJECTION TO APPOINTMENT OF SARAH AND ZECHARIAH AS RADIO OPERATORS IN MOSCOW EMBASSY?" he asked in a cable. Her daughter and fiancé, still in danger in the Negev, had become hostage to Golda's speedy recovery.

The onslaught from Tel Aviv, coupled with Golda's innate sense of indispensability, wore her down. Although her doctors strongly advised against it, she packed up her belongings and flew back to Tel Aviv, leaving herself with permanent leg problems.

When Golda and her staff gathered in Prague for their flight to Moscow, there was none of the giddy anticipation they might have felt if they'd been opening an embassy in Paris or Washington. Half the embassy staff had personal recollections of life inside what had become the Soviet Union, and several had relatives there. Golda herself was still haunted by memories of hunger and pogroms. And for her as a socialist, the Soviet Union was a graveyard of dreams of justice, egalitarianism, and freedom.

But in the schizophrenic world of war and diplomacy, the Soviet Union had proven a friend to Israel, an abrupt change of course for a county that had supported the Arabs, albeit halfheartedly, for three decades. The Soviets, however, were less interested in bolstering the Arabs than in ensuring that the British left the Middle East, and partition seemed the only strategy guaranteed to rid the region of Western powers.

So when the United Nations began debating the partition of Palestine, Soviet delegate Andrei Gromyko had set himself up as a staunch supporter of a Jewish homeland. "The heavy sacrifices of the Jewish people during the tyranny of Hitlerites in Europe emphasize the necessity and justify the demands of the Jews to create their own independent state in Palestine," he said.

The Soviet Union became the second country to grant official recognition to the provisional Israeli government, after Guatemala. And

Moscow had demonstrated its friendship in more concrete ways, arranging to sell the new state arms through Yugoslavia and Czechoslovakia while the United States maintained an embargo against arms sales in the region.

Although strongly pro-Western, Ben-Gurion and Sharett believed that Israel's survival depended on staying aloof from the Cold War. The military tide was turning, leaving Israel in firm control of the lower Galilee from Haifa to Lake Kinneret. Although they had been stalled in their attempts to retake Old Jerusalem, the IDF had stopped the Egyptian drive toward Tel Aviv. But the shaky truce negotiated in mid-July seemed unlikely to hold and they'd already fought off one attempt by the United Nations to reshape Israel into a small fragment of what they'd been promised in partition. So they couldn't afford to be ideologically picky about its friends. If Israel were to survive in a hostile neighborhood, they needed to build strategic bridges to a powerful country sending clear signals of amicability. Golda, they hoped, their most forceful emissary to the outside world, could be their architect.

So forty years after she and her family fled, Golda was returning with food in her belly, credentials as the minister plenipotentiary of a Jewish state, and a dizzyingly contradictory mission. On one hand, Israel was counting on her to build on those tentative signs of solidarity. On the other, she was to pry open the floodgates that would allow hundreds of thousands of Russian Jews to join the new exodus to the struggling state.

With no experience in diplomacy—an actual aversion to diplomacy—and only a halting knowledge of Russian, Golda was afraid that she wasn't up to the task. She needn't have worried. Stalin's seeming support for Israel had been little more than a ploy to remove Great Britain from a strategic region.

"The Jews in the civilized world are not a nation," Lenin taught. "The best Jews, those who are celebrated in world history, and have given the world foremost leaders of democracy and socialism, have never clamored against assimilation." Stalin had already decided that no matter their inclination, the Jews of the Soviet Union would be the Leninesque best Jews in the world.

*   *   *

The Soviet Foreign Ministry had ordered the Ministry of State Security to perform a background check on Golda, and after checking synagogue records and conducting interviews, the Ukrainian secret police gave her a clean bill of political health. Two of her aides were deemed to have "anti-Soviet propensities." Nonetheless, Foreign Minister Vyacheslav Mikhailovich Molotov approved their appointments, seemingly ignoring the State Security warning that the entire staff had been ordered "to establish contact with Jews in the Soviet Union and find the way to involve them in active pan-Zionist activity."

The Israeli Foreign Ministry had no budget yet, and the staff had all agreed to work for free, in exchange for housing, food, and cigarettes. Perhaps moved by that seeming idealism, or, more likely, by her own fervor, Golda had decided to run the embassy like a kibbutz.

But the Hotel Metropole, where they were delivered on September 2 and would remain until the Foreign Ministry found them a permanent home, looked nothing like a kibbutz. Reserved exclusively for foreigners, the posh art-nouveau edifice—the Tower of Babel, as Muscovites called the Metropole when it opened in 1901—dripped with thick Oriental carpets and huge cut-glass chandeliers, a stark contrast to the austerity of Israel. When Golda saw the first hotel bill, she panicked. The price of diplomacy was too high for a country without a functioning treasury. She grabbed Lou Kaddar, who'd been chosen to serve as her translator and aide, and went shopping for hot plates, butter, eggs, bread, and sausages. Borrowing crockery and cutlery from the hotel, since none was for sale in Moscow's half-empty shops, she instructed her staff to take only breakfast in the dining room and fend for themselves for lunch and dinner. Golda even set up a makeshift kitchen in her bathroom so she could cook Sabbath dinner for the single members of her staff.

Making do was the watchword in Moscow, late 1948. With Russia's farms and industry decimated by the German invasion and Stalin's collectivization, food was scarce, consumer goods nonexistent.

"What were they thinking, sending an old woman like Golda to a place like Russia after the war, to a devastated country?" thought Kaddar, who was thirty-five years old to Golda's fifty.

But the tiny diplomatic community took the Israelis under their wings, although not without causing some discomfort. The British consul in Jerusalem had warned his counterpart in Moscow that Golda was a "tough American Trades Union and Labour boss . . . not over intelligent but honest." When she appeared on the scene, the Moscow consulate wrote back that she lost no time "circularizing the whole of the Diplomatic Corps. This has caused some flutter amongst those of my diplomatic colleagues, and they are the majority, who have not recognized the State of Israel."

Still, the French told Kaddar where to find the best *kolkhozes*, farmers' markets, for fresh cabbage, onions, and potatoes. The British shared the addresses of the shops where bread was sold, at a premium, without endless lines. And the Americans offered the Israeli the handiest gift, the thick catalog of a Danish company that shipped in food and other necessities.

At Golda's request, Kaddar had purchased diplomacy books in Paris. But they offered no advice on Golda's first protocol crisis, what to wear for the formal presentation of her credentials. During her briefing at the Foreign Ministry, Golda asked what would be appropriate. Male diplomats, she was told, wore tails.

Having to worry about things like clothes was precisely the type of nonsense that made Golda despise diplomacy, and she was tempted to don the simple white cotton shift of a female kibbutznik and dub it their "national costume." But fiercely disciplined, Golda rarely gave in to petty temptations.

At 11 A.M. on September 8, she presented her credentials wearing a black evening dress she'd had made in Tel Aviv, a black velvet turban, and a $10 strand of fake pearls borrowed from her personal aide. The Foreign Ministry had requested that she deliver her remarks in English, but she refused, snapping, "I am not an Englishwoman or an American." Her ministry contact explained that they had no official Hebrew translator. Golda offered to bring one of her aides to interpret, a suggestion the Soviets declined. Nonetheless she spoke in Hebrew, knowing full well that the Russians present would have no idea what she was saying.

Settled in and officially credentialed, Golda was ready to work—to hammer out diplomatic agreements, open trade relations, test the waters about Jewish emigration, and make the rounds of Soviet offices to develop her contacts. But when she tried to reach out, she discovered that there were no contacts to be made, no diplomatic agreements in the offing, and no commercial relations in sight. The Soviets might have fought to receive a high-ranking Israeli ambassador, but they weren't interested in dealing with her. "From the moment they arrived in Moscow, Israeli diplomats were virtually unemployed," wrote Uri Bialin, an expert on Israeli foreign policy.

Twice, she met with Foreign Minister Molotov, but the conversations were diplomatic niceties, without substance. She put out feelers about loans or other financial assistance, but the Soviet Union was almost as broke as Israel. Mostly, she and her staff were ignored. Israel had nothing to bargain with but its neutrality, and Stalin was already convinced that Israel would become a lackey of the bourgeois imperialists.

"Golda had nothing to do but go from one cocktail party to another," said Kaddar, recalling the constant round of receptions.

The only relief she found at those dreary events was a petty game of diplomatic rebellion she played with Kaddar. Unaccustomed to speaking through an interpreter, Golda was impatient with Kaddar's insistence on translating even the most mundane of questions rather than simply answering them herself. The first time they attended a social event and Lou asked her, "How did you come to Moscow?" Golda snapped back, "What, you don't know?"

Over time, Golda began throwing out absurd answers. "We rode white donkeys," she responded one day. When a fellow diplomat asked, through Kaddar, "Where are you living?" Golda often said, politely, "In a tent," or "On the street."

Golda was exasperated at the utter boredom, and her annoyance was exacerbated by her sense of estrangement from events in Israel. Back home, the truce had collapsed, but the Israeli military had seized the entire Galilee and driven the Egyptians out of the Negev. While the Arab states were still refusing to accept defeat, Israel was on the cusp of the victory the world had assumed it could never achieve. Golda longed to be part of the moment.

Her tedium was compounded by a healthy streak of fear. Inside the hotel, a *smotryashaya po etazhu* sat by the elevator taking notes each time the Israelis left their rooms. When they left the hotel, the militia followed them. The Foreign Ministry sent them drivers and maids, whether Golda requested them or not. She'd brought a driver from Israel, but the Foreign Ministry refused to give him a driver's license.

Imagining that their rooms were searched when they left the hotel, they took to carrying their important papers with them. And warned that their rooms were bugged, Golda vowed to give the secret police nothing to hear. So if she wanted to have a private conversation with one person, they took a walk. When a large group had to meet, she convened the gathering in her enormous bathroom and turned on the taps, hoping that the noise would ruin any tapes. Every few minutes, Mordechai Namir, the legation counselor, yanked the chain on the old-fashioned toilet, just in case.

Golda could have handled both the monotony and the anxiety. Less sanguine about the possibilities of Soviet friendship than either Sharett or Ben-Gurion, she could even have handled being snubbed by officialdom. But Russia was home to three million Jews, her people, and she needed contact with them, and they with her, she believed, and she needed to open the door to their emigration to Israel.

Many clearly were anxious to leave. After the declaration of independence, the Soviet press had printed dozens of letters from Russian Jews congratulating the government for its support of the new Jewish state, and shortly after her arrival, Golda began hearing from them directly. "I would happily sacrifice my knowledge, experience, and if necessary, my life for the sake of reinforcing, developing, and strengthening the State of Israel," wrote one construction expert. A Jewish officer in the Red Army asked, "Can a Jew who fought four years against Fascism travel to his Homeland, to his loving country in order to be there at this hour with the finest of his people . . . ? Please print the answer to my question in a Soviet newspaper."

Promoting emigration from someone else's country, of course, isn't a very diplomatic mission, so Golda trod lightly. The first time she met with I. N. Bakulin, head of the Middle East Department of the Foreign Ministry, instead of broaching the possibility of Jewish emigration, she complained about British and American attempts to block the departure of Jews from the West. Bakulin, however, didn't rise to the bait. Rather, he sympathized with her plight, parroting the party line that emigration was necessary only where Jews lived with prejudice, the "capitalist countries."

She and Namir knocked on every possible official door, and the answer was always the same: a lecture on the evil influence of capitalism. "[It is] as though the problem of Jewish immigration to Israel exists in countries outside the 'East' bloc, and since [our] little state cannot absorb millions of Jews from the Diaspora, Jews have to fight for 'Socialism' in every country," Namir said in despair.

They had no more luck dealing with the Soviets about exceptional emigration requests. In late October, for example, Golda petitioned for exit permits for a small group of Israeli citizens who'd been trapped in Bessarabia and Bukovina by the outbreak of World War II. They received no answer.

One night a middle-aged Jew made it past security at the Metropole by waving his British Mandate passport. An ex-Communist who'd left Palestine to fight in the Spanish Civil War, he'd been wounded and transported to Russia for treatment. Unable to secure permission to leave, he appealed to Golda. But there was nothing she could do to help, just as there was nothing she could do for her second cousin from Pinsk, who showed up pleading for her intervention. "Not one Russian official would discuss Jewish emigration with me, not for individual Jews nor for the multitudes," she said.

Even meeting Soviet Jews proved impossible. Once word got out where the embassy was located, Jews gathered by the dozens in front of the hotel. But few dared enter. One afternoon Golda's daughter, Sarah, was stopped on the street by a man who introduced himself, in Hebrew, as a former member of Kibbutz Tel Yosef. Overjoyed to find another kibbutznik in Moscow, she urged him to come in and meet Golda. No, he replied. It's too dangerous.

The danger was torment for embassy staff members with family in Russia and for friends back home who'd asked them to make contact with relatives they hadn't heard from since the war. At first, Golda approached the authorities with requests for visits, but their coldness suggested how precarious the situation really was. Finally, she barred embassy personnel from seeing their family members. It was too risky, for all sides.

But obsessed with discovering whether Russia's Jews had clung to their historical identity after decades of repression, Golda had to do something. First, she threw Friday-night open houses. But while the reception room in her two-bedroom suite filled up with foreign reporters, a familiar cast of characters from the embassies, and the occasional visiting American, no Russian Jew ever dared attend.

Next, she sent information bulletins about Israel and the Israeli embassy to Jewish communities across Russia. Although she carefully avoided any anti-Soviet remarks, shortly after she submitted the first issue for printing, she was called to the Foreign Ministry and ordered to stop her attempts at illegal propaganda.

Finally, she took to attending the State Jewish Theater, standing conspicuously in the outer hall during intermission. Scores of Jews encircled her. An intrepid few sidled up with quick questions. But no one ever stayed long enough to converse.

Ironically, the only social contact Golda managed with a Russian Jew occurred at a reception celebrating the thirty-first anniversary of the October Revolution. It was one of those formal gatherings that Golda dreaded, hours of careful politesse among diplomats. But in the receiving line, she was introduced to the wife of Foreign Minister Molotov, Polina Zhemchuzhina, who approached her with a glass of vodka a few minutes later. "I have so many questions for you about Israel," Zhemchuzhina said in fluent Yiddish, grilling her about kibbutzim, the Negev, and Israeli youth movements.

Their conversation didn't last long because Molotov signaled to his wife that it was time to move on to other guests. But Zhemchuzhina remained long enough to offer Golda a bit of advice. "They told me that you're going to the very beautiful synagogue," she suggested, in fluent Yiddish. "Go, go there. The Jews want to see you."

Almost too stunned to respond, Golda asked where she'd learned to speak such perfect Yiddish. "*Ich bin a Yiddischer tochter,*" I'm a Jewish girl, replied Zhemchuzhina, née Pearl Karpovskaia. Then she took her leaving saying, "May things go well with you. If all goes well with you, things will go well with all the world's Jews."

Golda never saw Zhemchuzhina again. Within weeks of meeting Golda, she was arrested for treason and exiled to Kazakhstan.

If nothing else, Golda decided that she could at least be present, proudly and visibly present, not just as the ambassador from the State of Israel to the Soviet Union, but as the ambassador from the State of Israel to Moscow's Jews. Before she left Israel, Golda had instructed her male staff members to pack prayer shawls and prayer books. Then, during her first week at the Metropole, she sent one of them over to the Choral Synagogue, the only real functioning Jewish house of worship for Moscow's 500,000 Jews, to inform Rabbi Shloime Shliefer that the staff of the Israeli embassy would join him for services on Sabbath morning.

A secular Jew, Golda never went to synagogue, but that Saturday, September 11, she walked solemnly with her staff from the hotel to the Gorka, the Hill, as the Jews called the area around the synagogue, and climbed the steps into the women's balcony. Only a hundred or so old men had come to pray, the usual crowd in those grim days, and no one seemed to notice the solemn woman in the balcony or the young outsiders who were called up to read the Torah. Then, Rabbi Shliefer ended the service in an unusual way, not only with his standard prayer for the health of Stalin, but for the welfare of Golda Meyerson as well.

Suddenly, all eyes turned on the strangers, the Israelis. Still, no one dared approach. After the service, Golda stayed behind to talk to the rabbi, for whom she'd brought a Torah from Israel. Outside the synagogue, she lingered, hoping someone might dare, so caught up in tangle of emotions that she was unsure what street to take back to her hotel. An old man walked by her and mumbled in Yiddish, "Don't talk to me. I'll walk ahead. You follow." When they reached the Metropole, without looking back, he switched into Hebrew and recited the *Shehehiyanu*, the prayer for special occasions.

Praised are You, Eternal our God, Sovereign of the universe, for granting us life, for sustaining us, and for helping us to reach this day.

A savvy politician, Golda understood timing. So she waited two weeks to return to the synagogue, until Rosh Hashanah, the Jewish New Year. By then, the first warning had been issued, subtle but unambiguous in that climate of careful understatement and threats disguised as innuendo. The fact that it was written by Ilya Ehrenburg, Russia's most prominent Jewish journalist, a Bolshevik turned anti-Communist who'd transformed himself into a spokesman for Stalin, made the signal all the more dramatic. "Let there be no mistake about it," he wrote in *Pravda*. "The state of Israel has nothing to do with the Jews of the Soviet Union, where there is no Jewish problem and therefore no need for Israel. . . . And in any case, there is no such entity as the Jewish people. That is as ridiculous a concept as claiming that everybody who has red hair or a certain shape of nose belongs to one people."

About 2,000 Jews usually turned up for Rosh Hashanah, the rabbi had told Golda, so she wasn't surprised to find herself in a thin line of worshippers trickling into the immense synagogue on Arkhipovka Street. But that year the line never ended. Moscow's Jews kept coming, wizened old men and stolid babushkas, young couples carrying infants, teenagers who'd never heard a Hebrew prayer, children, army officers, the ragged, and the well-dressed filling the synagogue until no seats were left, overfilling the synagogue until there was no room to stand, squeezing into every corner until there was no space to yield. The women in the balcony sobbed as they approached Golda, reaching out to stroke her arm, to kiss her hands and the hem of her dress.

The outpouring was so unexpected, so improbable during that long winter of Stalin's supremacy, the welter of emotions—of pride and awe, longing and nationhood—almost too intense. Golda sat frozen to her chair, unable to speak, silently asking Russia's Jews for forgiveness for having doubted them. Then Israel's military attaché was called to read the Torah and proudly walked to the podium in his uniform, his yarmulke emblazoned with an Israeli flag. The service came to a halt as the congregation released the tension of its own ecstasy in a surge of tears.

When the shofar was blown for the final time, no one moved. Sensing that people needed to feel her—to feel Israel—move through their midst, Golda stood up and, without uttering a single word, made her way through the throng, suddenly desperate for space, for relief from the weight of her own emotion. But on the other side of the synagogue doors, as far as she could see, Moscow's Jews had gathered by the thousands—20,000; 30,000; as many as 50,000 by some estimates. When they saw her, they erupted in Hebrew, Russian, and Yiddish: *Golda shelanu; nasha Golda; Goldele, our Golda, our Golda, you must live.* From somewhere in the multitude, a voice sang out the opening strains of "Hatikvah."

Still, Golda couldn't find any words. Only after two of her staff members opened a path through the tumult and ushered her into a taxi did she find her voice. "*A danke eich vos ihr seit gebleiben Yidn,*" she cried out, leaning out of the window. "Thank you for having remained Jews."

The Israelis stayed away on the second day of Rosh Hashanah, too numb, too afraid for the city's Jews. When they appeared for Yom Kippur, the Day of Atonement, another enormous crowd awaited them. This time, so too did the militia, in force. But inside the synagogue, when the congregation intoned Yizkor, the prayer for the dead, Moscow's Jews prayed, too, for Israel's fallen.

Golda stayed all day, fasting for the first time in decades. Finally, Rabbi Shliefer arrived at the climax of the service, the Ne'ila. The shofar was blown for the last time and he chanted the closing phrase, *L'shanah ha'ba'ah b'Yerushalayim*, Next Year in Jerusalem.

"The words shook the sun as they looked up at me," Golda recalled. "It was the most passionate Zionist speech I ever heard."

No Russian newspaper mentioned the extraordinary events at the Choral Synagogue, the first unsanctioned public demonstration since the fall of 1927, when Trotskyites took to the streets and squares to protest Stalin's seizure of absolute power. A month later, the Yiddish newspaper *Eynikayt* was closed down, along with Yiddish theaters and the Yiddish publishing house Der Emes. Books by Jewish authors, including Boris Pasternak, Vasily Grossman, and Mikhail Svetlov, were removed from library shelves. And the government began to arrest leading Jews, including two winners of the Stalin Prize, Russia's leading Yiddish writers, a major general in the Red Army, and the head of the Soviet Information Bureau.

Stalin's campaign to stamp out all hints of a Jewish national consciousness was launched.

*   *   *

With Sharett and Ben-Gurion's hopes for friendship with the Soviets dashed, there was little point keeping someone of Golda's stature in Moscow. "One could send a dummy as Minister and it will have the same impact," Golda had reported. So on February 2, barely five months after Golda arrived, the government of Israel announced that Golda Meyerson would join the first cabinet, the only woman to be appointed, as minister of labor.

CHAPTER TEN

We do not rejoice in victories. We rejoice when a new kind of cotton is grown and when strawberries bloom in Israel.

On April 22, 1949, the day Golda moved home from Moscow, 2,463 bone-weary European refugees disembarked in Haifa and were led to a welcome center to be fed, deloused, clothed, and processed for Israeli identification cards. It wasn't a particularly frenzied day like the time when 15,000 new immigrants arrived all at once babbling in twenty-two languages with malnourished infants howling, their mothers pushing through a swarm backed up by the elderly, who could barely navigate the rickety gangplanks.

With independence, the floodgates had opened, and no one had the slightest idea where to put the patchwork of Polish professionals broken by years in concentration camps and illiterate Yemenites confused by electricity, of Hungarians confused by Libyan Jews who looked suspiciously like Arabs, and of Iraqis appalled at Czech women baring their legs against all the injunctions of the Torah.

The great Ingathering was one of those noble Zionist ideals that brought tears to the eyes of the Israelis. But that idyllic vision left no room for the messy reality that, unlike the early Zionist pioneers, the tens of thousands of immigrants who landed on the shores of the new country weren't starry-eyed romantics but refugees who'd survived pogroms, genocide, and internment. They needed homes, jobs, medical care, schools, language instruction, and vocational training. It was an emergency of staggering proportions and Israel, a beggar state surviving on the charity of increasingly tight-fisted philanthropists, had a meager national budget of less than $30 million.

A distracted idealist in matters other than defense, Ben-Gurion tossed out fanciful solutions—that every established family should take in some refugees or that they could build mud huts, Arab style. But even before the new state emptied the refugee camps on Cyprus of 24,500 inmates and the European DP camps of 239,000, before 250,000 Yemenite Jews lined up in Aden for Operation Magic Carpet and before Poland agreed to open the doors to allow 80,000 weary Jews to depart, Israel barely had enough housing for its residents.

Enter Golda, suffused with dreams of a Socialist Jewish state, a land of equality, justice, and full employment. The nature of the crisis, the nitty-gritty need for houses and jobs, was a perfect match for her literal style. Taut with pent-up energy after six months of enforced diplomatic ennui and driven by a need to prove that her lengthy isolation from the center of power was ill deserved, she threw herself into building a brand-new ministry from scratch until it became a domestic powerhouse and a political empire.

Three weeks after her return from Russia, just one month after Jordan signed an armistice agreement with Israel, Golda presented the Knesset, the new parliament, with her plan for building 30,000 new houses before the end of the year. Impossible! snapped Eliezer Kaplan, the minister of finance. The least expensive houses—300-square-foot boxes with outhouses—cost $1,300 to construct. Where was Israel going to get the money? I'll fly to America, Golda replied, always assuming that with a few weeks of schnorring—begging—she could shake loose the cash.

That promise didn't move the economists, who worried that spending on large-scale housing projects would destroy the current standard of living and pull money away from long-range development. "A Jewish state that aims at a high standard of living without unrestricted immigration . . . I for one don't see any need for it," Golda retorted. "I am always shocked by economists, themselves well fed, who concoct a theory according to which human beings cannot live with any dignity in the present."

Then city planners groused about the need for scientific data on the chosen sites, on soil and topography; mayors chimed in with their own ideas; and the Histadrut expected extensive consultations since they would supply both the labor and the cement.

The day before she was scheduled to leave for the United States to raise the funds she needed, Golda faced her biggest hurdle. The new state promised that Jews would finally have a voice, and the Knesset members were taking that promise literally in a chorus of monologues that routinely kept them in session long after midnight.

Ideologues on the right and the left flailed Golda for not depending more on private enterprise or for not moving more decidedly toward a planned economy. Shouldn't the houses be bigger and sturdier? others asked. Why are you building in the center of the country, which is already overcrowded? How much garden space will each unit have? What about schools and clinics? Hey, don't forget synagogues! representatives of religious parties insisted.

Golda had little patience for the endless prattling. There was no time for careful consideration, for negotiations with private companies looking for profits she couldn't afford, no time to develop mortgage financing programs, to order topographic surveys, to design ideal communities, or train skilled workers.

Would I prefer perfect small villages with shops, kindergartens, clinics, and public halls in ideal locations? she responded. Sure. But time is a luxury we don't have. For almost a year, the National Housing Department had been studying the problem, looking into alternative building techniques appropriate to a country without any lumber, steel, or plumbing industries, and the result was that 200,000 immigrants were living in tents or wooden shacks that flooded when it rained, baked in the torrid summer sun, and offered little protection in the damp cold of winter.

"We need cheap housing quickly" became her mantra, and Golda was willing to step on any toe that got in the way of her image of twelve or fifteen houses rising every day.

On May Day, she marched with 10,000 workers through the streets of Tel Aviv and from the podium expressed her hope that the following May Day would be celebrated in a Socialist state. Then she flew to the United States to ask Jewish capitalists to help her out of a bind. "I went to our Parliament two weeks ago . . . and presented a project for 30,000 housing units by the end of the year," she confessed, one of scores of guilt-inducing faux confessions she would make over the years. "Parliament approved it, and there was great joy in the country. But actually I did a strange thing: presented a project for which I don't have the money. It was an awful thing to do—to forge a signature to a check—but I have done it. . . . It is up to you either to keep these people in camps . . . or to . . . restore their dignity."

Leading Jewish fund-raisers were annoyed. Ben-Gurion was asking for contributions to the defense budget, the minister of social welfare was stumping for funds, and there was Golda claiming that her projects deserved the highest priority, in a year when giving was markedly down. But Golda's brand of candor, guilt inducement, and coquettishness never failed with American Jews. With the help of David Dubinsky, the president of the International Ladies Garment Workers Union, she kick-started a new approach to financing Israel's development, investment rather than charity, and created the Amun-Israeli Housing Corporation, which issued interest-bearing bonds backed by the Israeli government. "We pray for the day when we will no longer need philanthropic campaigns; we want to free ourselves of these kind and generous shackles," Golda declared, immediately selling $1 million of her new bonds to the ILGWU.

Back home, she moved on to her next drive, for employment. More than 5,000 people were registered for work at local employment exchanges, and most weeks half of them found nothing. Those numbers, she was quick to point out, didn't include elderly people who needed a bit of work to help make ends meet, women with children who would feel better if they were contributing to the social welfare, or those who'd simply given up. There was no shortage of work for the skilled. But most refugees, especially those from Arab countries, had neither trades nor professions, and their numbers doubled and tripled every month.

Golda's solution was a massive public works program of road and housing building, hospital construction, and tree planting that would guarantee every worker at least fifteen days of paid employment a month. For a tiny, impecunious state, the scope was staggering: a network of roads in the Galilee, the repair and broadening of roads in coastal areas, the rebuilding of the highway to Jerusalem, the opening of arteries to mines and remote kibbutzim in the Negev and to border settlements in the Judean Hills.

Again, the economists fought her, whipping out charts and graphs to show how much more economically productive it would be to channel the money into long-term economic development. "If we have to choose between making a road or vegetable growing, I'm for vegetables," she argued. "If we have to choose between a road and planning new orange groves, I'm for new orange groves. But the upshot is that we haven't one or the other, not because anyone has ulterior motives, heaven forbid, and not because anyone is trying to hide away from the difficulties." There just wasn't the money for both.

In the Knesset, the debate on Golda's programs turned into a vote of confidence on the government's record on the absorption of immigrants, and, predictably, Golda was heaped with ideological scorn from the left and the right and needled for every detail.

As usual, Golda prevailed, but, just as predictably, she had to find the money to finance her domestic revolution.

Returning to the American trough was becoming perilous. After more than a decade of raising enormous sums for Israel, Golda was equally weary—of the incessant begging and of Israel's dependence. She hit the wall when a man at a fund-raising event asked if some arrangement could be made to send care packages to Israel. Startled, she stopped cold on the stage. "Now this maybe seems funny to you, but it is not funny to me," she answered. "I am a citizen of Israel and I absolutely refuse to be classified as someone belonging to a people whose needs can be answered by packages."

The long-term solution, devised at a meeting of fifty American Jewish and Israeli leaders, was to expand on the approach Golda had developed for housing and convert Israel from a charity into an investment by issuing State of Israel Bonds. We "do not ask for any free ride to economic or political security," announced Golda, abruptly shifting her rhetoric. "Israel now stands on the threshold of an industrial revolution which must force the world to revise its concept of Israel as a poor country." All we need, she intoned, in another twenty cities, is $1.5 billion for a massive development program that will lead to Israel's "industrial revolution."

The major American Jewish fund-raisers grumbled that the new program would bleed money from their own campaigns and incur the wrath of the U.S. government. After Henry Morgenthau, secretary of the Treasury under Franklin Roosevelt, secured the approval of the Truman administration, the real concerns surfaced, that Israeli leaders like Golda didn't know how to manage money. "You had to be a little crazy, deciding to fund social insurance and free education when the State had no money and was surrounded by hostility," said Ralph Goldman, who worked in Ben-Gurion's office. "And the businessmen don't like crazy."

*   *   *

With the birth of Israel, the relationship between American Jewry and old Palestine had to change. "Every once in a while the Prime Minister of Israel or some other government member gets cables from good American friends who urge, 'For God's sake, don't do this!'" Golda complained. "You have taken a terrible decision. You are dependent upon the Jewish community of the United States. Now I don't say that every word of the

Prime Minister or any other cabinet minister is sacred or that there can be no mistakes, but have we no right to express our opinion without being punished immediately in dollars and cents? Is that the whip that is going to be held over the head of Israel forever? How much longer do you think can a community that has bought and bled in order to regain its dignity bear a situation of this kind?"

Golda had already ruffled some American feathers when she took over from Abba Eban as the chief Israeli delegate to the United Nations for the end of the seventh session of the General Assembly and used her bully pulpit to castigate the Soviet Union for its treatment of Jews. By what right does the State of Israel presume to speak in the name of a Jewish community outside its borders? many asked.

But no disagreement engendered more tension between America's Zionist leaders and Golda than Israel's economically risky decision to allow unfettered mass immigration into the new state. Israel's Law of Return extended citizenship to every Jew in the world, no matter his or her circumstances, and Golda took that piece of legislation literally. No matter how old, how ill, how crippled or poor, every Jew would be welcomed.

But the flood seemed unending. After the camps in Cyprus and Europe were emptied, almost all the remaining 37,000 Jews of Bulgaria and 8,000 more from Yugoslavia arrived. Another 20,000 made their way across the borders from Iran, 34,000 from Turkey, 32,000 from Libya, and 45,000 from Morocco, Tunisia, and Algeria. When the Yemenite Jews began streaming out of their mountain villages, Israeli planes had to fly eighty-nine trips a month to bring them to Tel Aviv.

Dozens of other Jewish communities were still in peril, and Golda fought tooth and nail to transfer them all to Israel: 110,000 Iraqi Jews, Kurdish Jews, the 106,000 survivors of the once proud Polish community.

"Mapai does not promise a land of plenty if it wins," said Golda on the eve of the 1951 Knesset election. "It promises only plenty of Jews and rationing to make that possible."

By the end of 1951, American Zionist leaders were appealing to the

Israelis to slow the pace or establish national quotas before decades of struggle were washed away in the tsunami. "You have heard that immigration will have to be curtailed," Golda responded at a UJA meeting in Chicago. "I don't believe it. Don't ask me how we are going to solve it if you don't help. But I know that there is not a single man or woman among us who would want the title of Minister or President or Prime Minister, who would want to live in the State of Israel, who would not rue the day the state had been established, if we have to reach the decision that there is a Jew anywhere to whom we have to deny admittance."

However, a growing of number of Israelis shared the American concern over the relentless wave that was, at once, both awe-inspiring and menacing. The face of Israel was changing at a dislocating pace, and European-born Israelis weren't sure what to make of the horde of dark-skinned, turbaned men and women who had never heard of Theodor Herzl, been inside a hospital, or flushed a toilet. Their food smelled different, their Hebrew lacked any twinge of *Yiddishkeit*, their prayers sounded foreign.

Making room for them, literally and figuratively, was jeopardizing everything Israelis had built. The economy buckled, the trade deficit sky-rocketed, inflation ran out of control, and the government kept demanding more belt-tightening. In August 1950, wartime rationing was extended to textiles and leather goods. By December 1951, the only unrationed products were bread, cream cheese, leben, and frozen fish filets.

In a new nation still teetering to find its political center, every election, national or local, turned into an anguished feud over "the crisis," both moral and practical. Every problem in the country, from loose cobbles on the pavement to the intermittence of telephone service, was blamed on the swarms of newcomers, and people began asking whether a tiny country living with a precarious peace could keep absorbing millions without upsetting its economic and social balance.

The answer was no, and Golda minced no words in hammering home that truth in a series of speeches that straddled the often fine line between dazzling leadership and public shaming. After shoes were rationed, she reduced the choice to the starkest of terms: "More immigrants or more shoes." When academics and professionals threatened to strike because they were earning the same wages as janitors, she met with them personally—to reprimand them for their greed. Israel had no millionaires and only one person with an income above 100,000 Israel pounds, roughly $275,000. Without any wealth to redistribute from the top, the only way to increase their salaries was to take it from the poor, Golda explained.

"I'm prepared to go not only from meeting to meeting but from door to door so that every single one of us may realize that when he sits at a table, when he puts on a suit, when he makes improvements on his house, does something for his child, he is doing so at the expense of all of us. Let's see how we'll enjoy life with this realization."

And month after month, she harangued workers about increasing productivity and lengthening the workweek. Disgusted with what she saw as the self-centeredness of trade union leaders, she abolished the trade union department in her ministry. "That is one job we certainly do not need to do," she sneered. The Histadrut, she insisted, would protect the workers while considering the needs of the state.

> The issue before us is simple and cruel. . . . Is there any connection between our talk in support of immigration and our deeds? . . . I am not worried about how soon a building worker will get the means to buy a refrigerator. I want to know by what means a Yemenite immigrant family will secure a roof.

*   *   *

When she returned from Moscow, Golda had been caught in the same housing crunch as Romanian and Yemenite refugees. The government hadn't yet arranged residences for its cabinet members and she was forced to beg for a home in Jerusalem, which Ben-Gurion had decreed the capital. Like many fortunate Jewish Jerusalemites, she wound up in an old Arab house in Talbieh abandoned by its owners in 1948, when Jewish residents of East Jerusalem fled to the Jewish West, and Arab residents of West Jerusalem took flight in the other direction.

The three-story Villa Harun al-Rashid, its name painted in Arabic on tiles inset above the second-floor balcony, was falling down. But the top floor had an amazing view over the city, and Golda, who was usually content with modest quarters, fixated on having an apartment there. The state engineers grumbled about the condition of the building and security officers worried about its proximity to the border, but no one balked at a cabinet member taking over an "abandoned" piece of Arab property. The government had already decided to allocate 400 such dwellings to senior officials.

All her life, Golda reminisced about her apartment at Villa Harun al-Rashid, about closing her door, brewing a cup of tea, and "feasting my eyes on Jerusalem's beauty," as she put it. But it was little more than the place that she scrubbed and tidied—her lifelong obsession since she admitted she thought best while cleaning—and a way station in her frenetic travels from Eilat to the Lebanese border, from New York and Los Angeles to London.

Overseeing a vast realm of housing projects, road building, refugee camps, and employment offices, Golda relentlessly visited every work site and development to keep up the tempo of transformation. Helping to scout out deserted Arab villages that could be quickly cleaned up for new immigrants, she then drove there to urge laborers to patch roofs more quickly. She planted the first trees in Eilat on the Red Sea, toured factories and industrial sites, and inspected model homes.

The bane of architects and builders alike, she examined every detail of construction, eviscerating them when their designs came up short. In Tiberias, she noticed that the kitchen windows above the sinks in a new development were well above eye level. "What, are you crazy?" she screamed at the foreman. "Are you idiots? A woman stands in the kitchen five hours cooking and you are forcing her to see the wall and not the Kinneret?" In Kiryat Shmona, the foundations were too high. "How would you like your wife to jump forty centimeters to the ground every time she wanted to empty the garbage pail?" she berated the engineers.

Invoking her mantra that "a progressive country cannot have islands of poverty," Golda established public works projects to increase Arab employment as well as Jewish. Knowing nothing about Arab society, but infused with her vision of an egalitarian nation, she reached out with particular concern to Arab women. "Arab women have to learn to work outside the home if they want to take their rightful place in Arab or Israeli society," she exhorted them at the opening of a handicraft center at the Jaffa sewing school, expressing the hope that an Arab woman would soon find a place in the Knesset.

When she attended the dedication of a new road into an Arab community, a local woman stopped her to offer a soft complaint. "To make us a road is very kind of you," the woman said. "But roads and cars are for men. We women still have to carry water pails on our heads from the floor of the valley up to the mountains." So Golda ordered new public works projects to link Arab villages up to main water lines.

Everywhere, she cracked the whip. Faster. Cheaper. Better. More gardens. More industry in the south. More productivity. More farms. More water in the Negev.

Scores of domestic problems wound up on her desk because no one else was willing to tackle them. When the Yemenites in the Misr refugee camp accused the brusque young Israeli workers there of treating them with disrespect by forcibly cutting off men's side locks and sending male physicians to treat women, it was Golda who had to find a way to deal with head lice and gynecology without offending religious sensibilities. After Yemenite Jews began wandering out of overcrowded refugee camps and offering to work at subminimum wage, Golda was called in. And when torrential rains turned the Sha'ar Ha'aliyah camp near Haifa into a swamp, Golda had to figure out where to move thousands of soggy refugees.

In between internal crises, she flew back and forth to the United States—to raise funds for the airlift of the Iraqi Jews, to sell Israel Bonds, for the Purim Victory Festival at Madison Square Garden, for meetings with U.S. government officials, and to lunch with Eleanor Roosevelt.

"Golda doesn't think in terms of hours of work or, 'I've finished my day's work,'" said Zvi Bar-Niv, her legal adviser. "She never closes the circle. . . . Working with her, we were all suicidal self-slave drivers."

Getting by on five hours of sleep a night, Golda was on a tear. There were too many fires to put out, too much cash to raise, too many people who didn't agree with her and needed to be pounded into submission. If the Ministry of Finance denied her money, she threatened to resign, and they always found a way to mollify her. At Passover in 1950, she grew obsessed with how little immigrants had for the holiday and demanded that Finance Minister Kaplan give the families what they needed for the seder. When he refused, she made herself comfortable and said, "I'm not leaving my office until I get the money. I won't prepare for the holiday as long as I know others can't."

Fighting with Golda, Kaplan knew, was an exercise in futility. Word around the Ministry of Finance was, "You better settle things with Zvi Bar-Niv (her legal adviser). If you don't, he'll go to the *balabusta* and she'll go to Ben-Gurion. Then it will cost more money."

There was a manic quality to Golda's tempo, beyond devotion to her cause or distress over the plight of the refugees, even beyond anxiety that she'd fail and be consigned to a trivial position as deputy prime minister or the head of what she called a "ministry of the superfluousness." Golda had no real life outside work to moor her, and she was battered by private blows that she bore with a phlegmatic stoicism widely seen as glacial detachment rather than as steely determination never to be seen as weak or vulnerable.

Although Golda always seemed strong, as she edged closer to the age of sixty, she was sick a great deal, a reality she studiously hid her from the public. Migraines continued to dog her, her gallbladder flared up repeatedly, and on a trip to St. Louis, she broke her shoulder. Finally, exhausted and overweight, in July 1954, she checked into Dr. Schlegel's clinic in Zurich, Switzerland, where she spent a week on a diet of juice and raw vegetables.

Things were strained with her daughter-in-law, Hannah, the daughter of old friends from America. Golda's delight at Hannah's marriage to Menachem had soured shortly after the couple moved to Zagreb, where Menachem was studying cello. Hannah got pregnant and returned to Jerusalem, where she believed the medical care would be more reliable. Furious that Hannah was ruining her son's career by forcing him to leave his studies, Golda pitched a fit. Hannah's estrangement from Menachem deepened, and she finally left him, turning Golda more bitterly against her. The birth of Meira Meyerson did nothing to calm the acrimony. Golda refused to have contact with her granddaughter, who was born with mild Down's syndrome, and insisted that the child be sent away to an institution. When Hannah decided to raise her daughter at home, Golda stopped talking to her and never so much as acknowledged the existence of Menachem's firstborn. "It was a terrible situation," recalled Ari Rath, the former editor of the *Jerusalem Post*. "Friends tried to mediate, but Golda was like a stone."

She saw Sarah and her grandchildren infrequently since visiting Revivim involved a long and difficult trip. And Golda was not the sort of mother who seemed to feel much need to spend intimate moments with her daughter or cuddle her children. "She was never the grandmother the kids ran up to hug or the grandmother with kids on her knees," said Rolf Kneller, a photojournalist who traveled with her regularly. "She was the grandmother who sat in an easy chair answering the questions of the children sitting on the floor."

Then, in May 1951, while she was in New York for a fund-raiser, Golda received a telegram that Morris, who had returned from Persia to be nearer his children and was living in her apartment while she was abroad, had had a fatal heart attack. Within weeks, David Remez, too, passed away. Although she and Morris had been separated for years and her relationship with Remez had grown more distant, for the first time in decades Golda was truly alone, detached even from distant intimacy.

Her father had died without seeing the foundation of the state, and her mother had languished, her memory gone, her eyesight fading. Clara was married and still living in Bridgeport, Connecticut. And Sheyna lived nearby, but she had never been known to offer much solace. Golda's superego, Sheyna nagged her incessantly—about her smoking, her children, her politics, and her soup.

Work—turning decades of dreams into the houses and jobs that would be the foundations of a Socialist state—became her life. Around her neck, Golda wore a brooch with an evil eye inscribed in Aramaic, "Let her own works praise her."

During her seven-year tenure as minister of labor, 120,000 refugee families moved into permanent homes, 80,000 received vocational training, and 400,000 began to work.

❋   ❋   ❋

In January 1952, Golda stood before the second Knesset and made her bid for posterity. For two years, she'd put out fires. It was time to demonstrate how serious she was about socialism.

> *It is a momentous occasion for any state when its legislative body opens its debate on a social insurance bill. May I be allowed to say that for our young country such an event is of seven-fold magnitude. . . . The aspiration for a just human society has characterized the Jewish people from its first appearance on the stage of history and inspired its prophets to fight for the cause of the poor and the widowed. These visions have left their imprint on the cultural development of mankind.*

She then laid out the first stage of a comprehensive national insurance plan for old-age pensions, widows' and orphans' benefits, maternity grants, and industrial accident insurance, deferring health, disability, and unemployment insurance until the nation had a functioning economy.

Half a century later, the program hardly sounds groundbreaking. But in 1952, few small countries—and not that many major ones—offered their citizenry the type of security that Golda sought to provide, or had already codified in her labor legislation. Israelis already had, by law, fixed hours of work with prescribed breaks, guaranteed overtime, a mandatory day off, and annual leave. Pregnant women received paid maternity leave from the state and could not be fired from their jobs

With the social security act, she moved on to address the problem of high infant mortality, especially among new immigrants and Arab women, with free medical care and hospitalization for pregnant women and direct subsidies to parents after the birth of a child. Men over the age of sixty-five and women over the age of sixty would receive full pensions. And provisions were made for the support of widows, widowers, orphans, and workers injured on the job.

It was Golda's bid to establish Israel as the sort of progressive social democracy she'd long envisioned, and she proved enough of a political shark to ram it through the Knesset despite the country's precarious economy. The Treasury department, under Finance Minister Levi Eshkol, predicted that the program, which would eat up more than 1 percent of the gross national product from day one, would foment economic disaster.

But it will be a social disaster if we don't do it, she told the Knesset. "The State of Israel will not tolerate within it poverty that shames human life."

*　　*　　*

Only once as minister of labor did Golda's zeal overstep her political acumen, but that gaffe dogged her for decades. No issue haunted Israel more than the prospect of *kulturkampf,* an open clash between the most religious of the state's Jews and its nonreligious majority. For ultra-Orthodox Israelis, the existence of a secular state was anathema, and the behavior of nonreligious Jews—women wearing skimpy bathing suits, men driving cars on Sabbath, kibbutzniks living out of wedlock—an abomination. Foreshadowing what would occur throughout the region in later years, they demanded that Talmudic law govern Israeli life. While the most rigid among them represented less than 10 percent of the population, the political parties that represented them constituted a vital part of Mapai's ruling coalition.

The fact of that coalition, of any coalition, was a deep disappointment to Ben-Gurion, who abhorred sharing power with any party, no matter its religiosity. But the first Knesset election had given Mapai only 46 of the Knesset's 120 seats. Despite deep popular misgivings, then, in exchange for their Knesset support, Ben-Gurion ceded moderate religious parties' control over marriage and divorce, persuading the public that in a time of economic uncertainty and political precariousness, Israel could not afford secular-religious confrontation.

But in 1953, Golda sparked precisely that type of conflict when she introduced the Bill for Compulsory Service for Women into the Knesset, an attempt to lift the blanket exemption from universal military service given to religious girls. Golda's plan would not require Orthodox women to spend time on active duty in the armed forces but to dedicate one year to work in hospitals, schools, or other community institutions under the supervision of the Ministry of Labor.

Golda, who'd been raised without any real religious training, had learned to tread lightly around the Orthodox after the storm over her appointment as head of the *yishuv*. So in drafting her military service bill, she had carefully consulted her moderate Orthodox political allies. But when the bill was announced, ultra-Orthodox rabbis stirred their community into a furor. Handbills distributed as far away as Brooklyn warned that Golda would send religious girls into brothels. Mobs of men and women marched on the Knesset, shrieking hysterically and attacking the police.

Furious at the reaction and believing that exempting religious girls from national service was an affront to their secular counterparts, Golda ignored the political consequences of open warfare with the Orthodoxy. "You will not force your way of life on us," she railed. "You will not terrorize a legislative body with demonstrations of hysterical women or by mob violence." After invoking the proud role Jewish women had played in the uprising of the Warsaw Ghetto, she asked, "If army life is degrading, why are they not concerned for the morals of their sons?"

Ultimately, the storm calmed and the bill passed, although it was largely ignored. But Golda's relationship with the religious community had been severely damaged, and the consequences were not long in coming.

As the 1955 elections approached, Ben-Gurion decided it was time for Mapai to take control over Tel Aviv, and that Golda should be his vehicle. Nothing was ever quite that simple in Israeli politics. Mayors, like prime ministers, were not elected directly, and Mapai did not win a majority of the City Council seats. The leadership turned, then, to their national coalition partners in the religious bloc for the two votes they needed to name a mayor.

Who do you want to install? the Orthodox councillors asked.

Golda?

They threw their votes to Chaim Levanon, the sitting mayor, leaving Tel Aviv out of Mapai's hands.

CHAPTER ELEVEN

*Internationalism doesn't mean the end of individual
nations—orchestras don't mean the end of violins.*

The torment of Moshe Sharett, Israel's foreign minister, began in the fall of 1953, when the brash young hawks of the Israel Defense Force stepped up Israel's response to terrorist attacks by moving beyond an eye for an eye to a tooth for twenty teeth. The armistice of 1949 had stilled the rumble of tanks and mortars belonging to national armies, but hundreds of Arab fedayeen who refused to renounce the struggle made their way across Israel's borders to blow up watertowers, murder villagers, and remind the Israelis of the high cost of victory. Between 1949 and 1954, more than one thousand incidents each month left Israelis intimidated in their own homeland.

Unable to prevent the incursions, Ben-Gurion had ordered deterrence through retaliation. Still the fedayeen wreaked nightly havoc on Israeli towns. Finally, Moshe Dayan, the new chief of operations of the Israel Defense Forces, urged ever more vicious retribution. On October 14, in reprisal for a fedayeen foray that killed a mother and two children, a special unit led by Ariel Sharon snuck across the Jordanian border to the town of Kibya, a fedayeen stronghold. When it left, sixty-nine civilians, half women and children, lay dead.

Horrified by the carnage, Sharett expected that Israel would issue a statement of regret. But ever protective of his military, Ben-Gurion refused and instead denied that any IDF unit was in the area that night.

Sharett and Ben-Gurion had been a tag team for two decades, the British-educated, multilingual diplomat the voice of prudent restraint against Ben-Gurion's daring. But the mercurial Ben-Gurion had fallen under the spell of the new generation of Israelis like Dayan, Sharon, and Shimon Peres, whose self-confidence bled unabashedly into arrogance. Teach the Arab governments a lesson so that they'll stop the fedayeen, they agitated. Ben-Gurion was enthralled.

Ben-Gurion began preparing for war, but Sharett kept finding ways to thwart his schemes to capture the Gaza Strip to clean out fedayeen camps, to help the Maronite Christians form their own state in southern Lebanon, even to scrap the 1949 armistice accords entirely. Livid, Ben-Gurion took to treating Sharett like a leper. When Sharett spoke, Ben-Gurion studiously read a newspaper or cut him off in mid-sentence. He stopped addressing or referring to Sharett by name. Finally, he stopped talking to him at all, reducing their communication to notes.

Sharett wasn't some minor politician who could be summarily dismissed. Having led the successful drive for partition at the United Nations, he was enormously well respected. But Ben-Gurion, who could be ruthless when thwarted, finally asked Golda what she thought of transferring Sharett from his cabinet post as foreign minister to the position of secretary-general of the party.

"But who will be foreign minister?" she replied.

"You," he responded.

No fan of diplomacy, Golda had long disparaged Sharett's job. "All a foreign minister does is talk and talk more," she told friends. But Golda was the most loyal of the Ben-Gurionists, admitting that if he asked her to jump out of a fifth-floor window, she would do so.

When rumors of Ben-Gurion's plan reached Sharett, the foreign minister was mortified. Golda had no education, no understanding of Arabs, no patience, and no appreciation of process, a sine qua non of diplomacy. "Does she know what it means to be a foreign minister?" he asked a colleague. "Does she think it's a cocktail party?"

But Ben-Gurion sent Sharett an ultimatum: Resign by 5 P.M. or I will. Ten days later, Golda Meir—having finally given in to Ben-Gurion's pressure to Hebraicize her name—was named Israel's second foreign minister.

Despite his mortification, for three days Sharett patiently briefed her, trying to impart decades of experience. "Golda didn't take a single note," said Sharett's son, Chaim. In fact, she dismissed her predecessor with a curt, "If I need something, I'll phone you." She never did.

The Foreign Ministry staff was horrified by this new mistress. Under Sharett, the ministry had been like a family; he trusted his employees and listened carefully to their arguments. Golda, however, had nothing but contempt for the Oxford/Cambridge crowd who saw three sides to every coin, and she did nothing to disguise it. She was a "hard woman," said Anne Marie Lambert, one of the deputy directors. "She had strong likes and dislikes and was not open to being convinced."

Everything about her style grated: her talent for simplification, her impatience with careful parsing of language, and, especially, her brusque style. "She liked to be the queen bee," recalled Esther Herlitz, a former Foreign Ministry official. "She didn't consult with foreign ministry people but with her political friends. . . . When people clocked in in the morning, she would stand there and watch to see if you were late."

Golda was no happier than her staff. When she interviewed Simcha Dinitz, then a junior ministry official, for a position on her staff, she asked him why he wanted the job. "Honestly, I really don't," he replied.

"That makes two of us because I don't like my job either," Golda said.

*   *   *

At the age of fifty-eight, Golda became the public face of Israel at the United Nations, in world capitals, and at international meetings. After spending time with her, Eleanor Roosevelt gushed, "She is a woman one cannot help but deeply respect and deeply love." She so charmed the notoriously diffident Charles de Gaulle that he deigned to cross the room to greet her, rather than forcing her to go to his side, as was his habit with other heads of state. During a tour of ten Latin American countries, she was awarded the National Order of the Southern Cross, Brazil's highest distinction for foreigners; the Order of the Great Liberator, Venezuela's greatest honor; and an honorary doctorate from the state university of Uruguay.

After Argentina appealed to the Security Council to condemn Israel for infringing on its sovereignty by kidnapping Adolf Eichmann, she brushed aside those who counseled her to pretend that Israel had not been involved and boldly admitted that Israel had violated the law, in the process delivering a lesson in the Holocaust that left delegates cringing in silence. "This is not a matter of revenge," she expounded. "As the Hebrew poet Bialik says, 'Revenge for a small child—the devil has not yet devised.'

"Will Eichmann's trial by the people to whose destruction he had dedicated all his energies constitute a threat to peace, even if the method of his apprehension in some way contravened Argentinean laws? Or didn't the threat to peace lie in the fact that Eichmann enjoyed freedom, Eichmann was not punished for his crimes. Eichmann was free to disseminate the position of his warped soul among the young generation."

The visitors in the gallery rose to their feet, applauding wildly.

Convinced that personal relationships trumped formal diplomacy, Golda spent months on the road, traveling across the United States and Canada, the Far East, Europe, Latin America, and Africa, one day dining with Emperor Hirohito, the next touring resettlement projects in northern Burma. Everywhere she went, Golda garnered a special measure of celebrity as the world's only female foreign minister. "Former Milwaukee Schoolteacher Is Israel's Iron-Willed Mother," trumpeted a headline in the *Montreal Gazette.* Before her arrival on a trip to Massachusetts, the *Boston Globe* announced, "Israeli FM 'Thinks Like a Man.'" The *New York Times* dubbed her the "Grandmother-Diplomat."

The arc of Golda's personal story—the impoverished Russian refugee who grew up in Milwaukee but gave up the comforts of America to build a Jewish homeland—captivated journalists. "A 20-Hour Workday," the *New York Tribune* declared, devoting more words to Golda's dark knitted dress and jacket, double strand of pearls, and silver brooch than to the politics of the Middle East.

Every fall, Golda appeared at the annual opening of the UN General Assembly, alternately playing Cassandra, the Trojan seeress who was never believed, and Poine, the goddess of vengeance. "What is the use or realism or justice of policies and attitudes based on the fiction that Israel is not there or will somehow disappear?" she asked after a Saudi delegate requested UN assistance in returning the 1.2 million Jews of Israel to their "countries of origin."

Little by little, with the simplest of rhetoric and the power of personality, Golda garnered enormous sympathy, for herself and thus for the country she represented.

"Like the Biblical Rachel sorrowing over her children, Israel's foreign minister Golda Meir stood in the rostrum of the United Nations this week calling for disarmament and direct negotiations with the Arab States for a treaty of peace," opined the *Telegram*, Toronto's afternoon newspaper.

＊　　＊　　＊

Outside of Israel, Golda radiated power, but that image camouflaged the reality that in the formulation of the major thrusts of Israeli foreign policy, she was a bit player. Ben-Gurion thought of the Foreign Ministry as little more than the PR wing of the Defense Ministry. So while Golda might have been the face of the nation globally, in all substantive matters, she was a marionette, with Ben-Gurion playing the puppet master.

The first hint of the role to which she'd been relegated was Ben-Gurion's attempt to appoint Abba Eban, Israel's ambassador to the United States and the United Nations, as the prime minister's chief adviser on foreign affairs, a sort of "watchdog over the new foreign minister," as

Eban described it. Extraordinarily eloquent and fluent in ten languages, Eban was everything Golda hated about diplomats: subtle, patient, and urbane. And Eban was no fan of Golda's truculence. "This is an infallible prescription for antagonistic explosions," Eban told Ben-Gurion when the prime minister suggested the new arrangement. At lunch at Golda's home, the two agreed that Eban should decline Ben-Gurion's proposition and that "she and I would be happy and creative in proportion to the geographic distance separating us from each other," Eban wrote.

But such easy accommodation with Shimon Peres, Ben-Gurion's fair-haired boy in the Defense Ministry, eluded her. An odd duck even in that swarm of idiosyncratic drakes, Peres was no Eban. The prime minister's personal shrewd operator, Peres was an ambitious technocrat, the architect of a dozen bold schemes that Ben-Gurion adored.

Golda's first collision with Peres, whose title of the deputy director general of the Ministry of Defense belied his importance, was sparked by Peres' convoluted negotiations with France, the only major power willing to sell Israel weapons. Well before Golda became foreign minister, Peres had become a familiar figure in Paris, sharing aperitifs with the guys at the French defense ministry and reporting back only to Ben-Gurion. Once Golda was appointed, Peres never bothered to introduce Golda to his contacts or bring her into his consultations. He studiously ignored her.

Four days after Golda took office, Peres and his alter ego, Moshe Dayan, flew to France for two days of meetings with Defense Minister Maurice Bourges-Manoury and emerged with a $100 million weapons deal. Neither Golda nor the Israeli ambassador in Paris was informed of the negotiation.

One of the first contacts Golda wanted to make was with Jean Monnet, the architect of the European Common Market, hoping to arrange preferential treatment for Israeli products. "Do you know Mr. Peres?" Monnet asked at their first meeting. She knew him, of course, but she hadn't been told that Monnet did.

"Who speaks for Israel, me or Peres?" Golda complained to Ben-Gurion, who soothed her bruised ego and promised to have Peres keep her in his loop. He never did.

"I told G[olda] that I am worried and regret her suspicions—which are completely groundless," Ben-Gurion wrote in his diary after Golda complained that she was being pushed aside by Peres. "I said that my work will be impossible if she goes around with such a feeling."

Peres was too valuable as the nation's behind-the-scenes arms wheeler-dealer and Golda as Israel's front woman for Ben-Gurion to choose between them. And firmly entrenched in a political culture in which dirty laundry was never aired in public, Golda lacked the internal clout to make an end run. All she could do, then, was threaten to resign, as she did regularly. But no one took her threats seriously.

"I want to quit," she moaned to her friend Amiel Najar, a former Israeli ambassador, while the two floated on a boat down the Seine. "I'm fed up."

Najar laughed. "You will never quit," he said.

Golda looked offended. "Why do you say that?"

Najar knew Golda well. "Because you adore what you are doing," he said.

*   *   *

In late July 1956, Gamal Abdel Nasser announced the nationalization of the Suez Canal Company, sending France and Britain into a tailspin. His action was yet another blow to their imperial pretensions, and their pique was heightened by concern over Nasser's growing alliance with the Soviet Union and the potential jeopardy it posed to the two-thirds of Europe's oil imported through the canal.

During Peres' next trip to Paris, the French defense minister briefed him on a nascent British-French stratagem to retake the canal by force and asked directly, "Can we work together?" Ben-Gurion was thrilled at the prospect. Not only had Nasser defied international law and United Nations resolutions by closing the canal to Israeli shipping; he was both harboring and encouraging the fedayeen whose forays against Israel had not let up. The proposed military action offered a perfect opportunity both to regain use of the waterway and to wipe out the terrorist camps.

Ben-Gurion decided to send a team to Paris to discuss the matter with Prime Minister Guy Mollet, Foreign Minister Christian Pineau, and Defense Minister Bourges-Manoury. Peres tried to exclude Golda from the negotiations but, for once, Ben-Gurion overruled him. So in late September, Golda, Peres, and Dayan boarded an old French Lancaster bomber sent secretly to fetch them. Few issues worried Golda more than Nasser's refusal to allow Israeli ships to navigate the canal and the nests of fedayeen he harbored in the Sinai Peninsula and the Gaza Strip. They were, she believed, the harbingers of war. But she was skeptical about Peres' rosy reports about French devotion to Israel and anxious to hear personally from the French prime minister.

But Prime Minister Mollet didn't show up as promised. And the French foreign minister suddenly complained that the British had cooled to the idea, or at least were seeking a postponement in deference to the opposition of the United States. Nonetheless, he said, France was ready to move in a joint Franco-Israeli action within two weeks. Golda concluded that Peres, who had a tendency to share only the most positive details of plans he advocated, had advertised French intentions dishonestly, although Ben-Gurion remained seduced by what he saw as a historic opportunity.

After a few weeks, the British came around, and final plans for the French-British-Israeli action were hatched at a meeting in Paris between Ben-Gurion, Mollet, and British foreign minister Anthony Eden, with Dayan and Peres as consultants. The scenario they developed was a delusionally eccentric plot to seize the canal while providing the British and French with diplomatic "cover." Israel would drop paratroopers behind Egyptian lines in the Sinai Desert and advance toward the canal in a cooked-up "reprisal" raid for fedayeen attacks. Feigning shock and dismay at the threat to the waterway, Britain and France would issue an ultimatum that both sides retire from the canal and allow an international force to secure it. When Egypt refused, the Anglo-French force would bomb Nasser's military airfields. Israel, in other words, would play the aggressor, leaving France and Britain in the role of the angels of peace.

In the late afternoon on October 29, 400 Israeli paratroopers dropped into the Sinai, as planned, and by the following morning were speeding across the desert toward the canal. The ultimatum was announced on schedule. Israel accepted; Egypt did not. Then the elaborate scheme fell apart.

The British and French delayed bombing for twelve hours, leaving the world convinced that Israel was acting alone. Before the Europeans could land any troops in Egypt, the international pressure for a cease-fire crashed down on Israel, including a chilling communiqué from Soviet premier Nikolai Bulganin threatening that his government was "at this moment taking steps to put an end to the war and to restrain the aggressors."

Nonetheless, the British pressured Israel to resist the outcry for a truce so that European forces could land on the Egyptian coast. But by the time the British and French airborne troops began scrambling for a foothold there, the war was over. In one hundred hours, Israel had captured the entire Sinai Peninsula, the Gaza Strip, and Sharm el-Sheikh, the promontory controlling the Straits of Tiran.

Ben-Gurion, Peres, and Dayan, however, had seriously overestimated the backbone of their French and British allies and the international fury they would provoke. When the wrath of the United States and the United Nations swept over them, Britain and France buckled, leaving Israel with none of the promised political cover. The General Assembly voted 95–1, Israel being the lone opposing vote, to order Israel to leave the Sinai and the Gaza Strip immediately. And President Dwight D. Eisenhower cautioned that failure to do so would "impair the friendly collaboration between our two countries."

Golda was dispatched to persuade world leaders that Israel had acted in self-defense and that retreat with no resolution of Egypt's defiance of international laws on the free movement of ships and encouragement of the fedayeen would do nothing more than sow the seed for the next conflict. In meetings with U.S. officials, in the UN delegates' lounge and from the rostrum at the Security Council, Golda pointed out that the aggressor was being confused for the aggressee.

> *Israel's people went into the desert or struck roots in stony hillsides to establish new villages, to build roads, houses, schools and hospitals, while Arab terrorists, entering from Egypt and Jordan, were sent in to kill and destroy. Israel dug wells, brought water in pipes from great distances; Egypt sent in fedayeen to blow up the wells and the pipes. . . . A comfortable division has been made. The Arab states were arbitrarily granted the right to make war; Israel was arbitrarily given the responsibility of keeping the peace. But belligerency is not a one-way street.*
>
> *Over and over again the Israeli government has held out its hand in peace to its neighbors. But to no avail. . . . What ought to be done now? Are we to go back to an armistice regime which has brought anything but peace and which Egypt has derisively flouted? Shall the Sinai Desert again breed nests of fedayeen . . . ?*

But with the United States taking out its ire at France and Britain on Israel, the Western powers squabbling, and the Soviet bloc enjoying the show, Golda's folksy simplicity struck no chord.

Unlike Dayan and Peres, Golda had no coherent strategy for dealing with Arab hostility. Although she was a champion hater, both globally and personally, she didn't hate the Arabs or consider them to be implacable enemies. Her oft-repeated criticism of Nasser for what she saw as his indifference to the plight of his own people wasn't mere platitudes. She was keenly aware that her fate was bound up with that of her neighbors.

But Golda had a limited repertoire of response to problems and approached peacemaking as she did the lack of housing and jobs: devise a solution and bludgeon anyone who opposed it into submission. Lacking any understanding of Arabs or any appreciation for the subtleties of diplomacy, in the wake of the international outrage, she reduced Israel's dilemma to the most simplistic of formulae: if Israel withdraws, the

Egyptians will send the fedayeen back across Israel's borders and again deny Israel an economically vital outlet to the Red Sea. The UN will say nothing about Egypt's behavior, reserving its condemnations for Israel should the nation act to defend itself. The only solution she could imagine was: don't withdraw unless Egypt agrees to behave itself.

"It wasn't just arrogance on Golda's part," said Herlitz. "All she could think of was to make us more secure, and she really thought that we could hold on to everything we'd captured, that it would endanger Israel not to do so. After all, she thought, other countries have conquered territory and kept it. Why should Israel be any different?"

Golda hadn't expected to make much headway at the United Nations; its demographic makeup was tilted heavily toward Arab nations and the Soviet bloc. But she still looked to the United States as the paragon of justice that would not turn its back on the Jewish people. Worried about open confrontation with the Soviet Union, however, Eisenhower showed no sympathy for Israel's predicament.

Those were terrible days for Golda and the Israelis at the United Nations as they worked to make their case to an impatient world. A night owl, Golda couldn't stand to be alone. When staff members showed up at the Essex House Hotel with the evening cables, they found her drinking coffee, and they couldn't escape her need to talk until 1 or 2 A.M. Golda was her own doctor, they joked, making her own medicine of Chesterfields and caffeine. "We finally established a rotation," recalled Herlitz. "Everyone would take a turn at a long night with Golda so that someone would be awake the next day at the UN."

In between negotiations, Golda went on the stump, convinced she could move Americans, especially American Jews, to pressure Eisenhower. "Had we waited for our Pearl Harbor, we wouldn't be here to tell the tale," she told an audience in Baltimore. Can you imagine what it's like to live with editorials in Arab newspapers saying that only Israel's disappearance can bring peace? What it's like to be a citizen of a small country without a single alliance when Soviet dignitaries regularly visit Cairo, Damascus, and Beirut?

Initially, Golda was articulating Ben-Gurion's position as well as her own. But when the French foreign minister mooted a compromise, Ben-Gurion backed down for fear of jeopardizing Israel's partnership with France. The scenario he and Eban devised in negotiation with the Americans was that Golda would deliver a speech to the UN announcing that Israel would withdraw from the Egyptian territories on three conditions: (a) that Israeli ships be allowed to pass through the Straits of Tiran, (b) that United Nations troops would monitor the cease-fire at Sharm el-Sheikh and administer the Gaza strip, and (c) that Israel would have the right to strike back if Egypt tried to prevent Israeli navigation or allowed fedayeen to cross its borders. As soon as she did so, Henry Cabot Lodge, the U.S. ambassador to the UN, would rise and pledge U.S. support for those understandings.

Eban hammered out the text of Golda's speech, with Secretary of State John Foster Dulles checking every syllable and comma. When Golda received it, she balked. She understood the dangers of defying world opinion. But withdrawal with no solution to the underlying problems, she believed, would lay the groundwork for another war, and she certainly saw no reason to trust Eisenhower or the United Nations.

"Let's just have it printed and distributed to all the delegates," she suggested to her staff—many of whom shared her sentiments—at a late meeting the night before she was to deliver the speech. "If you give that statement, you will be a traitor," warned her friend Amiel Najar. Eban, who'd patiently cut the deal and organized delegates to support it, walked out, slamming the door behind him.

Golda called Ben-Gurion and begged for a delay, but the deal had been sealed. The next morning she delivered the prepared remarks. The Americans, however, did not live up to their end of the bargain. As promised, Lodge guaranteed Israel's right to free passage through the Straits of Tiran. But he made no mention of Israel's right to retaliate or protecting Israel from the fedayeen in Gaza, saying only that the future of Gaza would have to be worked out.

When Golda left the UN that day, "she looked like someone who had

had the wind completely knocked out of her sails," said Meron Medzini, the son of Golda's old friend Regina. "She appeared to be completely losing it."

The next morning, a friend asked her what she'd done the night before. "I washed my underwear the entire night," Golda replied. "At least something should be clean around me."

*   *   *

Golda felt almost as sullied by Ben-Gurion's dealings with Germany. Although the prime minister offered up high-minded rhetoric about eschewing racism to justify his attempts at rapprochement with Bonn, his motivation was practical: reparations and arms from the New Germany, as he called it, could help keep Israel strong. Golda, however, was convinced that Germany was not all that "new," or that it could not be so long as the government and bureaucracy were laced with former Nazis.

The two had faced off over Germany while Golda was still minister of labor, when Ben-Gurion negotiated reparations to indemnify Israel for the cost of resettling Holocaust survivors. Israel exploded in an emotional gale the minute the $824 million deal was made public. Blood money! survivors protested. When the Knesset met to discuss the agreement, thousands of people gathered outside screaming "Treason!" They burned cars and threw rocks through the windows of the Knesset building. Inside, the tense debate was interrupted repeatedly by the wail of police cars and the explosion of gas grenades.

The next day, with calm restored but emotions still running high, the Knesset reconvened to vote. The chamber was packed, one Knesset member carried in on a stretcher from the hospital, where he was recovering from a heart attack. But Golda did not appear. Party discipline required that she vote with Ben-Gurion. Rather than do so, she stayed home sick.

As foreign minister, however, Golda couldn't easily evade the Germany question since Ben-Gurion was hell-bent on wringing everything he could out of German chancellor Konrad Adenauer while the guilt was still fresh. First, it was a few trade deals. Then, Shimon Peres took time out from seducing the French to play on German shame to secure weapons and technical assistance in missile development. Finally, Ben-Gurion himself met with Adenauer to negotiate a $500 million deal for military equipment and development aid.

Golda tried to ignore the relationship, the rumors that Germany had given Israel $1.5 million for an atom smasher, the deals for helicopters, torpedo boats, and antitank missiles. Then, her old friend Isser Harel, Israel's master spy, discovered that German scientists were helping the Egyptians build weapons. Nasser had long offered a comfortable asylum to Nazis on the run—propagandists, Gestapo officers, and concentration camp doctors. But in 1958, he went a step further, engaging the services of more than two dozen German rocket scientists to help develop his armaments industry.

German complicity with Egyptian military development was too much for Golda to swallow, so she and Harel confronted Ben-Gurion with evidence of the shape of his New Germany. Refusing to believe that the Bonn government was involved or to allow Golda to handle the matter herself, he sent Peres to ask the German defense minister to find a way to end the collaboration.

When the German explained that his government was powerless to control the employment of German citizens, Ben-Gurion and Peres were mollified. Golda was not, dubbing the German response "evasive and complicit." Desperate to avoid jeopardizing the weapons deals, Ben-Gurion began scheming to keep the news from going public. But Golda urged Harel to leak his evidence to newspaper editors in Israel and in Germany.

"There is no doubt that the motives of this evil crew are, on the one hand, the lust for gold and on the other, a Nazi inclination to hatred of Israel and the destruction of Jews," Golda told the Knesset in a direct slap in Ben-Gurion's face. Flinging around words like "weapons of mass destruction," she broke with party discipline and demanded—to no avail—that the German government stop the scientists.

Gradually, Germany became the third-largest consumer of Israeli exports, its second most generous supplier of weapons, and a major purchaser of Uzi submachine guns. Still, Golda steered clear of any involvement, leaving those negotiations to Peres and a small office in Cologne. But at every turn, she derided Ben-Gurion's depiction of the "new" Germany, where one-third of Adenauer's cabinet, one-quarter of the Bundestag members, and a huge percentage of civil servants, including eight ambassadors, were former members of the Nazi Party.

But over time, consumed with the economic and political challenges rocking Israel, Golda was worn down and swung into the camp of realpolitik and became an advocate of direct ties with Germany. Her new thinking was tested, however, when the first German ambassador to Israel was named. Dr. Rolf Friedmann Pauls was an experienced diplomat who had been central to the negotiation of reparations with Israel. But he was a former officer of the Wehrmacht, the Nazi armed forces, which the Nuremberg tribunal had nearly proscribed as a criminal organization. Why not someone younger, an intellectual, or a former member of the resistance? she asked.

Knesset members exhorted Golda to reject Pauls' appointment, accusing Bonn of attempting to use it to clear the name of the Wehrmacht. Rumors swirled around the country that Golda would not appear when Pauls presented his credentials to President Zalman Shazar.

But Golda saw Pauls' appointment as an opportunity. Unlike the intellectuals or resistance fighters with whom she'd be more comfortable, he had clout back home. Four days after his arrival in August 1965, Golda accompanied Pauls as he presented his credentials to Shazar. When the Israeli police band played "Deutschland über Alles," the German national anthem, an anguished roar erupted from the crowd of demonstrators, many elderly survivors and ghetto fighters carrying placards reading, ZKOR, ZKOR, REMEMBER, REMEMBER.

Demonstrating her customary indifference to popular sentiment, Golda was not fazed. "If we're serious about a special relationship with

Germany, we have to come to terms with Pauls' generation," she said
flatly, sounding suspiciously like one of the diplomats she'd long derided.

*   *   *

Golda's distance from the center of foreign policy making continued to
dog her. In early 1958, her ambassador in Paris was asked to explain why
an Israeli plane that made an emergency landing at a French military
airfield in Algeria was filled with arms destined for Latin America. She
could honestly report that she knew nothing about it. Providing weapons
to Anastasio Somoza, Nicaragua's dictator, was a Peres connivance to
which she wasn't privy.

Peres was a wellspring of wild machinations that Golda was regularly
forced to clean up. In early 1959, he asked the French if Israel might lease
French Guiana, an underpopulated colony rich in natural resources, "a
catastrophe, colonialism, imperialism," one of Golda's closest allies called
the notion. "Golda will let it pass over her dead body."

While Golda did manage to keep Ben-Gurion from embracing some
of Peres' more absurd ploys, including French Guiana, she never gained
the status that she craved. And shortly before the 1959 Knesset elections,
Ben-Gurion again urged Abba Eban to return home. This time, the
young diplomat, who'd long served as the Israeli ambassador both to the
United Nations and to Washington, didn't resist because Ben-Gurion offered him a position in his next cabinet. Eban assumed Golda was about
to step down and that he would be named foreign minister.

After Mapai scored its greatest electoral victory, Eban waited for Golda
to announce her resignation and make room for him at the top. Instead
she checked into a rest home outside Jerusalem, allowing rumors to fester, forcing her veteran comrades to make pilgrimages to her covered
porch to plead for her to remain in office. When she finally agreed, Eban
was appointed to serve as minister without portfolio, which gave him rank
without function.

"Wait your turn," Golda told him. "It will come soon enough."

Inside the Foreign Ministry, the disappointment was palpable. The

staff's distaste for Golda had only grown keener as she ignored and isolated them, routinely bringing in her own people as ambassadors and
mission heads. The staff union had even lodged a formal complaint about
her habit of reserving political plums for her cronies, but Golda had
quickly slapped down their leaders. "She was extremely vindictive," said
Benny Morris, whose father, Ya'akov, was head of the union. "He'd been
ambassador to New Zealand but she froze his career."

Having assumed that Golda would show special interest in their ca-
reers, the female members of the ministry were particularly frustrated
since Golda was never known for offering particular support to women.
"Many people thought that she hated women," said Herlitz. "I don't know
whether that was true, but she certainly wasn't a friend to us."

The men weren't much happier since Golda was notoriously quick to
ax subordinates whose behavior disappointed her standards. When she
discovered that one of her consul generals had been less than exact with
his per diem reports, she instructed her director general to demand the
offender's resignation. "If he doesn't want to leave, report him to the police," Golda said, allowing the man no opportunity to defend himself.

But her greatest wrath was reserved for womanizers, sexual harassers,
and men who had affairs while representing Israel overseas. "I don't care
about people's behavior at home," Golda said, after firing both Israel's
ambassador to Mexico and an Israeli embassy staff member in Brazil for
such indiscretions. "But abroad, they represent Israel and we cannot tolerate immoral people."

Her status within the Foreign Ministry was undermined as well by the
universal awareness of Golda's relative powerlessness. "If there was a
problem, she went to Ben-Gurion with her entourage and almost never
spoke," recalled Herlitz. "She simply took instructions. He was the foreign minister. She was just window dressing."

But no matter how much Ben-Gurion ignored or humiliated her,
Golda retained a passionate belief in his leadership. Their styles, however,
clashed violently. Perhaps the clearest example of that disjunction was
their long feud over Dimona, Israel's allegedly secret nuclear development

program, another intrigue cooked up by Peres. For Ben-Gurion, the nuclear program was part and parcel of his romance with science, his belief that nuclear energy could compensate for Israel's lack of oil, scarcity of water, dearth of minerals, and nightmarish defense situation. He was willing to spend almost any amount of money and risk any diplomatic brouhaha to build a nuclear facility in the Negev Desert. And Peres had convinced the French to help.

Israel's scientific establishment opposed the plan, calling it dangerous and irresponsible. And party leaders like Levi Eshkol and Pinhas Sapir worried about the expense, fearing that Ben-Gurion would sink millions down what they suspected would become a black hole. Golda didn't oppose the project on financial or ethical grounds, but she sensed that Israel would run into trouble over the elaborate web of lies Ben-Gurion and Peres wove to hide the project from view.

The first sign of danger occurred while excavators were digging the first massive holes in the desert for the complex of buildings. Isser Harel, the head of Mossad, Israel's external spy agency, learned that a Soviet satellite had overflown and photographed the site and that Foreign Minister Gromyko had gone to Washington shortly thereafter. For Harel and Golda, to whom he routinely reported such information, the conclusion was obvious: the Israelis had been caught and would pay in the most expensive of currencies, the goodwill of the United States.

Ben-Gurion was sufficiently alarmed by their report to call Peres home from Dakar, where he was attending the installation of Léopold Senghor as the first president of Senegal. "The situation is grave," Harel told him when his plane landed, intimating that Ben-Gurion himself, or at least Golda, would have to fly to the United States to preempt disaster. "So what?" asked Peres, sanguine as ever. "They have photographs of holes in the ground that could be for anything." Even if Harel was right, Peres urged Ben-Gurion to wait until the United States asked about the excavations.

His optimism about keeping such a massive secret was not, of course, warranted. The Americans quickly discovered that Israel was building a nuclear plant and sought a prompt explanation. Ben-Gurion stalled for time and then lied, writing, "The new Israeli reactor, now in the early stages of construction, is for peaceful purposes only." President John F. Kennedy didn't accept that story and insisted on sending American scientists on an inspection tour. Ben-Gurion agreed but delayed the visit for five months. When the scientists finally arrived, they were carefully guided around the facility and briefed on Israel's attempt to use nuclear power to bring its desert to bloom and desalinize water.

The inspectors swallowed that tale, which bought Ben-Gurion another two years. But no matter how far Ben-Gurion and Peres were willing to go in shading the truth, they could not keep the lid on what Eban called "an enormous alligator stranded on dry land." By June 1963, JFK was again pressuring Israel, and Golda was fed up with the deceptions. "Regarding Dimona, there is no need to stop the work . . . but . . . the issue is whether we should tell them the truth or not," she said in a policy meeting. "On this issue I had reservations from the very outset. . . . I was always of the opinion that we should tell them the truth and explain why. . . . If we deny that Dimona exists then it cannot be used as a source for bargaining because you cannot bargain over something that does not exist."

But Peres and Dayan didn't think of Dimona as a bargaining chip. They were determined not to trade away Israel's nuclear program.

*　　*　　*

Locked out of negotiations with France, sidelined by Ben-Gurion in the development of policy toward the Arab world, and bypassed in dealings with the United States by the prime minister's penchant for calling Abba Eban directly, Golda devised another outlet for both Israeli foreign policy and her own energies. In a world increasingly organized into blocs and alliances, of Cold War tensions and anticolonialist fervor, Israel was detached, invited to join none of the international confederations. Prompted by Socialist idealism, shrewd political calculation, and more than a dash of personal ambition, Golda strategized her own alliance, of Israel and the soon-to-emerge black African states.

In February 1958, during tough days of angry arguments with Egypt and the United Nations, Golda flew to West Africa for the opening gambit of her new game plan and was an immediate hit. In Liberia, the old Jewish woman was crowned a paramount chief at a mass rally of twelve hundred women, who decked Golda out in traditional robes. In Lagos, Nigeria's first president threw a garden party in her honor. In Dakar, she attended the opening of the Grand Council of African leaders of French West Africa and invited, it seemed, everyone she met to visit Israel. And on the eve of her departure from Ghana, at an enormous farewell cocktail party in her honor hosted by Kojo Botsio, President Kwame Nkrumah's closest colleague, she and a small group of Israelis taught 300 Nigerians the hora and then she stumbled through West Africa's most popular dance, the high life, with the foreign minister.

"Independence came to us, as it was coming to Africa, not served up on a silver platter but after years of struggle," Golda told quizzical Israeli journalists. "Like them, we had to shake off foreign rule; like them, we had to learn for ourselves how to reclaim the land, how to increase the yields of our crops, how to irrigate, how to raise poultry, how to live together and how to defend ourselves. . . . We have been forced to find solutions to problems that large, wealthy, powerful states had never encountered."

How can we help? Golda had asked Nkrumah during their first meeting. When he mentioned his desire for his own ships to reduce the cost of exporting cacao and importing manufactured goods, Golda arranged for Israel's Zim Lines to help him start Black Star Lines. For the construction support he needed, she charged Solel Boneh with creating a joint venture with the Ghana Industrial Development Corporation to build Accra's international airport, main highways, and main plaza. And she sent Israelis to set up demonstration farms, reorganize the Cooperative Bank, teach management courses, and help design both irrigation and drinking water systems.

To the consternation of her staff at the Foreign Ministry, Golda rarely drew on diplomatic professionals for African assignments, sending instead kibbutzniks, builders, union activists, and other hands-on specialists with instructions to "tell them about the mistakes we made so they won't repeat them."

Hanan Aynor, who Golda chose to run Mashav, the division of international cooperation she established, was awed by how involved the foreign minister became. "She mobilized top people through her personal appeal," he recalled. "And she mobilized a lot of money at a time when people in Israel were eating only one or two eggs a week. She was involved in everything, meeting with me twice a day or more, meeting with every Israeli who went out. If something didn't work, she intervened. Africans knew that she was sincere, and that broke Israel's circle of isolation."

Golda's early efforts were not without tragicomic relief. After she learned that West African farmers had an annual income below $100, she began promoting the idea of teaching them about poultry, perhaps a vestige of her own kibbutz experience. With a low investment, Golda reasoned, African farmers could raise a few chickens and eggs to supplement their meager incomes. But she didn't know—none of the Israelis knew—that neither the Hausa nor the Wolof ate white meat or eggs. "When I arrived as ambassador to Senegal, then, I was surprised to find an overproduction of eggs in the country," said Aynor. "The consumption was urban and europeanized and there was never much of a market."

In December 1958, Golda made her way into the heart of the African liberation struggle, courtesy of an invitation from George Padmore, the father of the pan-African movement, to meet in a special session with the 500 delegates to the first All-African People's Conference, including many of Africa's most prominent rebels, statesmen, and union leaders.

Through Padmore, Nkrumah, Julius Nyerere, and dozens of other anticolonial activists befriended by Israelis on the ground, Israel became intimately involved with liberation movements across the continent, establishing many of their intelligence schools; training men like General China, one of Kenya's most notorious Mau-Mau rebels; and financing the Accra office of Robert Mugabe after he was jailed for his opposition activities in Rhodesia.

Golda became a familiar figure across Africa, logging thousands of miles on annual, sometimes biannual, trips. During Liberian president William V. S. Tubman's fourth inauguration, she marched on the arm of the Liberian foreign minister. In 1960, she spent New Year's Eve celebrating the independence of Cameroon and was asked to head the jury for the selection of Miss Independence.

Every year, Israel's aid—although Golda banned the word, preferring "cooperation"—to Africa broadened and deepened. In Nigeria, Israelis built the house of parliament, hotels, roads, and bridges. In the Ivory Coast and Sierra Leone, they developed potable water projects, and in Uganda they paved the road to the Rwandan border and established a citrus industry. The Hebrew University–Hadassah Medical School trained hundreds of African health care professionals, while the Technion, Israel's institute of technology, prepared the first generation of engineers.

When famine struck Dahomey in 1962, Golda sent in a massive shipment of goods, and after the Belgians abruptly pulled out of the Congo, leaving the country with no physicians, it was to Golda that Dag Hammerskjöld and Patrice Lumumba turned for help. Her staff reported that there simply weren't enough Israeli doctors and nurses available to ship out at short notice, so Golda mobilized the Israeli army medical corps, which sent a team of forty-eight doctors and nurses on a special B-28 within three days.

"Back then, if you told a taxi driver in Conakry or Lagos that you were Israeli, he got excited," recalled Shlomo Hillel, former Israeli ambassador to Conakry and Abidjan. "We weren't just politically involved. It was a moral issue for us, and people knew it."

Africans became a common sight in Israel, and Golda's calendar was jammed with visits from a who's who of the continent. In June 1961 alone, she hosted Milton Obote, chairman of the Ugandan People's Congress; the consul general of Ethiopia; and the director of cultural relations of the Foreign Ministry of Mali. She sat for an interview with the editor of the *Accra Daily Graphic*, hosted a luncheon for the minister of natural resources of Sierra Leone, and gave the Tanganyikan minister of transport a tour of Jerusalem.

Golda never pretended that Israel was acting in a purely humanitarian vein. "We don't want to hide the fact that with Africa becoming independent and the countries taking their place in what is called the family of nations, naturally Israel was interested that all these peoples and countries should be friends of Israel and not enemies. . . . But . . . there is something much more basic about this, something instinctive. . . . A Jewish state that is not sensitive to discrimination and to the suffering of people on that basis would not be true to itself."

Israel was an almost ideal partner for African leaders anxious for help. Europe was hopelessly tainted by its colonial past and too obviously hungry to find new means of exploiting Third World wealth and strategic positions to be trusted. And the lessons taught by both Europe and the United States seemed too distant from the African experience to be irrelevant. Having just ousted Britain from Palestine and assimilated almost a million refugees, even while creating world-class universities, advanced agriculture, and an industrial infrastructure, Israel had a certain panache that offered hope, a glimmer of what Africa might accomplish.

"You have not tried to create us in your image," said President David Dacko of the Central African Republic during his 1962 visit to Israel. "Instead, Israel has contented itself with showing the new African nations its achievements, in helping them overcome their weaknesses, in assisting them in learning. In so doing, you have conquered Black Africa."

Ben-Gurion remained unconvinced by the growing involvement with Africa. Why spend so much money sending our people to Africa and bringing African dignitaries here? he asked. Golda responded impatiently. "Imagine that instead of coming to see us, all these presidents went to see Nasser."

Undaunted by Ben-Gurion's skepticism, Golda continued her African peregrinations. By the time Zambia achieved independence in the fall of 1964, she was an expected member of the gathering of the African clan among the likes of Kenneth Kaunda, Jomo Kenyatta, Julius Nyerere, and

Léopold Senghor. On that trip, her stock skyrocketed during a group visit to Victoria Falls. Rhodesian border guards attempted to separate the dignitaries crossing over from Zambia into separate lines for blacks and whites. "No, thank you," Golda said, turning back toward the Zambia border post. "I can do without the Falls."

The one African country she did not visit was South Africa, despite invitations from the local Jewish community, 100,000 strong. And in 1962, she provoked a crisis with them when she directed Israel's United Nations delegation to support antiapartheid resolutions at the General Assembly. Africans "have the right—and justly so—to expect Israel's support in their fight for liberty and freedom," she professed. A year later, she declined to replace Israel's representative in Pretoria, a move Prime Minister Hendrik Verwoerd labeled "a slap in the face of South African Jews."

Golda's undertakings did not go unnoticed by her enemies, of course. In the fall of 1960, the Arab League began sending Egyptian economic missions to Africa, and Radio Cairo reported that a special official had been appointed to direct an anti-Israel offensive there.

"The danger lurks under the glittering surface of trade and aid offered by Israel to some emerging African States," the League warned Africans. "Tel Aviv's offers have been, in reality, a facade for neo-colonialism."

The Arab counterattack made some headway. In January 1961, the Casablanca Conference, a meeting of senior government officials from Morocco, Egypt, Ghana, Libya, the Algerian Provisional Government, Guinea, and Mali, denounced Israel as "an instrument in the service of imperialism and neo-colonialism." Two months later, the All-African People's Conference, which had once embraced Golda, passed a resolution condemning Israel, along with the United States, West Germany, Britain, Belgium, the Netherlands, South Africa, and France, for neocolonialism.

Taking the long view, Golda nonetheless forged ahead, convinced that if she stayed out of the politics of Africa and put no pressure on African leaders to take sides in the Middle East conflict, Egypt's attempt to isolate

Israel from "the family of African peoples," as she called them, would fail. For a while, she succeeded. Despite Casablanca, at the 1963 Conference of Independent African States, Nasser didn't raise the Israel issue, having ascertained that he had little support. On the floor of the General Assembly, where Chad's vote weighed as heavily as England's, the growing number of African delegations insulated Israel, time and again, from Arab and Soviet hostility.

During a 1964 visit to Nigeria, Golda received a strong show of support against attacks stirred up by local Arab emissaries, who organized their wives to demonstrate against an invitation to Golda from the National Council of Women's Societies. "Diplomatic impudence and abuse of Nigerian hospitality," the Lagos *Sunday Express* branded the Arab action. The Nigerian government sent a strongly worded diplomatic note informing the Arab missions that they would tolerate no interference in internal affairs.

For Golda personally, the relationship with Africa, what *Newsweek* called "one of the strangest unofficial alliances in the world," rekindled an idealism that was beginning to flag. Israel had not turned out to be the Socialist country—the country without prisons, cheats, or criminals—that she'd fantasized back in Milwaukee, and Africa reawakened her oldest romantic spark. "You'll forgive me—but . . . I wouldn't dare—I admit my cowardice—to preach Socialism to a mother whose children are running around naked and barefoot, with little children, two, three, four years old running around with big bellies as a result of malaria," she confessed. "My Socialist duty is to see, to do everything I possibly can that those children should be fed and clothed and healthy and guaranteed a minimum education."

At a time when Golda was lonely, often sick, humiliated by Ben-Gurion, and treated with hostility at the United Nations, Africa provided her with a lavish dose of kinship, affection, even veneration. In Jerusalem, she tried to keep the elegant foreign minister's residence filled with friends and their children, but it was too stately for her puritanical tastes, too often empty of anyone other than Yehudit the cook and Esther the cleaning woman. "A woman so popular, so much appreciated with a following worldwide, the center of attraction, was lonely," recalled Zena Harmon, one of Golda's friends in that era. "She would call to chat and gossip. She would do nice things, like send flowers because she wanted a phone call to say thank you, someone to talk to."

But in Africa, she was always surrounded by tribesmen singing and revolutionary leaders paying her tribute. Her office was swamped with letters from ordinary Africans, letters of affection, letters asking for help, announcements that a female child in a village in Ghana had been named for her, or requests for her to serve as a godmother. On a continent where ancestors play a pivotal role in daily life, Golda became a sort of ancestor figure.

"At the opening of every General Assembly, Golda would be seated at Israel's table and the African delegates would all line up to shake her hand," said Yaakov Nitzan, who worked for her in the Foreign Ministry. "It wasn't politics. It was personal."

First known photograph of Golda
Meyerson, circa 1904.

Golda in Poale Zion pageant in
Milwaukee, 1919. PHOTOGRAPH COURTESY OF
THE JEWISH MUSEUM MILWAUKEE.

Golda with
her husband,
Morris
Meyerson,
1918.

Golda working in the fields at Kibbutz Merhavia, 1921.

Golda speaking at the Histadrut headquarters, 1946.

Golda as acting head of the Political Department of the Jewish Agency, 1947.

Golda shaking hands with Moshe Sharett, at the signing of Israel's Declaration of the Independence, 1948.

PHOTOGRAPH BY FRANK SHERSHEL, COURTESY OF GOVERNMENT PRESS OFFICE, STATE OF ISRAEL.

Golda with Eleanor Roosevelt at the United Nations, 1956.

PHOTOGRAPH COURTESY OF GOVERNMENT PRESS OFFICE, STATE OF ISRAEL.

Golda with Prime Minister David Ben Gurion, at the Mapai party convention, 1959.

PHOTOGRAPH BY HANS PINN, COURTESY OF GOVERNMENT PRESS OFFICE, STATE OF ISRAEL.

Golda greeting Rolf Pauls, first ambassador from Germany to Israel, at Beit Hanassi in Jerusalem, 1965.
PHOTOGRAPH COURTESY OF GOVERNMENT PRESS OFFICE, STATE OF ISRAEL..

Golda entering a state dinner with President Richard Nixon and his wife, Pat, at the White House, 1969.
PHOTOGRAPH BY MOSHE MILNER, COURTESY OF GOVERNMENT PRESS OFFICE, STATE OF ISRAEL.

Golda joking with Richard Nixon at the White House, 1969. PHOTOGRAPH BY MOSHE MILNER, COURTESY OF GOVERNMENT PRESS OFFICE, STATE OF ISRAEL.

Golda with Israel Defense Forces Chief of Staff Chaim Bar-Lev *(left)* and Minister of Defense Moshe Dayan at Beit Hanassi, Jerusalem, 1969.

Golda with Pope Paul VI, 1973.

Golda sharing a laugh with a veteran watchman at the 100 Years of Settlement Exhibition in Tel Aviv, 1973.

PHOTOGRAPH BY HERMAN CHANANIA, 1973.

Golda with troops on the Golan Heights during Yom Kippur War, 1973.

PHOTOGRAPH BY RON FRENKEL, COURTESY OF GOVERNMENT PRESS OFFICE, STATE OF ISRAEL.

Golda and Yisrael Galili (*right*) toasting the selection of Yitzhak Rabin to succeed her, 1974.

PHOTOGRAPH BY YAAKOV SAAR, COURTESY OF GOVERNMENT PRESS OFFICE, STATE OF ISRAEL.

Golda berating Henry Kissinger, U.S. secretary of state, outside her home in Ramat Aviv, 1975.

PHOTOGRAPHY BY YAAKOV SAAR, COURTESY OF GOVERNMENT PRESS OFFICE, STATE OF ISRAEL.

Golda with Mrs. Margaret Thatcher, leader of the British Conservative Party, in Tel Aviv, 1976.

PHOTOGRAPH BY MOSHE MILNER, COURTESY OF GOVERNMENT PRESS OFFICE, STATE OF ISRAEL.

Golda with Egyptian president Anwar Sadat at a meeting of the Labor alignment Knesset faction, Jerusalem, 1977.

PHOTOGRAPH FROM YAAKOV SAAR, COURTESY OF GOVERNMENT PRESS OFFICE, STATE OF ISRAEL.

Portrait of Golda Meir, 1976.

CHAPTER TWELVE

*It is easier to make a revolution than to uphold
the values for which it was made.*

Israel wasn't a decade old before Ben-Gurion began fearing for the health of his progeny. Democracy was running rampant, the system of proportional representation holding the nation captive to the whims of eighteen or more minor political parties. Even with Arab armies menacing them on all borders, veteran politicians seemed content to pontificate for hours. And someone—inside his party, outside his party, or in a foreign government—was always challenging him, whether Eisenhower griping about the tensions with Egypt or Golda threatening to quit.

Israel's leadership needed to be infused with new energy from younger men, he concluded, from *sabras*, native-born Israelis, who were, he thought, dazzlingly dauntless, free of ideology and the scars of the ghetto mentality. It wasn't that Ben-Gurion wasn't fond of the old guard, who'd followed him loyally for decades. But with the archetypical, or at least clichéd, narcissism of the Great Man in his own revolutionary indispensability, Ben-Gurion worried that Israel could not muddle on without him for more than a few days at a time. In his eyes, his old comrades were an aging clique of bickering weaklings who'd fall apart without his firm guidance. He needed to turn elsewhere to secure the nation's future.

In the run-up to the 1959 elections, then, Ben-Gurion announced to the party that he was going to bring "new blood" into the government. Golda knew what "new blood" meant. It meant sitting in the cabinet with Dayan and Peres as senior ministers, and she wasn't going to stand for it. She had worked too long and hard to be sidelined by impudent upstarts dripping with contempt for her and her ideals. Dayan was bad enough, with his puffed-up bluster and undisguised individualism. Peres, who was still undermining her at every opportunity, was unthinkable.

"It's not a question of age," she said to those who insinuated that the time had come for a changing of the guard. "A man has to have a birth certificate, but in public life it's unimportant, and it's not enough just to wave it about. I believe in certain values that do not change with time. Techniques change, priorities are altered, but underlying principles do not."

Ben-Gurion tried to reassure her that he wasn't trying to push her and her old comrades out. But they didn't believe him. He was acting too much like Jacob preparing to buy coats for his Josephs.

Golda's pique bled seamlessly between the personal and the political, but the root of her anger went well beyond ambition or a love of power. For Golda, Labor Zionism was Israel, Israel as it had to be if the Jewish homeland was to be more than just another nation. It wasn't just another ideology among many. It was *the* ideology, the wellspring of the rebirth of the Jewish people. In a near-primal way, then, she rejected any threat to Labor Zionism as a peril to the very soul of Israel, and the youngsters constituted such a menace. In Golda's eyes, for them, Mapai—the heart of Labor Zionism—was simply another party, Israel just another country, and socialism an inefficient economic system that did not maximize production.

Golda was not unique in her view, but she added her own personal penchant for inflexibility to the mix—"complete intolerance, complete disdain for any other opinion, a kind of primitiveness which was her strength," in the words of Uri Avneri, her greatest nemesis on the left. Levi Eshkol, one of Labor Zionism's leading figures, actually reduced Golda's character to three figures. When she was not displeased, he called her, in Yiddish, *de malke*, the queen. When she was annoyed, she turned into *de klafke*, the hag. When she was angry, she became *de mahsheve*, the witch, "mean, spiteful and with a very long memory of anyone who ever crossed her, from the ambassador's wife who wore a prettier dress than she did, to someone who spoke disparagingly about her, like Eban, to Sharett, who challenged her intellectual capability," commented Miriam Eshkol, Levi's widow.

"She was very sure of herself, so full of herself.... She was a *shlecte medercha Madonna*, a terrible, hysterical Madonna. She hated everyone but herself."

Fueled, then, by an overlapping blend of ideology, ambition, and spite, Golda launched her bid to undercut Ben-Gurion. As foreign minister, she might have been sidelined, but in domestic politics she was a power-house. Two other political heavyweights joined her crusade against the new generation: Zalman Aranne, the minister of education, and Pinhas Sapir, minister of trade and Mapai's kingmaker and wheeler-dealer. The troika, Israelis called them.

The opening shot in the public battle of the generations was fired by Dayan, who'd just resigned from the army, a clear prelude to his entry into the political arena. At the Mapai convention in May 1958, he let loose at the Histadrut, demanding that the labor confederation's power be curtailed, wages frozen, and massive layoffs ordered to streamline the ailing economy. And why do we need to mobilize the youth move-ment to build the new Ein Gedi–Sodom road? he asked. Bulldozers are more efficient. A shiver ran through the room at the clear implication that Dayan had no commitment to pioneering values.

But Dayan wasn't done. At a meeting of the students' club in Tel Aviv, he tore into almost every institution in Israeli society, from the kibbutzim and the economists to the politicians and the bureaucrats. "These men of the last generation have reached an age where they can no longer carry out revolutions," he said. "These men look back proudly on their achievements of 1902. . . . But we are interested in 1962."

Irate, Golda dropped a warning shot across Ben-Gurion's bow: she would not serve in the cabinet, the Knesset, or any other body of the new government. Knowing that he had little hope of forming a coalition without his old comrades, he tried to pacify her by openly lecturing Dayan about his inappropriate behavior. She was not mollified. A week before the elections, he finally bowed to Golda's political clout by announcing that Dayan would serve in the minor position of minister of agriculture and Peres in the noncabinet position of deputy minister of defense in his new government.

But before the tension could ease, five years of buried resentment over a secret intrigue popularly called the Mishap erupted, tearing at the fabric of Israel's political establishment. The seeds of the scandal were sown in 1954, when senior officials of the Defense Ministry, worried that the pending evacuation of British troops from the Suez Canal might signal a buildup to war, began batting around ideas for forestalling the British departure, including a plan to commit some minor sabotage that could be blamed on nonexistent Egyptian hotheads in order to provoke serious misgivings about Nasser's responsibility in Washington and London.

In July, two groups of young Egyptian Jews did just that, planting incendiary devices at a mailbox in Alexandria's central post office, in the reading room of the American libraries in Cairo and Alexandria, in movie theaters in both cities, and at Cairo's train station. The explosions caused no harm, to passing Egyptians or to Egyptian relations with the West. But the Egyptian government managed to capture the culprits, executed two of them, and sentenced the others to lengthy jail sentences. A new cloud of suspicion descended over Egypt's Jews.

Before the Israeli government could mount an official investigation into who was responsible for the Mishap, Peres began spreading rumors that Pinhas Lavon, the defense minister, had given the order. Peres' allegations fell under immediate suspicion since everyone knew how much he despised his superior. A charismatic labor leader and well-known dove who'd been handpicked by Ben-Gurion to run the Defense Ministry, Lavon was an odd choice since he had no military experience and was notoriously emotionally volatile. Golda and other old colleagues had cautioned Ben-Gurion against him, but both Peres and Dayan had supported Lavon's appointment. It was an appointment the two men quickly regretted when Lavon refused to share power with them. Soon they were openly complaining that Lavon was too rash, too disrespectful, and too ambitious.

Peres' rumors about Lavon's responsibility for the Egyptian fiasco weren't easily dispelled because the two-man commission convened to quietly sort out what had happened in Egypt found the mess impossible to unravel. The director of military intelligence had clearly ordered the action, but he insisted that he had acted on direct instructions from Lavon, who protested his innocence equally vociferously.

The matter should have ended there since it was the type of security scandal that could have disappeared in an easy cover-up in the days before the media became hungry. But it became entangled in generational rivalries, intraparty feuding, personal vendettas, and Ben-Gurion's unique brand of quixotism.

Convinced that Peres had tried to set him up, Lavon tried to fire him. When Dayan threatened to quit if his closest ally was ousted, Lavon himself was forced out, left to lick his wounds and wait for an opportunity to clear his name. That opportunity arose in the midst of Golda's fight with Ben-Gurion over his promotion of Dayan and Peres. Discovering that a key witness against him in the investigation into the Mishap had admitted that he'd committed perjury, in 1960, Lavon petitioned for public exoneration.

Ben-Gurion refused and attempted to prevent the opening of a new inquiry, but Lavon appealed to the Knesset, and the battle was joined, the younger generation lining up behind Ben-Gurion, the troika and most of the veterans behind Lavon. Journalists had a field day as the two sides leaked classified information, old rumors, details of Golda's ill health, anything and everything to tarnish the reputations of their opponents.

Levi Eshkol, the minister of finance and the most even-tempered of the political elite, tried to mediate. But Ben-Gurion suddenly demanded a full judicial inquiry into the Mishap, raising the specter of a long, drawn-out process that would inevitably divide and undermine Mapai. Golda tried to reason with the Old Man, but he thundered about the sanctity of democracy and the separation of powers, although, curiously, he never used his authority to establish such an inquiry himself. Ben-Gurion's motivations were never clear: his acolytes insisted that he was standing on principle, while his detractors suggested, alternately, that he was protecting Peres, that he was using the affair to reexert control over his increasingly rebellious party, and, even, that he was losing his grip on reality.

After days of consultation and table pounding, Eshkol offered a compromise, a committee of inquiry led by the justice minister instead of a full, public investigation. The cabinet acquiesced despite Ben-Gurion's opposition, the first time in Israeli history that his entire senior cabinet lined up against him. But Ben-Gurion never accepted defeat graciously. Before the committee's findings could be presented to the cabinet, Ben-Gurion demanded that its report be rejected and that a judicial inquiry be launched. When the cabinet ignored him and voted to accept the committee's conclusion that Lavon had not issued the fatal order, Ben-Gurion gathered up his papers and stormed out.

*   *   *

Angry and hurt at Ben-Gurion's indifference to the political chaos he was creating, Golda informed the prime minister's aide that if Ben-Gurion continued to press for an inquiry, she'd quit the government. It never occurred to him to take his loyal warhorse seriously.

But Golda had gamed out every move. She assumed that Ben-Gurion would resign over the cabinet's disobedience, which would bring down the entire cabinet. Free to form a new and more pliant one, he could then disown the ministerial committee's findings. To forestall such a power play, she handed in her own letter of resignation. The ultimatum was clear: if you want to re-form the government, you'll have to do so without me or the troika. Ben-Gurion understood that the Knesset would not approve such a government.

Misjudging Golda's fire, Ben-Gurion's followers tried to trump her by proposing a series of resolutions, all of which boiled down to the same theme: there can be no government without Ben-Gurion. "Stating that no party member [but Ben-Gurion] shall be prime minister had a discordant ring," she said in what became the opening cry of a palace coup. "Is it permissible to surmise that a man—even the most revered amongst us—is capable of error, or is it sacrilegious to think it? If it is permissible to think it, our colleagues are permitted to think that Ben-Gurion is proposing something on which he errs, and they differ."

Like any good divorce, the final act was messy and drawn out, with reconciliation achieved and shattered repeatedly and not an ounce of poignancy. By 1962, power in the party had begun to shift from the prime minister to the Gush, the bloc of party functionaries Ben-Gurion himself had created in the 1940s to preserve the veteran leadership. It was a bloc that Golda dominated utterly.

The denouement was precipitated by the public outcry over the issue of the German scientists working in Egypt. Harel, Israel's security chief, resigned over Ben-Gurion's lackadaisical attitude, with Golda threatening to follow him. Ben-Gurion tried his own ploy and resigned himself. "There are personal needs," he said. "I propose to keep them to myself." It was the fourth time Ben-Gurion had quit as prime minister, a post he had held for all but fourteen months of Israel's fifteen-year existence, and few believed he was serious. "If there is a crisis everything can change," he told the media, prompting deep skepticism as to his intentions.

He was not out of office two weeks before he made the shape of his "personal needs" clear by delivering a speech about the importance of reopening the Lavon investigation. Arranging with a newspaper reporter to delve into Defense Ministry files for new evidence, he immersed himself in the hundreds of documents and transcripts already gathered. The troika implored him to desist, to consider the consequences for his chosen replacement, Eshkol. But Ben-Gurion was beyond caring what others thought and started firing off scathing salvos against Eshkol and the troika, calling them liars, incompetents, and menaces.

A man given naturally to moderation, Eshkol was in agony and repeatedly implored Ben-Gurion to give up his crusade to reopen the Mishap investigation. But fixated on proving that he was correct and hurt that the nation, finding him too strident and authoritarian, had moved on, Ben-Gurion proved unyielding. He might have resigned voluntarily, but he blamed Eshkol for his political demise.

With Eshkol constitutionally incapable of waging the final battle to dethrone Ben-Gurion permanently, it fell to Golda to play Ben-Gurion's nemesis. She was an unlikely combatant. In 1964, at the age of sixty-six, she'd applied for old age benefits, and she was constantly ill. After decades of heart problems, circulatory difficulties, phlebitis, herpes zoster, kidney stones, and migraines, she'd been diagnosed with lymphoma, cancer of the lymphatic system, and was undergoing debilitating weekly treatments while still smoking three packs of cigarettes and drinking dozens of cups of coffee each day.

But weary of what she called Ben-Gurion's "erratic" behavior, Golda moved into the breach and orchestrated his political demise. Her first step was to bolster Mapai's strength in the Knesset through an alliance with Ahdut HaAvoda, another Labor Zionist Party, which had split from Mapai in 1944. Led by activists of the middle generation, between Golda and Dayan, Ahdut was marginally to the left of the Mapai mainstream and thus had few members pushing to turn Israel's back on socialism. Golda's strategy was obvious: the added Knesset seats and clout of Ahdut's leaders would diminish the prospects of the Dayan-Peres crowd.

Sensing that his own power was also in jeopardy, Ben-Gurion prepared himself carefully for the tenth Mapai convention in February 1965, scheming to regain control. But Golda had already decided that he'd thrown down the gauntlet one time too many, telling friends that he was "daft" and a "wild man."

"I don't know what the people want," Ben-Gurion cried out after rising to the podium to a thunderous ovation. "But I know what it SHOULD want—that truth and justice shall reign in our land!" To achieve both, he called on the party to disavow the findings of the ministerial committee that investigated the Mishap and endorse the creation of a judicial panel.

"A judicial committee?" cried Eshkol in exasperation. "God in Heaven, what are we doing here trying to restore the Roman forum? I say to Ben-Gurion: Give me a chance, give me room."

From a wheelchair, Sharett, five months from death, released almost a decade of pent-up indignation over his ouster from the Foreign Ministry. In a feeble voice, he excoriated his old partner for making decisions without considering the consequences for the party or the country. "A leader cannot subjugate the movement, stop it from thinking, merely by some aristocratic right," he proclaimed. "I hope and believe that the party will close ranks on this point, and once and for all shake off this nightmare, exorcise this dybbuk."

When he finished, Golda, who'd never shown an ounce of guilt at having allowed herself to be used in Sharett's removal, leaned over and kissed the dying man. The Ben-Gurionites were stunned. For years, they mentioned that kiss in conversation, in articles, in their memoirs as the evening's coup de grâce.

But having orchestrated the Night of the Long Knives, as that evening has gone down in Israeli history, Golda wasn't finished. How dare Ben-Gurion humiliate Eshkol by calling him a liar unfit to lead the country! she cried out. If Eshkol is "a liar, we are all liars." The hall sank into utter silence as Golda mocked the "hypocrite" who speaks of justice yet "accuses and . . . judges from the outset."

Wilted by the onslaught, Ben-Gurion walked out of the hall, muttering, "Everything is finished." When friends stopped by later, they found him obsessing about Golda, a woman he'd long loved and admired, the woman who was always seated to his left. "He always said that she had a unique quality of getting to the heart of things," recalled Yitzhak Navon, then Ben-Gurion's aide, later the president of Israel. "He made excuses for her, saying that she'd had a hard childhood."

He could talk of nothing but Golda. "What does she want from me?" he asked. "What did I do to her? It was sad to hear her speak like this, spouting venom." In his diary, he wrote, "I don't know where all this poison comes from."

Within days, however, Ben-Gurion had recovered and he went on the offensive to oust Eshkol from the leadership of Mapai and regain control over the government. But at a Central Committee meeting in May, Golda summarily submitted that Ben-Gurion's time had passed. "There is now a personal war, aimed at liquidating certain comrades," she warned, referring to the rumors being spread by Ben-Gurion's allies. "What do they mean by raising the question of Ben-Gurion as head of the list and prime minister? . . . What is the present premier supposed to do until then—and will other countries wish to deal with him in the meanwhile?"

Like an aged volcano that defied extinction, as *Newsweek* put it, Ben-Gurion erupted once more. Quitting Mapai, he gathered the new generation around him and announced the formation of a new political party, Rafi.

The election campaign was the ugliest in Israeli history, an open contest between the Old Man and Golda. Ben-Gurion told reporters that she had visited him ten months earlier and said, "The Party is Tammany Hall, not the Tammany Hall of today but of the 19th century. . . . I will not remain in this party. There is no other party I can join. I will become independent."

"LIAR!" shouted Golda. "How is it that all of a sudden we have all become cheats, as Ben-Gurion claims, while for the first fifteen years of the State we were such upstanding citizens?"

Thinking himself a modern Moses, Ben-Gurion assumed that the Israelites would flock to his side. But when the votes were counted in November, Rafi won only ten seats to the forty-five of the new Mapai/Ahdut alignment. Israel's personality cult had ceded to the political prowess of Golda and the cult of Mapai.

Her work completed, Golda resigned as foreign minister while retaining her seat in the Knesset. Donating the statues, books, jewelry, and other mementos she'd collected to museums, she moved home to the modest duplex house she shared with her son, Menachem, and his second wife, Aya, in suburban Tel Aviv. She wanted time to sew bathrobes for their three sons, to make them pancakes on Saturday morning, to attend concerts, enjoy weekends with Sarah and her children at Revivim, cook and wash like a normal person, she said. Her sister Sheyna was growing feeble, losing her memory to Alzheimer's. It was time to retire from public life.

No one was fooled. Everyone knew that the troika and the other members of the Gush would drop by to share the children's pancakes and that drinking coffee in Golda's kitchen would remain an Israeli political institution. Nonetheless, Golda kept up the pretense.

On the afternoon of January 26, 1966, the Knesset chamber filled early for the first session of the new government. Golda walked in smiling and headed for her old chair, the foreign minister's seat at the long ministerial table. Suddenly, as if it had just struck her that she'd resigned, she turned toward the backbenches.

"Oy vey," she said. "It's hard to teach an old horse the way to a new stall."

*   *   *

Golda whipped up a few Saturday breakfasts for her grandsons, arranged a visit to Revivim, called up old friends, and delighted in the public reaction when she was spotted riding the public bus—although the drivers often veered off their routes to drop her at her front door. But her retirement didn't last three weeks before the guys came calling, Sapir and Aranne, Eshkol and Yisrael Galili, the new minister of information and one of Golda's firmest allies in the Ahdut camp, asking her to give up her relaxation to become secretary-general of Mapai.

It was a position of enormous power without any glory, which fit Golda like a glove. Worried that the divisions within Mapai might lead to an opposition victory, tantamount to a Zionist apocalypse in Golda's lexicon, she agreed to try her hand at reuniting Rafi, Ahdut, and perhaps Mapam, the more Marxist party, into a single Labor Party and reignite popular enthusiasm for a labor movement sorely challenged by a deep recession and skyrocketing unemployment. The task, she admitted, was daunting, a "job of bridge building, of reconciling the various points of view and personalities, of dealing with old wounds without making fresh ones," all polite euphemisms for tense negotiations, arm-twisting, and head-bashing.

Although Eshkol was prime minister, Golda became the leading political figure in the country, one part kingmaker, another conscience, and, mostly, an old-fashioned party boss. The locus of power shifted from the cabinet to the party, from the prime minister's office to Golda's kitchen. In meetings, she openly berated Eshkol as if he were a dunce. When he wanted to see Golda, he didn't summon her to Jerusalem; he made the trip to her office or house in Tel Aviv.

"She was such an Amazon, though without the swords or spears," said Miriam Eshkol. "She dwarfed them, like Snow White with the seven dwarfs."

* * *

Nasser disrupted Golda's plans for party reunification. On the morning of May 16, 1967, the chief of staff of the Egyptian army instructed the commander of the United Nations Emergency Force to remove his troops immediately. The presence of the UN contingent in the Sinai Peninsula had been the price Egypt paid for Israel's withdrawal after the 1956 war, and Dag Hammarskjöld had pledged that those troops would never be pulled out without the consent of the General Assembly.

While Israel was still absorbing the implications of that request, Egypt began massing troops in the Sinai and Gaza, thousands of men with tanks, armored personnel carriers, Soviet howitzers, and Katyusha rockets, backed up by Soviet MiG fighters. "The existence of Israel has continued too long," declaimed Radio Cairo. "The battle has come in which we shall destroy Israel."

The Soviets had informed the Egyptians that Israel was preparing to invade Syria, which was also moving troops within sight of Israeli settlements. There was no truth to the allegation, as the Israelis offered to prove to the Soviet ambassador with a quick trip to the border. He declined, leaving many to suspect that the Russians were trying to provoke war.

Israeli diplomats around the world scurried to reassure both the Egyptians and the Syrians that Israel was not about to invade either nation, sending messages through the British, the Americans, and the United Nations to correct what they hoped was a misunderstanding. They invited the chief UN observer in the region, General Odd Bull, to tour northern Israel to verify the absence of Israeli troop concentrations.

At 7 A.M. on the morning of May 18, Golda was woken up by a phone call from one of Eshkol's aides asking her to attend a meeting. The news was grim: U Thant, who'd succeeded Hammarskjöld as UN general secretary, had ignored his predecessor's pledge. Without consulting the General Assembly, Israel, or the great powers, he had agreed to remove the UN Emergency Force, arguing that the UN force would not legitimately remain in Egypt without the consent of the government in Cairo.

During the decade that those units—4,500 troops scattered in forty-one observation posts—had been stationed in Egypt, shipping had passed peacefully through the Straits of Tiran, no fedayeen had infiltrated Israel from Gaza, and the Egyptians and Israelis had not faced off in a single clash. Their exodus from Egypt, Israel feared, would lead to another war, the very war Golda had predicted before she gave her fateful Israeli withdrawal speech to the United Nations in 1957.

Even Golda, who was boundlessly cynical about the UN, was shocked by what she called "U Thant's ludicrous surrender to Nasser. It was against all rhyme or reason for a force that had come into existence for the sole purpose of supervising the cease fire between Egypt and Israel to be removed at the request of one of the combatants the very moment that the cease fire was seriously threatened," she wrote of that day.

Eban, who'd replaced Golda as foreign minister, offered to make the rounds of the international capitals, but the cabinet preferred to send Golda. "She'll be tougher," Eshkol said. Humiliated, Eban fought for his right to represent Israel, and Golda, who had the "flu," her euphemism for a tough bout of cancer treatment, wasn't fit to travel in any case. So the foreign minister caught the first flight to Europe, hoping to find someone to mobilize the international community to save the Middle East from another war.

Israeli military intelligence couldn't decide whether Nasser was trying to lure the Israeli army into the Sinai Desert or hoping to seize part of the Negev. But Eshkol suspected that he was looking for prestige more than conflict. So while he waited to see how the major powers reacted, he kept quiet to avoid escalating the rhetoric, provoking speculation that he was wallowing in indecision.

Suddenly, a well-orchestrated outcry flared out of the camps of Rafi and Herut, the primary opposition party, led by Menachem Begin: This is a dangerous time and Eshkol has no military experience. Establish a National Unity government of all the political parties and bring Ben-Gurion back to lead it.

Eshkol dismissed the agitation. "These two horses cannot be hitched to the same carriage," he said of a proposal that would have him work for Ben-Gurion. And with Golda's hearty support, he also rejected the creation of a National Unity government.

On May 22, perhaps emboldened by U Thant's easy agreement, Nasser closed the Straits of Tiran to vessels bound for Israel, again defying international law and the United Nations. Still Eban found little outrage in the Western capitals, only stern admonitions against a preemptive strike. President Lyndon Johnson made some noise about organizing an international flotilla to sail through the straits with an Israeli ship. "Nice gesture," Golda called it, remembering all of Eisenhower's promises about Israel's right to defend itself if Egypt ever again closed the international waterway. But what would happen after the international flotilla sailed away?

Israel "will not stand alone" as long as it does not act alone, Johnson promised Eshkol. In response, Eshkol asked Johnson what he wanted Israel to do when faced with such direct aggression. LBJ reassured, pressured, and cajoled. But he did not answer Eshkol's question. Israel's leaders knew what that meant: you're on your own.

With thousands of Egyptian and Syrian troops on her border, UN troops decamping, and UN observers being kept off Syria's Golan Heights, Yitzhak Rabin, the military chief of staff, knew that more than free passage through the canal was at stake. "It is now a question of our national survival," he said,

So began *Hahamtana*, the Waiting. The reserves were called up—every adult under the age of fifty-five years—and those left behind started clearing out basements and restocking them as emergency shelters. Private vehicles were commandeered for military use, factories closed to allow workers to join their reserve units, and the elderly began delivering the mail. Although not a member of the cabinet, Golda was too embedded in the inner circle not to be pulled into the planning, so she made the regular trek to Jerusalem and kept the coffeepot hot for the ministers who dropped by her house.

Rabin urged a preemptive strike. But caught between the clear danger from the Egyptian-Syrian alliance and the equally clear warnings he was receiving from Washington, Eshkol hesitated, worried about a repeat of the firestorm of criticism that rained down on Israel in 1956.

Mistaking Eshkol's desire to calm the tension for dithering, the public grew increasingly susceptible to the organized drumbeat for him to step aside, especially once Moshe Dayan emerged as the new, Rafi-inspired savior. "Eshkol's a coward," they chanted, now demanding that he form a National Unity government with Dayan replacing him as minister of defense. Dayan, Dayan, Dayan. The chant crescendoed into a chorus, as if the cocky general were Israel's only possible salvation.

As the anti-Eshkol demonstrations continued, the Mapai secretariat, the inner circle, convened. Suspecting that the youngsters were trying to use the war frenzy to undo the political damage she'd inflicted on them, Golda protested that any change in the government was unnecessary, harmful, and equal to a vote of no confidence. "A whispering campaign has been started to the effect that there will be additions and changes in the country's leadership and I wish to make the following proposal," she said. "The party thinks the Government in its present form is fine and opposes the participation of the opposition on a permanent basis."

Realizing that the nation needed to hear from him directly, Eshkol went on television to allay public fears. But a poor speaker at his best, that night Eshkol was exhausted. As an Israel in need of reassurance watched and listened, he fumbled, lost his place, and stuttered his way through a dry, legalistic speech, shattering the national morale.

Bringing the opposition into the government in a wall-to-wall national coalition became inevitable, but convinced that the anti-Eshkol movement was the result of an elaborate political theater, Golda resisted. "A one-woman stumbling block" of destructive political fanaticism obstructing national unity, *Ha'aretz* called her.

Hoping to deflate the growing hysteria, Golda finally threw her support to the creation of a national war council, one step short of a full National Unity government. But that didn't mollify Gahal or Rafi, which refused to cooperate with any scheme that didn't include Dayan's appointment as minister of defense. "Over my dead body," she told a meeting of the Mapai and Ahdut Knesset caucus.

"As long as the Old Lady is there, she controls the party and she won't let me in the cabinet," Dayan complained.

Golda tried to sideline Dayan into the position of minister without portfolio, but he declined. Eshkol tried to tempt him with the post of deputy prime minister. To no avail.

In the midst of the political meltdown, King Hussein of Jordan signed a defense pact with Egypt, putting his own troops under Cairo's command, raising the specter of a war on three borders. Still, Golda and Eshkol tried everything to keep Dayan out of the leadership, Golda understanding that he'd try to use the defense portfolio to reinvigorate his image as a national war hero, Eshkol hurt by the assumption that he could not run the military. As Eshkol reminded the few who would listen, he had been the modernizer of the Israeli Defense Forces, and Dayan's track record was less than stellar. Israel's military genius had predicted that the Egyptians wouldn't be ready for war until the 1970s. So why were 465,000 Arab troops, more than 2,800 tanks, and 800 aircraft ringing Israel?

To appease those sincerely worried about Eshkol's ability to conduct a war, Golda suggested that Yigal Allon be appointed as minister of defense. A founder and commander of the Palmach, Allon had led most of the decisive operations during the War of Independence and was, by far, the most experienced field officer in the country. But Rafi wasn't really worried about the war; they were making their bid for power. So they refused Golda's compromise, leaving her and Eshkol consulting nonstop with opposition leaders while one hundred female demonstrators stood outside her office with signs reading, GOLDA—ENOUGH HATRED.

Finally, Golda relented. Elevating Dayan would be politically expensive, she knew. But the alternative was simply too dangerous.

With the National Unity government, a wall-to-wall coalition of all Israel's political parties, in place, the government debated how to respond to the mounting threats and whether to launch the preemptive attack that the military urged. "I understand the Arabs wanting to wipe us out, but do they really expect us to cooperate?" asked Golda, who held no official position in the government but took her place at the cabinet table as if by right. Then, looking from Eshkol to Rabin, she added, "I don't see how war can be avoided. Nobody is going to help us."

At 7:14 A.M. on June 4, the Israeli Air Force took to the air, leaving only twelve fighters to defend the national airspace. In less than two hours, while the Arab pilots were still eating breakfast, they destroyed 300 Egyptian aircraft on the ground. Turning north, they then hit Jordanian and Syrian airfields. By the end of the day, the Egyptian, Jordanian, and half of the Syrian air force had been decimated.

On the ground, the IDF seized the Golan Heights from Syria, ending the hail of destruction that had been falling on the settlements below for years, and recaptured the Sinai and the Gaza Strip, the same territory Golda had publicly yielded ten years earlier. Three days later, on June 7, Israeli paratroopers captured the Old City of Jerusalem.

The following morning, Golda drove to the Western Wall that had meant so little to her when she and Morris lived in Jerusalem. Stunned by the sight of the most spiritually and historically valuable piece of real estate in Judaism, hundreds of young soldiers stood there in awe, their weapons slung across their shoulders as they undertook the almost visceral Jewish ritual of wedging notes of hope and prayer into the cracking mortar of centuries-old stones. Not recognizing Golda but seeking female comfort, several soldiers threw themselves into her embrace.

The antithesis of a romantic, Golda was never one to indulge nostalgia. She didn't pray or weep over the reunification of the ancient city, then; she flew to the United States to raise money for a national treasury getting dangerously empty.

"Again, we won a war—the third in a very short history of independence," Golda told the crowd of 18,000 at a rally in Madison Square Garden two days later. "A wonderful people these Israelis! They win wars every ten years whatever the odds. And they have done it again. Fantastic! Now that they have won this round, let them go back where they came from so that Syrian gunners on the Golan Heights can again shoot into the kibbutzim, so that Jordanian Legionnaires on the towers of the Old City can again shell at will, so that the Gaza Strip can again be a nest for terrorists, so that the Sinai Desert can again become the staging ground for Nasser's divisions. . . . Is there anybody who can honestly bid the Israelis to go home before a real peace? Is there anyone who dares us to begin training our ten-year-olds for the next war?"

The crowd swelled in a chant, "NO, NO, NO!"

* * *

Israel threw itself into blind euphoria over the latest evidence of its prowess, and Golda went back to work to stitch together a new Labor Party. If all the fragments that had broken off from Mapai over the years—Ahdut, Rafi, and Mapam—reunited, Labor could rule supreme without the pesky necessities of appeasing coalition partners, Golda hoped. Such an alignment was not without danger, from Golda's point of view. Bringing Rafi back into the fold would position Dayan to succeed Eshkol as prime minister, and Golda was firm in her belief that Dayan would be a disastrous leader. Unlike her feelings for Peres, her opposition to Dayan wasn't personal; it was a matter of ideology and loyalty, neither of which he was known to possess. Quirky and intensely individualistic, Dayan wasn't even loyal to his own beliefs, which swung wildly from day to day, often in dramatic contradiction.

Peres was determined to lead Rafi back into the Mapai fold to give himself and Dayan a shot at power. But no matter how hard he pushed, Golda wasn't about to reunite with Rafi until Ahdut committed itself to the merger. Ahdut's members, with a firm Socialist ideology close to Golda's, would serve as her bulwark against the Dayans and Pereses. And unlike Rafi, which was little more than a vanity party for a handful of young Turks, Ahdut had a strong base in kibbutzim and an influential newspaper. Most important, perhaps, it had as its star Golda's anti-Dayan, the man she saw as the next prime minister, Yigal Allon.

Golda might have been a sick old woman, but she ran the merger negotiations essential to achieve her vision of a labor-dominated future for Israel with immense political skill. Riding roughshod over anyone who tried to thwart her, she blithely dismissed Rafi's appeals for electoral reform, soothed decades of grudges, and faced down Mapai veterans like Eshkol, who feared that Ahdut might overwhelm Mapai. More than ideology, the division of key positions, or the new party structure, Golda battled the towering egos endemic among all three parties, overwhelming them with her own.

How did she do it? "She comes clumping along with that sad, suffering face drawn with pain from her varicose veins and God knows what-all," one old-timer said. "You rush to help her to your seat. She thanks you kindly. The next thing you know you're dead."

With opponents, she was alternately ruthless and accommodating, throwing them off balance. The younger men, even those who despised her, found themselves unable to resist her perfectly pitched amalgam of guilt, motherhood, historical privilege, and ruthless application of conscience.

"Golda knew what power was and how to wield it," said Abraham Harmon, former Israeli ambassador to the United States. "She was a hard politician. Either she liked you or she didn't. And woe betide you if she didn't."

No one found Golda easy to battle. "She was the heir to a long Jewish tradition of argumentation," remarked Abba Eban. "Whenever she saw or heard some kind of dogmatic assertion she would rise up against it in an effort to reveal its superficialities and its weaknesses."

The Charter of Unity establishing a new Labor alignment was finally signed in Jerusalem in January 1968, but the harmony implied in the founding document was fictive, at best. All the old factional disputes continued, the long-standing rivalries, the ideological divisions. At the first sign of discord, Golda tried imposing harmony through blackmail. "You people don't really want me," she complained, announcing that she would not serve as secretary-general. "The last two years have been the most miserable in my life. I have not been properly appreciated."

It was a classic Golda dramatic ploy to maintain control and followed the script she would act out again and again. Golda threatened to quit. Wringing their hands, party leaders appeared at her house to beg her to reconsider. Golda refused, complaining that she was tired of the infighting and the lack of gratitude. Everyone showered her with compliments, soothed her bruised ego, and promised to obey, at which point she relented, her power to move forward with her own program bolstered.

When she abruptly resigned again in July, then, no one took her all that seriously. But this time, when the leadership made the usual trek to her kitchen, Golda did not relent. A wave of anxiety spread across the country at the prospect of Israeli politics without Golda, the nation's towering party boss, and the media threw themselves into an orgy of speculation: Is she really serious? Is she sick? Is she fighting with Eshkol, fed up with Rafi, manipulating the new Labor into an early alignment with Mapam?

All the speculation was correct. Threatened by Golda's overt attempts to position Allon as the heir apparent, both Rafi and Eshkol were challenging her and her courting of Mapam was ruffling Rafi's prickly feathers. Golda said none of this in public, of course. "I want to be able to live without a crowded engagement book," she explained. "I want to be able to read a book without feeling guilty for taking time from other matters and to go to a concert when I like. . . . At seventy, one is entitled to some of the pleasures of life."

Israelis were stunned by the pending disappearance of the last public face of the old *yishuv*, a woman who one newspaper called "something more than a political leader: a symbol of the new Israel, of its courage, its strength and its boundless devotion."

With a modesty that was, at once, both genuine and formulaic, Golda dismissed the Sturm und Drang. "No one is indispensable anywhere," she professed. "You will manage without me."

But it wasn't clear how they would. No one else seemed to have the fortitude or moral authority to keep Labor's warring factions and monumental egos in check. No one else seemed to have what journalists called her "granite-like strength."

It seemed the end of Israel's first era. Breaking from Rafi after party reunification, Ben-Gurion had formed a new mini-party, but he dwelled in a political wilderness. Sharett had died not long after the Night of the Long Knives. Israel's first two presidents were gone, as were half of the signers of Israel's declaration of independence. On August 1, 1968, the woman who had loomed over *yishuv* and Israeli politics for more than three decades, who'd raised hundreds of millions of dollars to build the nation, spoken as Israel's international voice, and maintained a perennial political open house (after which she stayed up half the night to wash up the cups and mop the floors), left her office on the third floor of Labor Party headquarters on Rehov Hayarkon in Tel Aviv and went home.

With Menachem and Aya studying in Connecticut, and Sarah and her family still at Kibbutz Revivim, Golda said she expected to travel, perhaps take a holiday in Switzerland, and indulge her favorite pastime, cleaning. For once, Golda wasn't being disingenuous. One part of her had always longed for a "normal" life.

But she never denied the other part of her character. "I am not going into a political wilderness," she promised. "I do not intend to retire to a political nunnery."

CHAPTER THIRTEEN

*Authority poisons everybody
who takes authority on himself.*

At the time, the Six-Day War victory seemed the dawn of a new era of blessed serenity, of life without the tension of military call-ups, air-raid sirens, and the agonizing din of Nasser's incessant threats. But with the Eshkol government mired in indecision about what to do with its new bargaining chips, the West Bank, Gaza, East Jerusalem, Sinai, and the Golan Heights, Israel could find no way to translate its military triumph into peace.

The cabinet quickly came to something approaching unanimity on the question of Jerusalem and the Golan Heights; the former was too emotionally and historically momentous to be returned to Jordan, which, in any event, had no more legal claim to the city than did Israel, and the latter too vital to the safety of the settlers who lived on the plains below. But they were hopelessly divided about what to do with the remaining real estate.

To Dayan, the solution was obvious: the West Bank should be econom-ically integrated into Israel while leaving the citizenry vassals of Jordan.

Cabinet doves, like Golda's close comrade Pinhas Sapir, objected vehemently to any de facto annexation, certain it would prove an insurmountable stumbling block to any resolution of the conflict. Trapped between
his two ministers, Eshkol vacillated.

The future of the territories wasn't the only issue about which Sapir
and Dayan, Eshkol and Dayan, Allon and Dayan bickered. Israeli politics
was a bacchanalia of squabbling, quibbling, and open brawling, egged on
by a scrappy press regularly fed juicy leaks by all sides. Behind the weighty
issues dogging the country loomed the ever-present jockeying for power,
and as Israel moved toward elections scheduled for November 1969, the
volume and the vitriol grew deafening.

Eshkol tried to protect his flanks against both Allon and Dayan, the
leading pretenders to the throne. But the more he rebuked Dayan for his
cowboy approach to government, the more popular the general became.
Meanwhile, Allon griped incessantly at being excluded from key meetings and crucial decision making. And Israelis followed every tiff, battle,
and complaint with the fascination of viewers of a gripping soap opera.

Toward the end of 1968, Eshkol collapsed. The government announced that the prime minister had fainted from a high fever brought
on by bronchitis, but, in truth, at the age of seventy-three, he had suffered
a heart attack. Fearing that Allon, the deputy prime minister forced on
him by Golda, would push him out of office, Eshkol refused admission to
the hospital. For months, oxygen tanks were delivered to his residence
stealthily to conceal how ill he really was. Finally, on February 26, Eshkol suffered four more heart attacks within a few hours and died at home
in bed.

Israelis anxiously awaited the resolution of the long Mexican standoff
between Dayan and Allon, and no one expected either to cede graciously.
While Dayan's bravado had earned him the affection of thousands of Israelis who saw their own increasingly macho élan reflected in his swagger, his party leaders alternately despised and feared him.

Although he evoked little popular enthusiasm, Allon was the party's
chosen one. But the choice was a political nightmare. If Allon were se-

lected, Dayan would likely bolt, and perhaps create a new party. If Dayan was tapped, Labor's popular support might soar, but the party itself would be imperiled by Dayan's indifference to its health, and, many believed, the state too would suffer from a maverick prime minister who disdained the sort of consultation essential in a democracy.

Well before Eshkol's death, Sapir, who'd replaced Golda as secretary-general of the new Labor alignment, set out to forestall such an internecine struggle. Two months earlier, at an early morning meeting at the barbershop at the Essex House Hotel in New York, he had informed his protégé Yossi Sarid that Eshkol was dying. "Golda is going to replace him," Sapir announced. "It's final and decided. Only she can hold the party together."

From New York, Sapir had flown to Switzerland to sound out Golda, who was resting at the Hotel Dolder, an elegant old spa in Hottingen, on the edge of Zurich. "You realize you'll be next in line," he told her after disclosing that Eshkol was failing. "You have to step in to avoid a suicidal clash between Allon and Dayan." Golda dismissed him snippily. "As long as Eshkol lives, what do you want from me?" But she didn't refuse.

A masterful political strategist, Sapir then went home to get his ducks in a row, playing the Allon supporters against the Dayan fanatics, emphasizing that Golda would serve for only six months, until the November election.

So long as Eshkol remained functioning, Sapir didn't meet much resistance. But the moment the prime minister died, all hell broke loose. Sapir arrived at the prime minister's residence just hours after Eshkol's demise and found Allon and Dayan there, quarreling about where to bury him. Eshkol had expressed a desire to be interred at his kibbutz. But Jordan had been shelling that area, and Dayan, worried that a large gathering would attract more fire, insisted he be laid to rest on Mount Herzl in Jerusalem.

Later in the morning, Golda showed up to offer her condolences to Eshkol's wife, although Miriam Eshkol so despised Golda that in 2006, she still kept a crudely made effigy of her nemesis hanging over her kitchen stove so that she could fantasize burning it. Once Golda was seated in the center of a large sofa in the living room, all debate stopped. "He will be buried in Jerusalem," Golda decreed, and no one dared challenge her. Sapir smiled in satisfaction. That decisiveness, and the deference it evoked, was precisely why he had decided that Golda should be named prime minister.

At a meeting in the Knesset dining room several hours later, Sapir, Dayan, Allon, Galili, Golda, and the minister of justice planned Eshkol's funeral and agreed that during the period of mourning, Allon, as deputy prime minister, would head the interim government. Only then would a prime minister be chosen.

But the political wrangling didn't wait for the mourning to end. Galili tried to enlist Sapir to rally support for Allon. Sapir agreed, but he also began planting doubt, conjuring up visions of Rafi's defection. Painting the same scenario to Allon, Sapir extracted a muttered agreement that the interim appointment of Golda would be wise. Then he returned to Galili to report that Golda was Allon's choice.

Allon and Dayan were both too ambitious to acquiesce meekly to any postponement of a final decision. Their attempts to thwart Sapir were conducted through surrogates, a group of veteran politicians lobbying for Allon, most of the old Rafi members campaigning for Dayan. But neither man had the savvy to stop Golda, with her aura of folksy motherliness and cold political skills. On March 3, with the mourning period about to expire, the party ministers voted to ask her to step into the breach, Dayan alone abstaining. Later that day, the party leadership confirmed that decision, although Golda coolly encouraged Allon and Dayan to submit their names to the Central Committee for consideration, knowing full well that neither would dare do so.

Dayan quickly proved Golda correct, announcing that he had no interest "in being a Don Quixote." And lacking the killer instinct necessary to face down the party bosses, Allon, too, stepped aside.

Golda had been dreaming of becoming prime minister since Ben-Gurion first resigned in 1953 and had been keenly disappointed when both Sharett and Eshkol were chosen instead. In July 1968, the Israeli press speculated that she was angling to replace Eshkol. And in his final days, the prime minister muttered in Yiddish about the "shrew who sits and waits for me to die."

Golda might have pretended that she harbored no ambitions, but her comrades knew better. "She couldn't wait," said David HaCohen, a long-time member of the Knesset. "It was clear that she'd kick anyone who got in her way."

But with the prize finally before her, Golda couldn't ignore either her age or her health. "Seventy is not a sin," she quipped in public. But to Senta Josephtal, a leader of the kibbutz movement, she expressed trepidation that she'd grow senile in office and that everyone would sense her disability except her. "Don't worry," Senta told her. "I will tell you when you are senile."

When she asked her physician, the head of the Oncology Department at the Hadassah Hospital, what he thought her life expectancy was, he told her, "Ten years." In ten years, Golda would be eighty-one years old. More than enough time, she concluded

The final vote by the party Central Committee was scheduled for the morning of March 7, and despite his public withdrawal, Dayan still had not given up. "I am so divided against the party leadership, the procedures, that psychologically speaking I would not have much difficulty in leaving it," Dayan had told a group of students a few days earlier, a clear hint of the price Labor might pay for selecting Golda.

The latest poll showed that only 3 percent of Israelis favored Golda as prime minister—to Dayan's 45 percent and Allon's 32 percent. Nonetheless, when the four hundred members of the party Central Committee voted, not a single ballot was cast against Golda, although forty-five Rafi diehards abstained.

When the results were announced, Golda, sitting in the front row of Tel Aviv's Ohel Auditorium in a simple dark suit, froze. At Eshkol's funeral, she'd tearily told her colleagues, "I have always carried out the missions the state placed on me, but they have always been accompanied by a feeling of terror. The terror exists now." No one had believed her because Golda's toughness so thoroughly masked her lingering insecurity. But her mother's voice still echoed inside her, reminding her that she wasn't good enough or wasn't smart enough.

Finally, Golda rose and walked slowly to the platform, tears streaming down her face. As she reached the podium, she was greeted by a thunderous ovation and cries of "Golda, Golda, Golda shelanu, our Golda."

The response outside the party was less enthusiastic. "The people of Israel have the right to expect that the helm will be given to a younger person whose power of action will not be restricted by age or health," opined *Ha'aretz*. And while the Sephardic chief rabbi ruled that a female prime minister, unlike a female president, was acceptable given the historical precedent of Deborah the prophet, many ultra-Orthodox Israelis believed her appointment to be a clear violation of Jewish law.

Ordinary Israelis celebrated their progressiveness in having the first modern female chief of state who owed nothing to the "appendage syndrome" that had brought Sirimavo Bandaranaike to the premiership of Ceylon in the wake her husband's assassination and Indira Gandhi to office not long after her father's death. But they weren't sure what to think about their new prime minister, a woman whose *New York Times* obituary had already been written. With none of the maverick panache of Dayan, or military aplomb of Allon, had the country just inherited an aging political hack as its leader?

After all, not even Golda claimed to be the most appropriate choice. "I became prime minister because that was how it was, in the same way that my milkman became an officer in command of an outpost on Mount Hermon," she later admitted.

*   *   *

Golda would never have admitted it publicly, but she wasn't overly fond of the Israel whose mantle she inherited. After two decades, the country hadn't matured into the Socialist paradise she'd envisioned back in Milwaukee, a land without thieves or murderers, a gallimaufry of free Jews with hands calloused from tilling their own soil in a nation that radiated Jewish justice and integrity.

Instead, Israel was a raucous and argumentative country that could make a New Yorker feel mellow. No issue, no matter how mundane, was resolved without lengthy, passionate debate. More than a dozen political parties competed for power, each fractured by a dozen caucuses and egos too large to be contained within a single political organism.

The kibbutzim, the heartland of the Zionist pioneers, were on the wane, the zeal for self-sacrifice supplanted by modern self-indulgence. Female soldiers were shortening their skirts, showing entirely too much knee for Golda's puritanical tastes. Her beloved "New Jews" were shunning the hora and folk songs for the Beatles, whose appearance in the country had been banned by the ministry of education as "a bad influence on the young generation."

In Tel Aviv's cafés, disputations about how to turn a fast buck had replaced debates about the dialectic, and consumption overwhelmed egalitarianism.

Golda gritted her teeth as she watched her old dreams unravel into the nitty-gritty hallmarks of a class society—expensive shops and eateries, ostentatious homes, fancy clothes, and luxury cars. "They've got some chutzpah!" she sniffed with disgust one afternoon when she discovered the price of the fish she'd just eaten at lunch with Zalman Aranne, her minister of education, and his assistant. Cohen's, where they'd dined, was hardly Jerusalem's most expensive restaurant, and the bill was not exorbitant. But for Golda, anything more than modest was inappropriate. "It was just a piece of fish," she seethed.

Most painful to a woman inspired by the concept of Jewish labor, a flood of inexpensive Arab workers from the occupied territories was pushing Jews out of physical work, and unlike in the old days, the displaced workers were applauding since the booming economy created opportunities for them that didn't demand backbreaking manual labor.

Golda had nurtured her naive girlhood fantasy about the shape of a Jewish homeland well into retirement. But like so many aging revolutionaries who trade idealism for pragmatism, strand by strand, Golda let go of the social justice aspect of her old dream, taking comfort in its other facet, the sight of Jewish policemen keeping the streets safe, of Jewish pilots flying airplanes emblazoned with the Jewish star, and Jewish judges handing down fair treatment.

Israel wasn't supposed to be just another country, but just another country was better than another 5,000 years of Jewish exile. As prime minister, then, no matter her ambition or her need to prove that Bluma had not been right, after all, everything boiled down to that single reality: it was now her job to maintain a safe harbor for children terrorized by anti-Semites and for adults scarred by some version of a swastika.

It was an ironic historic mission for a woman whose passion had always been for domestic matters and whose political romance was with the Histadrut, not the Israel Defense Forces. But the Egyptians had launched the War of Attrition, one of the deadliest border conflicts in modern history. Almost daily artillery barrages along the Suez Canal turned the Israeli front in the Sinai into an armed camp. And fedayeen newly equipped with Katyusha rockets continued to sneak in from Lebanon and Jordan to wreak havoc in settlements and markets. Defense was consuming 40 percent of the national budget.

Israelis had become accustomed to the violence; it was part of their daily routine. A year earlier, depression had hit with the realization that they had won the war only to lose the peace. But the country had moved on, accepting the reality that another clash was inevitable. The struggle was to hold at bay the fumbling efforts of the United States to impose a settlement not to Israel's liking while appeasing Washington sufficiently to keep arms supplies flowing. That battle would define Golda's premiership.

But Golda couldn't turn her full attention to the security challenge, or any other quandary, until she established firm control over her obstreperous and pugnacious government. Eshkol's wall-to-wall coalition of leftist idealists and militant rightists, a holdover from the run-up to the war, was a nightmare of quarrels and power struggles. No matter the issue—the place of religion in Israeli society, taxation, investment policy, or security—the cabinet could never achieve consensus. The public might have been enthralled with the National Unity coalition, giving it 90 percent approval ratings, but Golda found herself running a government by gridlock.

While Eshkol was prime minister, no conundrum had consumed more days of discussion while yielding fewer decisions than the occupied territories, a much-heralded bargaining chip and security perimeter that had become a millstone around Israel's neck. Golda saw no way to resolve the problem, so she simply decreed an end to squabbling. "I don't see why we have to quarrel among ourselves," she said. "As long as the Arabs are still at war with us, I am opposed to a Jewish war."

She brought that same authoritarian hand to every decision and to every meeting. The international media painted Golda as a doting grandmother "past the biblical threescore and ten," but she was more the prophet Deborah than the gentle matriarch in the fashion of Rebekah, wife of Isaac. Strong, decisive, overbearing, and intolerant, she ran roughshod over her ministers and unified the government in short order. How did she do it? pundits and journalists asked in awe. "Golda told them all to shut up, so they shut up," one government official explained.

Ever inclined toward the puritanical, Golda kept firm control on spending, even demanding that her ministers seek her permission before traveling abroad. When Yigal Allon served as minister of education, he always hesitated before making any such request, fearing the humiliation of rejection. "She held them on a short leash," recalled Eliezer Shmueli, former director-general of that ministry.

Like a schoolmarm with a classroom of incorrigible pupils, if a cabinet member began leafing through a newspaper during a meeting, she barked, "This is not a reading room." When Dayan continued to waver in his commitment to the party, she pointedly admonished him, "Conditional party membership is not acceptable." Frustrated that her cabinet of warring egos could get nothing done at their Sunday meetings, she convened her own mini-cabinet—the kitchen cabinet—on Saturday nights to make the serious decisions.

"If you could only watch her come into the cabinet chamber, I swear she grows an inch every time she walks through the door," one government minister told *Newsweek*.

Her imperiousness made Golda less than universally popular among other party officials. Like Ben-Gurion, she frequently bypassed the cabinet in making decisions, and her ministers cowered in fear of annoying her. Eban thought so little of her that he regularly quipped that she chose to use only 200 words although her vocabulary extended to 500. Peres called her a member of "a self-perpetuating oligarchy with a powerful sense of self-pity."

But no matter how much they denigrated Golda, they bent to her formidable will, not merely because she wielded prime ministerial and party power but also because she manipulated them with aplomb. At one time or another, almost all of Israel's leaders, including Ben-Gurion, remarked that she reminded them of their own mothers. When Ben-Gurion did so, Golda rebuked him. But with the younger men, she used her resemblance to their mothers to her own ends. Being berated by one's prime minister was difficult enough, after all. Being berated by a prime minister who felt like your mother was intolerable.

"There was never a mother who reproached a son in a sterner manner," said Simcha Dinitz, who became her ambassador to Washington.

Even with her closest advisers, she could be incredibly sharp, yelling and throwing ashtrays across the room. Lou Kaddar, who'd served with her in Moscow and became her confidante, scheduler, and purchaser of appropriate clothing, long remembered with uncharacteristic bitterness the afternoon Golda went to airport to pick up her sister Clara, who was arriving from the United States, and realized how much else she had to accomplish that day. "What are you doing to me?" she screamed at Lou and Dinitz, her steely gray eyes blazing. "I've never seen a more stupid schedule than the one you've organized!"

Although she radiated age, becoming prime minister was a tonic that infused new life into Golda. An average day plucked from her appointment calendar began with hard-boiled eggs, toast, and coffee at seven A.M., when she'd peruse the press with her longtime aide Lou Kaddar, always homing in on the articles that annoyed her. At the office, Golda looked through the incoming cables and doled them out to her aides and ministers. The long reports that stacked up on her desk were rarely touched. Never much of a reader, she preferred briefings to sloughing through written documents.

Between consultations with department heads, discussions with party leaders, and appointments with foreign ambassadors, she tried to slip home for lunch, but she rarely managed it. There was always another speech to give, another helicopter flight to a remote corner of the country, another foreign donor to woo. Her door was perpetually open, to Soviet Jews, to the mothers of dead soldiers, to friends of friends from America, to the beseeching and the carping.

Her day never seemed to end because Golda still hated being alone. If she had no evening commitments, she heated up whatever her maid had left her to eat and then called a friend to stop by to chat, took cake and coffee to her security guards, or made tea for her old friend Zalman Shazar, sipping her brew through imported sugar cubes in her cheeks. Only when she was desperate did she sit alone to watch her favorite television programs, detective stories like *Ironside*.

Her pace was so grueling that her aides lost all semblance of personal life. Although Lou had agreed to join her when Golda pledged that she would serve only six months, Golda couldn't get along without her. Simcha Dinitz, who had a family, took on the job of political adviser, thinking that he would be able to leave the office with her in the late afternoon. But Dinitz was excellent company, and Golda always wanted him to stop by in the evening or on Sabbath, and he found it impossible to escape.

In a country like England, where the role of head of state requires more formal mien, Golda could not have survived politically. But she was an ideal fit for Israel. She wore her office easily, more naturally than as a right. Although she was suddenly flying around the world in private planes and staying in fancy hotel suites, she still assumed everyone would call her Golda and that no one would bow or scrape, and her modesty allowed Israelis to indulge their illusions about their own fading idealism.

Everyone knew that Golda's maid and driver ate lunch at her table, that she abolished the tradition of official pomp when she arrived at the airport, and that she hated her inevitable body guards. Greeting Katherine Graham, publisher of the *Washington Post*, on a visit to Israel, Golda began by asking, as she so often did, "So, how's your mother?"

When Senator Frank Church came to town, Golda invited him for breakfast, then promptly disappeared into the kitchen. Unsure what was happening, Church followed and found her cooking. "You came to help?" Golda asked. Church shrugged his shoulders. "Okay, why not?" he replied. No, no, go into the living room and I'll bring you breakfast, she urged, then added, "But there are a few things you can help me with—Phantoms, land-to-air missiles, and help Russian Jews to emigrate."

The Israeli public became enthralled with the colorful Old Lady who clumped around the country in her heavy lace-up shoes—*na'alei Golda*, Golda shoes, as they're called in Hebrew—smoking incessantly, quipping acidly. How's your health, journalists asked. "Nothing serious," she'd respond. "A touch of cancer here, a little tuberculosis there." Her *shtick* was legendary, not just her sarcasm and bluntness, but her regular boasting about her cooking—which anyone who sampled it knew was likely to include an unappetizing selection of overbaked cakes, peppered gefilte fish, and odd creamed mushroom casseroles—her selection of dowdy clothes, and her ubiquitous outmoded handbag. (At one point, Miriam and Levi Eshkol could no longer abide her purses and bought her a stylish Hermès bag in Paris. Golda never once carried it.)

In less than a month after her selection, Golda's approval ratings had reached 61 percent and there was little doubt that the interim prime minister would stand as the Labor candidate for prime minister. "I have never had any political ambitions, but I'll bow to the will of my comrades," she declared, as usual. By July, her approval rating was 89.9 percent of Israelis.

Only .33 of the populace expressed dissatisfaction with their new prime minister.

*   *   *

Golda took Israel's helm at a moment of sweeping change. The tide of Arab labor that poured in from the occupied territories fed an orgy of building, and the economy began a dramatic transition to high technology. The ultra-Orthodox found their political footing, the gap between rich and poor widened at a frightening pace, and a new vocabulary of individual rights began edging out old rhetoric about the collective good.

Golda had risen in the labor movement because of her political skills, her loyalty, and her prowess as a fund-raiser, not because she was inventive or imaginative. "She wasn't there to be a revolutionary," said Yossi Beilin, a longtime Labor activist, but to continue the policies of the past.

Each time a problem grew into a crisis, then, her first reaction was to stick an old Band-Aid onto a newly gaping wound.

Golda's inability to consider new ways of coping with changing realities was demonstrated in no arena more clearly than in stormy Israel's labor relations, which hit a new low point during her tenure. Israel looked fat and happy. A postwar euphoria had swept the Diaspora, and overseas Jews contributed an astonishing $1.2 billion to Israel Bonds and the United Jewish Appeal, to Israeli hospitals, schools, and universities. And the new trade between the West Bank and Jordan and the new pool of cheap labor within Israel sparked a massive surge in the economy, which grew by 35 percent from 1969 to 1973, with a concomitant rise in inflation.

Bound by contracts—with little protection against the soaring cost of living—workers rebelled, and the country was plagued by a rolling series of labor disputes by postal workers and hospital administrators, port workers, physicians, and stevedores. Service workers and administrators shut down hospitals and clinics, leaving them without clean linens, food, or sterile equipment. Electrical workers walked off the job, shutting down both power and water. Civil aviation workers closed Israel's air link to the rest of the world. Every day a new group—from the assembly line at the Elite chocolate factory, from canneries and mines—joined the rising national chorus complaining about wages.

Although covered by union contracts, Israel's workers had little say about what was bargained on their behalf since their unions were all part of the Histadrut, which alone had the power to negotiate. A multinational conglomerate of banks and building companies, factories, financial institutions, and other businesses, as well as a labor federation, the Histadrut even bargained on behalf of workers employed in its own companies, just as it represented public employees, although its leadership overlapped with the management of the government.

Golda saw no contradictions or conflict of interest in this system. Israeli workers, in her view, didn't need much protection since theirs was a Socialist state led by men and women committed to justice. But in an era of rising expectations and revolts against paternalism of all sorts, that approach was anachronistic, at best. So workers began clamoring for wage increases in the middle of contracts. Inured to appeals to egalitarianism, skilled workers, like physicians, insisted on being paid in accordance with their level of education, the demand for their talents, or, Golda's horror, their "importance" to society.

As impatient with what she saw as blatant self-interest as she'd been when she faced down Jewish workers in the *yishuv,* and still clinging to her utopian concept of "from each according to his ability, to each according to his need," Golda haughtily dismissed them. "These aren't the lowest-paid workers who are striking, but the highest-paid," she complained angrily. "Organized labor has not only rights but obligations.

"I will not spend my last days at the head of a government that only considers how to make some people millionaires."

When lectures and entreaties failed to stem the tide of strikes, Golda didn't look for a new approach to socialism; she borrowed a page out of the handbooks written by strikebreaking governments all across the capitalist world. After physicians at public hospitals began a slowdown, she threatened to send in volunteers—scabs, in fact. That threat fell on deaf ears, so she instructed her minister of health to issue an emergency order forcing them back to work. When he refused to take such a drastic measure, a virtual declaration of war on organized labor, Golda signed the order herself.

Three months later, she introduced legislation criminalizing strikes "in defiance of proper union authority." To the squall of outrage, she said only, "I simply cannot stand by and watch everything deteriorate into a political and moral catastrophe." No new raises during contract periods, she decreed. No slowdowns. No walkouts.

As she turned on labor, Golda ran into the formidable figure of Yitzhak Ben-Aharon, the secretary-general of the Histadrut, an iconic figure in Socialist Zionist circles. Refusing to be subservient to the Labor Party, he encouraged strikes instead of helping to stop them and accused Golda of acting like a corporate boss.

Golda struggled to isolate Ben-Aharon politically, but a man of fierce determination and immense moral authority, he was immune to intimidation. Finally, Golda denounced Ben-Aharon for ignoring the skyrocketing cost of defense, implying that he was jeopardizing national security. Nonsense! blared Ben-Aharon, declaring that there was plenty of fat in the public coffers. "From where will the money be taken?" asked Golda sardonically. "By reducing tea drinking, perhaps?" When Ben-Aharon pointed to the rising salaries of public administrators, Golda scoffed, unwilling to entertain any discussion of corruption within her own ranks.

Golda won her battle against Ben-Aharon, but in the process helped to undermine the very institution she was attempting to save. The Histadrut was paralyzed not just by strikes and the changing face of the country but by the Labor Party's refusal to allow it any semblance of independence. The labor federation she had long thought of as the heart of Israeli socialism collapsed entirely within a decade, victim, in large measure, of her inflexibility.

Golda proved equally rigid and clumsy in confronting the growing rift between secular and religious Israel, the legacy of Ben-Gurion's curious coalition between old socialists and Orthodox Jews. For twenty years, secular Israelis, the overwhelming majority of the population, had lived in the limbo of quasi theocracy, with public transportation, movies, and restaurants closed on Sabbath and no civil marriage or divorce, the price of the National Religious Party's alliance with Labor.

At the very moment when the population was beginning to chafe under the religious bit, the growing clout of the ultra-Orthodox parties pushed the NRP toward greater and greater rabbinical control over daily life. The most secular of Jews, Golda blithely smoked and drove on Sabbath, and ate frequently at the Mandarin, her favorite nonkosher Chinese restaurant. But having learned her political lessons about the dangers of offending the Orthodoxy, she would do nothing that might anger the NRP.

Her first attempt to balance the power of the religious establishment against the desires of the populace was sparked by one of those absurd brouhahas that make no sense except as part of a wider cultural war. After years of deferring to religious sentiment against televising on Sabbath, the Israel Broadcasting Authority decided to extend its programming to seven nights a week. But five days before the first Friday-night broadcast, the head of one of the religious parties prevailed upon Golda to stop the sacrilege and, deferring, Golda asked the IBA to change its decision.

Rather than acquiesce, the director of the IBA threatened to resign, and Golda invoked an obscure clause in the regulations of the public body governing broadcasting to countermand his decision. But a single viewer in Tel Aviv appealed to the courts and found a judge willing to order Golda to show cause why the IBA should not go on the air, and she could find no grounds.

"It was the day the majority rebelled and succeeded, for a fleeing moment, in imposing its will on the minority," Ephraim Kishon, Israel's most biting satirist, wrote in his regular column, suggesting that the anniversary of the first Friday-night television broadcast should be made a national holiday.

Nonsense, said Golda dismissively, who'd watched the Friday-night chapter of *The Forsyte Saga* from Kibbutz Revivim. "For this they had to make a lot of noise?"

The secular majority, however, won few such victories while Golda was prime minister, and the conflict reached its breaking point in an esoteric squabble over the minutiae of Jewish religious law as embodied in the halacha. The case involved a brother and sister barred by the religious courts from marrying other Jews because their mother had left her first husband, a convert, and remarried without a religious divorce. Under the halacha, Miriam and Hanoch Langer were thus *mamzerim*, bastards, and ineligible to marry Jews who were not also *mamzerim* or converts.

The specter of two young Israelis, one a burly army sergeant-major, the other recently mustered out of the military, blocked from true love by elderly rabbis citing abstruse religious injunctions, inflamed the Israeli media. And the resulting explosion of frustration with clerical control threatened to topple Golda's government when some Knesset members, partners in her coalition, introduced bills to permit civil marriage.

With her limited repertoire of political responses, Golda tried to solve the dilemma by bullying the religious court. "If God has to be compassionate, then rabbis also have to be compassionate! she barked angrily against the backdrop of a demonstration outside the Knesset by supporters of civil marriage.

"I cannot accept the statement that's the way it's written in the halacha," she told a gathering of students. "That's not the answer to everything."

But alarmed at the threat to her alliance with the NRP, she then turned around and used all of her political skills to block any legislation that would open the door to civil marriage, even rejecting a compromise that would allow civil marriage only for those barred from religious ceremonies under the halacha. "Such a law will create discrimination against the children of those married in such a way," she declared, an odd bit of logic under the circumstances.

"I am not religious, but we must honor the religious," she coaxed, reflecting both political exigency and the sympathy she had developed for religion as she aged. Falling back on her oldest rhetorical device, she dredged up a bit of her personal history to make her point, recalling how she had tried to avoid the use of a chuppah when she married. "Then I realized, 'You idiot!'" she exclaimed. "'You are making your mother miserable. It's not as if you won't be able to live happily ever after if you stand under a chuppah.'"

Religion kept the Jewish people unified in exile, she argued. "If I had remained in the Soviet Union, I might have become Orthodox because synagogue was the only place Soviet Jews could express their Jewishness."

Ultimately, Golda resolved the Langer issue with a masterful stroke of backroom politics around the selection of new chief rabbis that led to the appointment of a religious leader willing to find a way to allow the Langers to marry. But, as in her dealings with the Histadrut, she missed the point. Product of a paternalistic collectivist political culture, Golda was blind to concerns for individual autonomy, individual identity, individual liberty. But caught up in the international fervor for individual rights, a growing number of Israelis were sick of being controlled by a theocracy.

The fundamental shift from a collective social model to individualism continued to elude her, alienating a growing bloc of liberal support in the process. "In Golda, you had a combination of ignorance and self-righteousness, which means that she didn't understand the other side and didn't even try," said Shulamit Aloni, the godmother of Israel's civil rights movement. "When I started talking about human rights, she accused me of advocating bourgeois, egoistic values, and those were bad words in those days. But what I remember most was how she responded when I stood up at a meeting and started to talk about something and began by saying, 'I think.' Golda interrupted and chastised me. 'There is no I think,' she said. 'There is only WE think.'"

*   *   *

"We" was the most powerful word in Golda's personal vocabulary—if the "We" referred to Labor Zionists or to Zionists, Jews, or Socialists. She gave short shrift to almost every other group identity, and the one she dismissed with the greatest disdain was women.

Given the era and Golda's position as the first Western female to head a state, it was inevitable that she would become a feminist icon, and, initially, the lionizing puzzled her. The first time she saw a women's movement poster with her photograph and a caption reading, "But can she type?" Golda laughed. "You know, I never did learn to type," she remarked.

Golda did have her moments of feminist sensibility. Shortly after the first Knesset was formed, journalists asked how members should be addressed, by their first names, last names, or both? David Remez said that he thought the public should call them by their given names followed by their father's. Why not their mothers' names? Golda asked.

When one cabinet minister suggested a curfew on women to deal with an outbreak of assaults on women at night, Golda countered, "But it's the men who are attacking the women. If there's to be a curfew," she decreed, "let the men stay at home."

But special considerations for women rankled her, as they had when she was active in the Women's Council before the creation of the state. When Israel's nascent feminist movement asked that 25 percent of the Labor Party's seats on the Histadrut Executive be reserved for women, Golda agreed only grudgingly. "In a free, egalitarian society, there should be no need for a legal defense of women's position, which should be achieved on merit only, irrespective of gender," she said. "But reality not being what it should be, better to be ashamed that we have to pass such a ruling than not to pass it."

Organized feminism, however, struck no chord with her, and she did nothing to conceal her hostility to the movement. "What do they have to be liberated from?" she asked. "Are they bored?" Queried about handicaps she might have felt because of her gender, she responded scornfully, "I don't know—I've never tried to be a man." And when asked about the Women's Liberation movement, she sneered. "Do you mean those crazy women who burn their bras and go around all disheveled and hate men?" she asked. "They're crazy. Crazy."

Golda enjoyed traditional femininity too much to identify with feminism. Despite her old-fashioned clothes and antipathy to makeup, she was a vain woman. Her nails were always manicured, and she never went to bed without washing and brushing her hair. "I'm a realist and I suppose if I believed that I could go to a beauty parlor and I would really become beautiful, I suppose I would have done it," she said. "But I knew that it wouldn't help."

Flirting with men was a standard part of her repertoire, both personally and politically. At parties and receptions, she carefully checked out the clothes of the other women, gossiping cattily about their style and fashionability. And she bragged about her emotionalism, women's emotionalism. "It's no accident many accuse me of conducting public affairs with my heart instead of my head," she once said. "Well, what if I do? . . . Those who don't know how to weep with their whole heart don't know how to laugh either."

Thinking of herself as a Jewish pathbreaker rather than a feminist pioneer, she conceded only that career women faced a tougher road than men. "After all, it's the woman who gives birth. It's the woman who raises the children. . . . When a woman also wants to work, to be somebody . . . well, it's hard. . . . More difficult, more tiring, more painful.

"But not necessarily through the fault of men—for biological reasons."

*   *   *

During the struggle against the British, Golda became steeped in a political culture of clandestine planning and tight-lipped discipline, and, like so many other rebels-turned-rulers, she never made a comfortable transition from the essential furtiveness and obedience of the underground to the openness necessary in the halls of power of a democracy. The specter of leaks, breaches of discipline, and loss of control haunted her as if she were still coordinating insurrection against the authorities rather than wielding the authority herself. By the time she became prime minister, she had become adept at dealing with challenges from within the political system. But meeting challenges from outside it eluded her. A dynamic democracy, after all, requires deftness and subtlety; Golda possessed neither.

Israel's increasingly aggressive press became her bête noire with its constant questions and perennial skepticism, which felt like treason to her. How dare newspapers publish whatever they want? she declaimed, as if the role of the media, like that of the workers, was to serve the state. Faced with leaks, she snippily informed journalists that giving out details of closed meetings was theft and reporters who published such details guilty of "receiving stolen property."

"If you can't control the press, run from it" was her motto. "There is one thing they won't say after my death, that newspapermen were delighted with me," Golda boasted. "I'm prepared to suffer unpopularity."

Golda didn't shy away from foreign journalists, assuming she could manipulate them. Her favorite tactic when they confronted her about the peace process was to shut them down by pulling out the *Gvilei Eish (Parchment of Fire)*, four volumes filled with the names and photographs of every Israeli who had fallen in battle. She'd read pieces of poetry left behind by Israel's heroes, leaving most foreign reporters too emotionally battered to question her with much vigor.

But Israeli journalists learned to steer clear of her, knowing that she would object to virtually anything they wrote. "She was very thin-skinned," remarked Naftali Lavi, a journalist who later became Dayan's spokesman. "She could be really sarcastic with journalists. If you wrote something she didn't like, you knew it from her infamous nasty look."

Rolf Kneller, who photographed her often, still recalls how small she made him feel. "In front of Golda, you felt yourself put in your place. She could be charming as hell with a coquettish smile. But many, many people were afraid of her."

Challenges from the media were hard enough for Golda to handle, but when they spread to the public, she was lost, confounded by citizens' growing penchant for questioning government actions and for refusing to accept them. In the summer of 1972, she confronted her first open act of civil disobedience when the villagers of Bir'im and Ikrit, Israeli Arabs who'd been forcibly evacuated in 1948 "until the security situation allows their return," returned home although the government had denied them permission to do so.

For Golda, the matter was simple: the IDF had concluded that the villagers' presence almost on the Lebanese border was a security hazard, so the Arab peasants should be removed. But their plight became a cause célèbre in intellectual circles, and twenty writers called on Golda to justify the government's behavior. "Security concerns," she explained, expecting that the matter would end there. After all, what respectable Israeli would dare question security? The writers did, however, and demanded a meeting with the prime minister. Never one to shrink from such gatherings, Golda agreed.

For nearly seven hours, with her daughter, Sarah, serving tea, Golda haggled with the writers, citing the dangerous precedent the return of the villagers might establish with Palestinian refugees. "They're Israeli nationals whose sons have served in the IDF," countered Haim Hefer, one of Israel's most prominent songwriters and filmmakers. "A grave injustice has been done."

But Golda wouldn't talk about justice, even when the writers threatened to form a permanent opposition bloc. "She couldn't see justice in any position but her own," recalled Hefer. "She liked people who kissed her shoes, and she didn't know how to deal with people who didn't."

Golda was uniquely unprepared, then, when Israel's first popular uprising erupted on her doorstep in March 1971. The first sign of trouble was a flood of leaflets and crudely drawn posters announcing a demonstration at the square by Jerusalem's City Hall. ENOUGH, they read in bold letters. "Enough of unemployment. Enough of watching apartments being built for new immigrants while we have to sleep ten persons in one room. Enough of government promises that are never kept. Enough of police brutality. Enough of exploitation. Enough of discrimination." They were signed by the Panthereim Shechorim, the Black Panthers.

A politically astute group of North African youth had seized on the name of the bad boys of the American civil rights movement to force the government to address the widespread discrimination against the *mizrachim*, Jews of North African and Middle Eastern descent. Charges of inequality were a touchy issue in Israel, especially with Golda, but the distance between poverty and wealth in Israel was more often than not measured in ethnicity. For every dollar an Eastern European Jewish family earned, a *mizrachi* family brought in just 69 cents. Although more than two-thirds of Israel's elementary school children were *mizrachi*, that percentage plummeted to 18 percent in high schools and 5 percent at the university. And in the halls of power, the dominance of Eastern European Jews was overwhelming.

"One percent in the Knesset, 96 percent in jails," a Black Panther pamphlet proclaimed.

The Panthers reserved particular contempt for Golda because of her long commitment to Soviet Jews. No matter her official position, after her stay in Moscow, Golda kept the international spotlight on Soviet Jewry at rallies and meetings. Moved by the courage of men and women who were fired from their jobs, expelled from school, or arrested simply for asking for permission to emigrate, Golda never had a conversation with an American president or senator, with her comrades at the Socialist International, or with Diaspora leaders in which she failed to make a plea for world pressure on the Soviets to allow Russia's Jews to leave.

By the time Golda became prime minister, the gates had begun to crack open. And unlike the North African and Middle Eastern Jews, who'd been offered tents or tiny houses when they'd arrived, the Soviet Jews were being welcomed with spacious garden-style apartments. Carried away by the liberation of *her* people—and there was no doubt that Golda felt that special historical bond—at one meeting she'd waxed nostalgically that any Jew who didn't speak Yiddish wasn't a real Jew.

"Imagine how we felt as we watched all this," said Kochavi Shemesh, one of the original Panthers and now an attorney in Jerusalem. "Our families were still living ten people crammed into miserable old houses, and Golda was busy talking about us as if we weren't 'real Jews.'

"She was our special target. We came up with the name Black Panthers to frighten her."

Golda had the Black Panther leaders thrown into preventive detention to forestall their first demonstration, but she showed little sign of such fear. The young protesters were nothing but a gang of school dropouts and military rejects who lived on the margins of society, after all. When they threatened to go on a hunger strike at the Western Wall unless Golda agreed to meet with them, she didn't hesitate to schedule time for them—although she offered them little satisfaction.

"We felt like we were meeting with a social worker, not a prime minister," recalled Shemesh. "We demanded slum clearance, more money for education, youth rehabilitation and housing. She offered us historical rationales, pages from Zionist history, and a lecture about the expense of security.

"All we got out of the meeting was cigarettes. We'd forgotten to bring ours, so we helped ourselves to the box on her table."

But the following week, Golda felt the first stirrings of foreboding. On the Night of the Panthers, as Israelis came to call it, what began as a routine public meeting about the plight of the *mizrachim* turned into a raucous march up Jaffa Street in downtown Jerusalem, hundreds of people chanting, "Golda, teach us Yiddish" and "Either the cake will be shared by all or there won't be any cake!"

As the demonstrators converged on Zion Square, their numbers swelled to more than 5,000, and hundreds of helmeted police backed up by water cannons ordered them to disperse. When they refused, the police attacked and the young men fought back, breaking shop windows and beating up policemen. All night, the battle raged. By the morning, 150 panthers had been arrested and their families were demonstrating outside of Golda's house.

The violence enraged Golda, and when the president of the Moroccan Immigrants Association called the Panthers our "sometimes nice friends" a few days later, she exploded. "They are not nice boys," she exclaimed. "How can a Jew throw a Molotov cocktail at another Jew or at a

Jewish building. Are these nice people? Perhaps they were good once and I hope they will be good in the future, but they certainly are not good now."

Her comment inflamed not only *mizrachim* but also many Eastern European Jews. And every time Golda opened her mouth, she made things worse. At one point, she offered her explanation for why the Eastern Jews were lagging behind their European counterparts. They "brought deprivation and discrimination with them in their baggage from their countries of origin. . . . Fate caused them to live in countries that did not develop—in terms of education, industry, culture, etc.—and therefore they lack intellectual capabilities. They have it inside, but it was suppressed so that they haven't developed it as did Jews from Europe or America."

What, to Golda, sounded like a reasonable statement touched off another wave of fury at what was heard by Iraqi lawyers, Iranian intellectuals, and Algerian businessmen as not-so-subtle racism. And she kept digging herself deeper into that same hole. "What has happened to us?" she asked on the floor of the Knesset, her deep voice rising in frustration. "What has happened to our understanding? To our self-discipline? We are behaving as if there were no danger ahead of us, as if we had already achieved the peace we long for."

Insulted at the implication that they did not care about Israel, the Panthers tacked up wanted posters with Golda's visage: "Wanted for violence against the panthers . . . for robbing poor families . . . for character assassination."

At that point, Golda began having trouble sleeping, and with good reason. Panther groups had sprung up all across the country asking why the government was spending millions of dollars on marble floors for Jerusalem theaters rather than on playgrounds for poor children. In July, they streamed into Jerusalem for the Quiet Demonstration, a well-controlled protest of more than three thousand Israelis denouncing racism, seeking strong government action against poverty, and calling for Golda's resignation. A month later, they returned to Zion Square and burned Golda in effigy. "A state in which half the population are kings and the other half are treated as exploited slaves—we will burn it down!" they shouted.

Responding like the well-schooled bureaucrat, Golda appointed a special commission, the Prime Minister's Committee Concerning Children and Youth in Distress, to make recommendations for streamlining Israel's response to poverty and called for volunteers to help underprivileged children with their homework. Although she began using her bully pulpit to acknowledge the problems of poverty, she always gave the government a bit of an out. "A nation at war cannot be better off during the war than before it," she propounded patiently, although few took her seriously.

Predictably, her rhetoric did nothing to cool the Panthers' ardor. Although the ragtag group splintered, they kept up the pressure. A year after they first surfaced, they began nightly Robin Hood forays into Golda's neighborhood, which was largely European, stealing bottles of milk and leaving tags reading, "The children in the poverty-stricken neighborhoods do not find the milk that they need at their doorstep each morning. In contrast, there are cats and dogs in rich neighborhoods that get plenty of milk, day in–day out." And they invaded a Histadrut convention, silently raising placards reading GOLDA, GO HOME and 300,000 CHILDREN ARE HUNGRY.

For the *mizrachim*, it was a defining moment, especially in their relationship with the Labor Party. Month after month, they listened as Golda explained, in utter sincerity, how defense spending left little money for slum clearance and why charges of discrimination were baseless. They listened. But they never once heard her admit to the racism that everyone knew existed or watched her treat their plight as an emergency.

It fact, it never was to Golda. Not that she was indifferent to poverty. Nor was she blind to *mizrachim* and their problems. But the charge of discrimination cut too deep in a woman whose vision was clouded by her own progressive self-image and by her utopian fantasy of a Jewish homeland in which all the scars and ruptures left by 2,000 years of exile would vanish in a paroxysm of brotherhood. Ultimately, the *mizrachim* were not part of Golda's Israel, which, like her old neighborhood in Milwaukee, was a modernized shtetl of Yiddish-speakers and secular Jews.

In June 1973, more than two years after the first Black Panthers were thrown into preventive detention, the Prime Minister's Commission presented the government of Israel with its final report, a 750-page document laying out a broad program of reform—a negative income tax, recalculation of children's allowances and old age pensions, the revamping of the educational system, the funding of informal educational facilities, and new initiatives in housing.

The next government, Golda promised, the one that would be elected in October 1973, would deal vigorously with the committee's recommendations.

CHAPTER FOURTEEN

*Moses dragged us for forty years through the desert to bring us to
the one place in the Middle East where there was no oil.*

Golda hadn't been prime minister for a month before the telephone
by her bed started ringing in the middle of the night with the news
that another soldier had been killed in the Egyptian shelling of Israeli
positions in the Sinai. Some nights, the reports weren't quite so awful,
two troops lightly wounded or a bunker exploded. Once, an operator
called at three A.M. to inform her that twenty-six sheep had been killed in
an attack, and then a few minutes later to correct the information: the
number was really twenty-five.

Golda had left orders that she be informed about every skirmish, no
matter how minor, and it was fortunate that she didn't need much sleep. In
May, June, and July 1969, after Nasser launched the War of Attrition, the
Egyptian bombardment came nightly, killing 47 Israeli soldiers, wound-
ing another 157. The fedayeen in Jordan lobbed mortar shells onto Israeli
settlements an average of every third day. During one ten-hour period of
intense night fighting, Israel Lior, her military attaché, phoned eight
separate times.

On off nights, when the phone didn't ring, Golda still heard the jarring tones in her recurring nightmares. "Suddenly all the telephones in my home start ringing," she told her friend Yaakov Hazan. "There are a lot of phones, located in every corner of the house, and they don't stop. I know what the ringing means, and I'm afraid to pick up all the receivers. I wake up covered in a cold sweat. It's quiet in the house. I breathe a sigh of relief, but I can't get back to sleep. I know that if I fall back to sleep, the dream will return."

Even when the dream did not, the stark reality haunted Golda's days, magnified by the din of hostile rhetoric emanating from Cairo and Damascus, Beirut and Baghdad. In November 1969, Nasser decreed that Egypt would forge "a road toward our objective, violently and by force, over a sea of blood and a horizon of fire." Six months later, Yasir Arafat, the chairman of the Palestine Liberation Organization, fulminated, "We shall never stop until we can go back home and Israel is destroyed. . . . We don't want peace, we want victory. Peace for us means Israel's destruction and nothing else." All the leading Arab figures weighed in, with King Faisal of Saudi Arabia declaiming, "All countries should wage war against the Zionists, who are there to destroy all human organizations and to destroy civilization and the work which good people are trying to do."

*   *   *

While Israelis cocooned themselves in an illusory euphoria over the depth of their borders and their military prowess, Golda lived in terror that twenty years of existence could be wiped out in an instant if she made the wrong decision or took the wrong bargaining stance. Apprehensive about her own security savvy, she compensated with steely determination not to swerve from the formula she believed was the key to peace: Israel and her neighbors needed to negotiate directly and sign a peace treaty. Then Israel would withdraw to the "secure borders" agreed upon.

Golda did everything she could to instigate such talks, although there wasn't much she could do beyond announcing repeatedly that she was ready to go to Cairo, "on an hour's notice—in the time it takes to pack a suitcase." But the Arabs had a different formula for peace: the Israelis had to withdraw to their pre-1967 lines before the warring parties could begin to negotiate, indirectly, for a nonbelligerency pact or some other sort of agreement short of a treaty.

Since Golda showed no inclination to accept that blueprint, Nasser, without whom no other Arab leader would dare make a move, ignored her and the Arab world mocked her offer. "Mrs. Meir is prepared to go to Cairo to hold discussions with President Nasser but, to her sorrow, has not been invited," opined one Jordanian newspaper. "She believes that one fine day a world without guns will emerge in the Middle East. Golda Meir is behaving like a grandmother telling bedtime stories to her grandchildren."

Stymied by the fundamental difference in approach, and convinced that the Arabs weren't serious in any case, Golda concluded that Israel could do nothing but sit tight on the occupied territories, Israel's bargaining chip. "Why should we talk about giving back these territories? There is no one to give them to, even if we wanted. . . . We can't send them to Nasser by parcel post."

The rest of the world wasn't inclined to wait for Golda to receive an invitation to Cairo and dozens of diplomats, thinkers, and diplomatic wannabes offered up peace plans, although few planned Israel's peace without some wider agenda. Since the United Nations had proven embarrassingly inept in dealing with the Middle East, the Big Two—the United States and the Soviet Union—held their own Middle East peace talks, as did the Big Four, the wider circle including France and Great Britain.

None of those countries could lay claim to being a neutral arbiter. The military quartermasters of the Arabs, the Soviets had designs on military expansion in the eastern Mediterranean, and the French were desperate to repair their relationship with oil-rich countries still suspicious of Algeria's colonial master. Britain's history in the region was too fraught with resentment to give London any moral authority with either side.

And the United States was torn by so many competing, often conflicting, agendas that it was often difficult to tell whose side Washington really was on. Hungry to build a solid foreign policy legacy, Richard Nixon threw the United States full bore into resolving the long Middle East conflict, but his primary goal was to use the regional conflict as a wedge toward détente with the Soviet Union. Less optimistic about Nixon's grand design for linking regional wars to the East-West contest, the State Department struggled to shore up U.S.-Arab relations and protect American oil interests.

While American Jewish leaders used their political clout to enlist scores of senators and congressmen to a pro-Israel foreign policy, business and industry leaders pushed in the opposite direction. Meanwhile, the bumbling but genial William Rogers, the U.S. secretary of state, who had not a whit of foreign experience, took pains to prove that he could outnegotiate Henry Kissinger, Nixon's national security adviser, who knew as little about the Middle East as he did.

No matter the messenger, Golda received the same tidings from all the would-be peacemakers: You can't base your security on conquered territory. What can we base it on, then? Golda asked curtly. That's where most negotiations broke down, since no one offered her any realistic countersuggestions.

The United States, Britain, France, and the Soviet Union had boundless faith in Big Power talks, international guarantees, and UN peacekeepers, and Golda scoffed at them, having developed a strong aversion to international guarantees and international peacekeepers since they had proven worthless in preventing the 1967 war. And the notion of Big Power talks infuriated her, intimating that the Jewish homeland was an adolescent in need of parental supervision.

"We refuse to be a protectorate of the four powers or the six powers or twelve powers, or even twenty-nine powers of the United Nations," she averred at a United Jewish Appeal dinner in 1971. "We have had UN observers. . . . Demilitarized zones, we have had. . . . And we said, 'Never again.'"

In the face of united Arab insistence that Israel withdraw from every inch of the land it captured in 1967, the United States pleaded with

Golda to at least announce which, if any, pieces of the occupied territories Israel hoped to retain. An old hand at shopping in markets, Golda knew that the first rule of bargaining was never to give away your bottom line. "It isn't a question of, let's say, territory . . . or something concrete where you can discuss it and a mediator can try to bring about some compromise," she remarked on *Meet the Press*. "The Arabs, to our great sorrow, just don't want us to be. They want a destruction of the state. . . . There is no compromise on a question of that kind. . . . The question involved is, are they prepared to acquiesce to our lives and presence in the areas? . . . For that it is necessary that we meet them and negotiate with them and if they are not prepared to meet us then they certainly are not prepared to live in peace with us."

Even if Golda had been inclined to flexibility, she had little room to maneuver politically. After the October 1969 elections, it took her thirty-five days to form a cabinet, and what emerged was an unwieldy National Unity government split between the Greater Israel crowd, determined to annex every square inch of what Israel had conquered, and peaceniks willing to trade away virtually everything, with a dozen ambitious and ambiguous strong-willed personalities in the middle.

As unimaginative in her approach to peace as she was to domestic problems, Golda saw no reason to jeopardize the stability of her government when she didn't have a peace partner. So in the face of a morass of competing and colliding agendas, both foreign and domestic, she dug in, and the Israeli people, as distrustful of the outside world she was, applauded her obduracy.

One of the most popular songs during Golda's tenure was a ditty with a chorus that ran, "The whole world is against us. . . . If the whole world is against us, we don't give a damn. If the whole world is against us, let the whole world go to hell."

*   *   *

Golda was still struggling to stitch together her coalition government when a cluster of peace initiatives fell from Washington. Flattered by

Charles de Gaulle's enthusiasm for diplomatic cooperation, his bid to restore France's waning international clout, President Richard Nixon agreed
to his idea for a Big Four Middle East peace effort to end the fighting
between Egypt and Israel.

Simultaneously, Henry Kissinger plotted a secret Soviet-American dialogue as part of his wider vision of using "linkages" between regional
conflicts and Soviet-American relations to forge détente: Moscow and
Washington would develop a joint plan for peace without consulting either Cairo or Jerusalem and they would then muscle their respective vas-
sal states into accepting. It was a quintessentially Kissingerian concept,
built on what turned out to be the dangerously misguided assumption
that the Soviets would abdicate power in a contested region in return for
friendship with the United States.

Both sets of negotiations were launched, but the former were stillborn,
and the secret talks with the Russians bogged down because Kissinger
had seriously underestimated the finesse with which the Soviets would
dangle cooperation under his nose while shipping massive quantities of
armaments to Egypt and Syria.

Then, Secretary of State Rogers moved in with a State Department
initiative. In one of his prototypical moments of manipulative theatrical-
ity, Kissinger decided that Golda should be kept in the dark about the
Rogers plan initiative, although she was due in Washington to discuss
arms purchases on the eve of its announcement. So when she tried to
engage in negotiations with the president, he seemed to be loath to discuss anything but baseball. "Milwaukee lost the Braves but they got you
back," he said in their meeting in the Roosevelt Room, apropos of Golda's
scheduled trip to her old hometown. "As a matter of fact, the Braves could
use you as a pinch hitter right now."

Suspecting that Nixon's reticence was a harbinger, Golda returned
home and waited for the bomb to drop from Washington. Two months
later, in December 1969, it fell with Rogers' announcement that he would
attempt to negotiate an agreement for an Egyptian promise of nonbelligerency in exchange for Israeli withdrawal from all the occupied territories

with "insubstantial" border changes. To diplomatic professionals, Rogers' proposal was a masterful exercise in vagueness, spelling out no details about the meaning of either "insubstantial" or "nonbelligerency." All that, Rogers said, could be worked out once Israel and Egypt had agreed to the basic principle.

At heart, the Rogers plan was a minor revision of UN Security Council Resolution 242, passed in the aftermath of the 1967 war, shortly after the Arab states rejected Levi Eshkol's offer to trade all the occupied territories in exchange for peace treaties. At a summit in Khartoum, they had issued the famous Three Nos: no recognition of Israel, no negotiations with Israel, no peace with Israel.

The holy grail of Middle East peace, 242 called for "withdrawal of Israeli armed forces from territories occupied in the recent conflict," an end of the state of belligerency, mutual recognition of all established states, and "secure and recognized boundaries." At the insistence of the United States and Great Britain, the word "the" had been omitted from the phrase about territories. Assuming that "territories" implied the word "the," the Arabs used 242 to demand that Israel give up every square inch of what they'd conquered. Parsing the resolution more closely, Israel insisted that it left room for negotiations. Rogers' offering did not. Nor did it include Golda's conditions for peace, direct negotiations between Arabs and Israelis, and a peace treaty, not some vague promise that they would stop shooting.

In fact, Golda simply didn't believe all the Arab rhetoric about the occupied territories presenting an insurmountable stumbling block to peace. "Now they say we should go back to the '67 borders, but that's where we were, so why was there a war?" she protested. "And we had '47 borders. . . . We didn't like them very much, but we said yes to them. But there still was a war. And after the '48 war they said we should go back to the '47 borders. But that's where we were . . . and that's where they wanted to get us out from. . . . They still nurture a hope that at some time we'll disappear."

Golda summarily rejected Rogers' plan and instructed Yitzhak Rabin, her ambassador to Washington, to mount a full American Jewish assault against it. "We didn't survive three wars in order to commit suicide so that the Russians can celebrate a victory for Nasser," she said. "Israel will not be sacrificed by any power policy and will reject any attempt to impose a forced solution on it.

"It's appeasement!"

Over the years, Golda had refined her message to American Jews until few doubted that they and Israel were one. "It's our Israel that's the most important thing, and since it is OUR ISRAEL so together we see to it . . . that it should be safe and that it should develop and it should be a decent Israel in every way," she preached on her regular trips to the United States. "So we work together. It is not just a figure of speech when we say we have a partnership. It is. . . . And we won't let you people down. That we promise."

Calling on the devotion to Israel she'd engendered, as well as the contacts she'd developed among America's most powerful, she and Rabin brought a firestorm of criticism down on Rogers. Cardinal Cushing of Boston urged Nixon to hold out for direct Arab-Israeli negotiations. Seventy members of the U.S. Senate and 280 members of the House of Representatives seconded that call. And George Meany, president of the AFL-CIO, denounced the secretary of state for appeasing Soviet and Arab dictators.

A novice at international diplomacy, Rogers was stunned, both at the gall of the prime minister of tiny Israel and at the intensity of the reaction she was able to arouse. His plan—all the U.S. peace plans—was based on the assumption that Israeli dependence on the United States for military aid gave Golda few options but to succumb to American pressure. Perhaps deluded by her lined visage and gray bun, he had severely miscalculated.

If Golda's initial reaction wasn't sufficient to drive that point home, her next decision should have convinced Washington that she intended to run Israel's security affairs without regard for their sensibilities. Weary of the human and economic toll of the War of Attrition, she ordered the

Israeli air force to begin deep penetration bombing of the Egyptian interior with the triple goal of relieving the pressure on Israel's troops, bringing the war home to the Egyptian people, and forcing Nasser to respect the cease-fire that he'd broken by launching the border conflict.

Aghast, world leaders denounced the Israeli action as a threat to international peace. Soviet premier Alexei Kosygin threatened to increase military supplies to Egypt. And the Arab world accused Golda of trying to topple Nasser.

That's not my goal, she explained, "although I wouldn't shed any tears if it had that impact."

The State Department pressed Nixon to punish Israel by withholding the Phantom jet fighters and Skyhawk attack planes that Golda had requested. But the Soviets were already sending the Egyptians advanced MiG and Sukhoi bombers, installing surface-to-air missile batteries, and flying air cover missions for them. Could the United States allow the delicate Middle East balance to be upset?

Nixon knew that he could not, so he dithered, sensing that Kissinger's "linkages" weren't working, but unwilling to give up on that high concept. Privately, he sent Moscow an expression of "grave concern" over the quickening pace of arms delivery to Egypt, repeating a long-standing plea for a joint arms embargo, although publicly he promised that Israel would be sold whatever military equipment was necessary for self-defense. The Soviets responded by pouring in more SAMs as part of a new integrated air defense system they were building that would make Egyptian skies lethal to the Israeli air force.

But Moscow played Kissinger and Nixon with the skill of Vladimir Horowitz. Every time the United States grew fed up and seemed on the verge of selling Israel more aircraft, the Russians coaxed Washington into another round of talks, leading Nixon to hold Golda's increasingly anxious armament requests in abeyance. By June 1970, however, the balance was tipping dangerously in favor of Egypt. U.S. intelligence estimated that twelve thousand Soviet military personnel were working in Egypt, training combat troops and manning SAM-2 and SAM-3 missile batteries. And the Soviets were introducing SAM-6 mobile missiles, their most advanced MiG fighters, as well as Tu-16 high-speed jet bombers with long-range air-to-surface missiles.

Rogers decided on one final gambit to cool down the situation and proposed a temporary, ninety-day cease-fire and military standstill to give the United Nations mediator, Gunnar Jarring, time to negotiate with the Israelis and Egyptians. Nixon implored Golda not to be the first to reject the suggestion and pledged that, in return, he wouldn't pressure her to accept Rogers' original blueprint. Her first response was an emphatic no, both out of pique over the withholding of military equipment and out of fear that any truce would give the Egyptians time to finish their missile shield unimpeded. Understanding Golda's temper, Rabin, however, did not transmit that rejection to Rogers and waited in the hope that Golda would cool down and change her mind.

During the last week in July, however, the Egyptians forced her hand by giving in to Soviet pressure and agreeing to the new Rogers plan. Then Nixon sweetened the deal by promising Golda increased economic and military assistance. Still terrified that Nasser would use the ninety-day truce to consolidate his new air defense system rather than to negotiate, Golda sought reassurances from Washington that the cease-fire would include a military standstill in the thirty-two-mile truce zone to which the Soviets would also agree. Caught up in the euphoria of pending success, Rogers giddily promised that all sides had accepted those conditions. Trust us, he importuned her.

Still, Golda held out. What about a verification process? asked Golda. Why isn't the standstill mentioned in the truce documents? Precisely what did the Russians promise?

From his second White House in San Clemente, Nixon appeared on national television and tried to mollify her. "It is an integral part of our cease-fire proposal that neither side is to use the cease-fire period to improve its military position," he announced.

Golda yielded to the pressure and the promises. But during the first day of the cease-fire, Israeli intelligence discovered that the Egyptians were moving more SAM batteries into the truce zone. Furious, Golda complained to the Americans and demanded that the new batteries be withdrawn. The Americans refused to believe the Israeli evidence, just as they refused to admit that Rogers wasn't entirely sure what Soviet ambassador Dobrynin had actually agreed to in terms of a standstill and that the United States didn't yet have reconnaissance flights flying over the area. Obsessed with his diplomatic coup, Rogers had neither secured the promised Soviet commitment nor made the necessary surveillance arrangements.

When the Americans denied what she was seeing in aerial photographs—the movement of battery after battery of Russian SAM-2 and SAM-3 missiles—Golda threatened to cancel all negotiations with Jarring, and some members of her cabinet urged her to take more aggressive action by ordering a preemptive strike against those batteries. For three weeks, the United States resisted admitting what was happening, arguing that it had found "no conclusive evidence of violation." By the time the CIA acknowledged that their evidence, too, showed that the Egyptians had moved fifteen new missile batteries into the truce zone, Nixon knew that it was too late to demand their withdrawal. So he mollified Golda by rushing specialized electronic countermeasures equipment and Shrike air-to-ground missiles to Israel, a token contribution that hardly corrected the military balance. But before the end of the month, he promised her another eighteen Phantom jets as well as some Skyhawks and M-60 tanks.

It was the beginning of a new cycle in Israeli-American relations. The United States would push too hard, or too clumsily, in its quest to link détente or its relationship with the Arab world to the future of Israel. Golda would freak out and either phone Washington directly or call in the American ambassador and lecture him about pogroms in Russia. (After the tenth time hearing the same story, Ambassador Walworth Barbour could no longer take it and said, "I didn't mind the first few times, but every time I want to find out where Israeli troops are in a raid, we have to start in Russia.") To pacify her, Nixon would release a few planes or tanks, and Golda would calm down until the next mini-crisis.

*   *   *

Against the backdrop of a dozen cease-fire violations, the breach of the military standstill was a small matter. For Golda, however, it was a defining moment. Peace initiatives, after all, presuppose a certain level of faith, or at least trust. Hardly a trusting person, Golda was left even more suspicious, particularly of the American efforts and of international guarantees.

"Everybody tells us that there is no such thing in the world today as secure borders," she declared at a rally in London. "Yet I haven't seen any people stepping back from their borders and saying, 'We don't care what our borders are.' . . . They tell us there will be international guarantees. . . . Does any other people depend on international guarantees? No. But we should. And if we don't, we're stubborn or intransigent or not accommodating or we don't care about public relations."

The points of reference Golda brought to the peace process, as to almost everything else in her life, traced back to her past, and they didn't encourage optimism or risk taking. When Golda heard Arab threats, she flashed back to her father's helplessness in the Ukraine, to Poland and Germany, and instinctively became haunted by the specter of a new Holocaust, an Arab Holocaust.

Her experience in the Middle East had only reinforced those old phantasms. Golda had lived through three major series of Arab riots, three wars, economic embargoes, boycotts, and the closure of international waterways. Arab leaders had not renounced the Khartoum NOs, and their radio stations and newspapers printed a steady stream of anti-Jewish and anti-Israeli invective. The Beirut daily *Al Moharrer,* for example, demanded that Barbra Streisand films be banned, as had been Omar Sharif's films after the Egyptian actor kissed Streisand in *Funny Girl,* a clear sign that he had become too close to Jews.

And every time the world's would-be peacemakers berated her for failing to produce a peace map, Golda flashed back to the Peel Commission and a dozen other British attempts to extricate themselves from Palestine that ended in an argument over borders instead of wider issues.

When Golda added up what she had seen and experienced, what she was still seeing and experiencing, she saw no reason to believe that the Arabs were interested in peace. "If the Arabs felt that they could have a war and win, we would have had another war a year ago and six months ago and yesterday," she reasoned. "They aren't deterred because they don't like war. . . . If they don't attack us in an outright war it is because they know exactly what is going to happen."

So no matter how often her sister Clara, who brought an American leftist's perspective to Middle East affairs, asked her to try to empathize with the Arabs, and she did so regularly, or how frequently Eban and others told her not to confuse rhetoric with reality, Golda couldn't hear the menacing diatribes as bombast. For her, they were the rantings of a new group of anti-Semites plotting to murder Jews.

When the self-styled peacemakers preached to her about Nasser's warnings that her rigidity would lead to war, she fell back on images straight from the shtetl, especially the tale of the man in a czarist village who always knew when horses were about to be stolen—because he was the gonif, the thief. "When Nasser warns that there's going to be a war with Israel, how does he know?" she asked. "Because he's the gonif."

Nasser wasn't Golda's only target. Knowing that Nixon was meeting with oil company executives and Kissinger was dreaming of détente, Golda lashed out at American pressure for Israel to withdraw to its pre-1967 boundaries. "This is not the border of the USA but of the Jewish people," she preached. "Why should we be the ones to serve as guinea pigs for borders? . . . Why should we be the only country in the world that agrees to become a protectorate surrounded by a framework people by Americans, Russians, Yugoslavs, and Indians? . . . Some say it is bad for our image if we stand firm. What did Czechoslovakia lack in the way of a good image, yet Soviet tanks rode into Prague?"

In the halls of diplomacy, Golda was being asked to trade land for vaguely worded Egyptian promises of nonbelligerency, and she was intent on forcing the Americans and Europeans to understand that the land in question protected Israeli settlers in the north from the potshots of Syrian gunners, the nation's commerce from the strangulation of trade through the Straits of Tiran, and the center of the country from Egyptian columns massing on a border within hours of downtown Tel Aviv.

It wasn't just the substance of the proposals that dismayed her but the idea of having an agreement foisted upon Israel by outsiders. She had an almost religious belief in the value of direct negotiations, of enemies sitting together to bury their swords. "It would be a fatal error to try to explain Israel's stand in psychological terms such as stubbornness, suspicion, and the like while disregarding our balanced attitude," she cautioned several months after taking office. "In principle, we do not hold a situation whereby powers arrogate to themselves the right to discuss the destinies of nations and countries without the participation of those concerned, and in lieu of immediate colloquy between the nations themselves."

As Golda entrenched herself in a sort of one-woman siege mentality, a mythology took root, Golda "the intransigent," as her opponents called her, and her sharp tongue and caustic rhetoric left them plenty of ammunition. In response to entreaties that she take the Arab mentality more seriously and leave Nasser room to pluck honor from his shame, she said, "We're told over and over again that Nasser is humiliated. Humiliation as a result of what? He wanted to destroy us, and, poor man, he failed. . . . What can we do for him except to say that despite the fact that you attacked us, you wanted to destroy us, we won the war, now we want peace."

Golda wasn't taking her cues only from her own instincts but from the Israeli military leadership. Shortly before she assumed office, a reporter asked her to comment on a military exercise they had recently witnessed. "Do you think that an old lady like myself would have anything to say about things like this?" she responded.

Thus while she kept a tight rein on Dayan politically, understanding that she was dangerously out of her depth on military affairs, she deferred to the minister of defense she'd inherited from Eshkol, just as he began deferring to her. It was the beginning of an odd relationship, the younger man intimidated by the older woman of extraordinary strength, the older woman almost enamored of her swashbuckling general, the two engaged in a strange dance of flirting that was a sort of political courtship. If they had a fight, Dayan inevitably was the first to concede, asking, "Do you still love me, Golda?"

But it was anything but a relationship among equals. Dayan's biographers characterized Golda's treatment of Dayan as "slavish dependence," an absurd allegation since she frequently put Dayan in his place and he did nothing to stop her. Golda knew that caught up in his public image as the brave military hero, Dayan was too inconsistent in his own thinking to be an entirely reliable adviser. In the early days after the 1967 war, he had been Israel's leading empire builder. The 1949 armistice agreements, he remarked offhandedly, were no longer "sacred law. . . . If the Arabs want a change, they should phone up." Obsessed with Israeli access to the Red Sea, he exhorted, "Better Sharm el-Sheikh without peace than peace without Sharm el-Sheikh."

Yet Dayan regularly advocated opposing positions within hours, and then changed his mind once more. He was the first to advocate a swift response to Egypt's violation of the standstill agreement, threatening to resign if Israel did not force the issue, and the first to change his mind once he had calmed down. Golda, then, routinely included the chief of staff and senior IDF officers in her kitchen cabinet to broaden her view.

The job of the military, of course, was defense, not translating victory into peace; that was the task of the diplomats, and Golda never learned to trust diplomats. As Ben-Gurion did to her when she was foreign minister, Golda regularly cut Eban out of the loop, dealing directly with the White House and with Yitzhak Rabin or Simcha Dinitz, her ambassadors in Washington. But diplomacy was never her strong suit. She kept her eyes clearly focused on how to defend Israel, then, with little consideration given to what was needed to make peace.

*   *   *

Golda's old adversary Nasser was as intransigent as she was, if not more so, and equally boxed in by political pressures. But in September 1970, he suffered a massive heart attack and was replaced by his vice president, Anwar Sadat. In world capitals, new hope became epidemic. Maybe without Nasser . . .

While lacking in Nasserian braggadocio, Sadat's first statements were anything but reassuring. "We shall liberate our land," he vowed to a meeting of top officials. "It is more honorable to die while defending our land than to live on our knees in surrender." In an interview with the *New York Times*, he declared that negotiations were impossible until Israel withdrew from "every inch" of the occupied territories. And under no circumstances would he establish diplomatic relations with Israel. "Never, never, never."

So prospects for peace looked dim, and international efforts did nothing to brighten them. Golda returned to the bargaining table with Jarring. But despite a flurry of diplomatic shuttles and optimistic verbiage emanating from UN headquarters, that mission was abortive. So too was another attempt for a comprehensive peace by the indefatigable Rogers, who couldn't get past his belief that international peacekeepers and Egyptian promises would sway Golda.

The only glimmer of hope was an idea first broached by Dayan, who urged Israel to launch its own peace initiative by offering a partial settlement in exchange for partial peace. The partial settlement he had in mind was a twenty-mile Israeli pullback from the Suez Canal and the partial peace, an Egyptian promise of nonbelligerency. It was a shocking suggestion since Israel's slogan had long been "no territorial concessions without a final peace treaty." But the logic was compelling. If Israeli troops moved back from the canal, Sadat would gain some breathing room. The canal could be reopened, bringing Sadat much-needed revenue, and the Egyptians who lived on the west side of the waterway could return to their normal lives. If all went well, Sadat might feel calm enough to negotiate a peace treaty. Even if he did not, he'd have an enormous incentive not to resume hostilities.

Golda worried that any withdrawal would become a slippery slope of retreat, and she knew that her government would pay a high political price for relinquishing any territory, especially if Israel received nothing concrete in return. Nonetheless, she began to consider more limited withdrawal, of six miles instead of twenty, although assuming that Sadat would reject it.

Then, shortly after Dayan outlined his plan on *60 Minutes*, Sadat announced his own New Initiative: if Israel agrees to withdraw to the Sinai passes, and allows Egypt to reopen the Suez Canal and post its troops on the east bank, Egypt would extend the temporary cease-fire by six months and move toward a full peace with Israel.

The mention of peace electrified the Western diplomatic community. But the passes to which Sadat referred weren't six or twenty miles from the canal but halfway across the Sinai, 115 miles from the canal. Sadat's insistence on placing Egyptian troops on the side of the canal abandoned by the Israelis would leave no physical obstacle between the Egyptian army and the Israeli heartland. And Sadat was quick to add that he saw the pullback as the first step toward Israel's total withdrawal from all occupied territories.

In retrospect, it is clear that Sadat wasn't trying to forge a settlement so much as to manipulate the Russians into providing him with more military equipment. Certain that Golda could order up Phantoms, Skyhawks, and other advanced military equipment from the United States as easily as a New Yorker could order pizza, he was frustrated that the Soviets wouldn't simply open their warehouses to him and angry at a new condition they'd attached to their military largesse: that Egypt consult them before using Soviet weapons against Israel. Chafing at such control, Sadat had plotted a shrewdly intricate ploy to play the United States off against the Soviets as he prepared for war.

Golda rejected Sadat's initiative, but she made some tentative moves of her own. Most important, for the first time, she drew a sketchy map of the future borders she envisioned. Israel must keep the Golan Heights and Sharm el-Sheikh to protect Israeli settlements in the north and Israeli transit through the Straits of Tiran and would continue to rule over a united Jerusalem, she announced. Although she left her view of Israel's future borders with Jordan fuzzy, she made it clear that no Arab troops would be allowed to cross the Jordan River. Finally, she expressed the hope that Israel could withdraw from the Sinai Peninsula, which would then be demilitarized. But she vowed that the Gaza Strip, which was never Egyptian in the first place, would not be returned to Cairo but become, perhaps, part of Jordan with an access corridor across Israel.

Golda's outline of Israel's future borders sparked a revolt in the Knesset, the "not one inch" crowd led by Menachem Begin introducing a vote of no confidence. She squeaked through with 62 of 120 votes, but the message was clear: not a centimeter more.

Ten days later, however, she pushed the Israeli government further by raising the issue of a pullback from the canal in a cabinet meeting. In a critical shift, her ministers effectively renounced the long-held principle, established during the government of Eshkol, of no withdrawal except in the context of a contractual peace settlement.

It might have been the beginning of a new era, but the National Unity cabinet deadlocked when they tried to define how much land they were willing to give up for a partial peace. Still, they might have found a way to break the stalemate had William Rogers not, again, bumbled onto the scene to push for a partial settlement. At Sadat's invitation, Rogers flew to Cairo, the first trip there by any U.S. secretary of state since 1953. Rogers smelled a diplomatic coup in the offing, and by the time he reached Jerusalem, he couldn't contain his enthusiasm for Sadat's sincerity and the prospect of Israel's withdrawal from the occupied territories.

Golda, whose memory was long and grudge-carrying legendary, refused to share Rogers' optimism, and their exchanges grew bitter. When she asked Rogers to fly over the Golan Heights to see, firsthand, the danger that Israeli settlements faced, the secretary refused. Astonishingly ill-informed about Israeli politics, he made matters worse by trying an end run around Golda, asking to meet with the Knesset Foreign Affairs Committee, a request Golda could hardly deny given the frequency with which she met with members of the U.S. Congress. There the discussions were even more acrimonious than they had been with the prime minister, since the committee included her opposition on the right, especially Begin, only hardening resolve against any partial settlement agreement.

Ultimately, all the talk about such an agreement was probably irrelevant since Sadat, by his own admission, was engaged in a more complicated dance than peacemaking. His invitation to Rogers had sparked the harsh response he'd expected from the Soviets, and, in turn, he arrested or ousted the members of his government most firmly allied with Moscow. As Sadat expected they would, the Soviets panicked and sent Russian president Nikolai Podgorny to Cairo to smooth things over with a treaty of friendship and the promise of "all the weapons you have asked for," in Sadat's words. Between September 1970 and July 1971, he received 100 MiG-21 jet fighters and 80 Mi-8 troop-carrying helicopters.

But having promised that 1971 would be the "year of decision," Sadat wanted more, especially more aircraft. So when the promised weapons did not arrive promptly, he called the Soviet bluff and ordered the expulsion of all 15,000 Soviet troops in Egypt. Washington celebrated what it considered to be Sadat's decision to scale back his military activities. "That interpretation made me happy," Sadat later wrote. "It was precisely what I wanted them to think." The West's leading diplomats never considered Sadat's real reason: "No war could be fought while Soviet experts worked in Egypt."

Sadat's expulsion of the Soviets reinvigorated American attempts to arrange an Israeli pullback from the canal. But this time, Washington's messenger was Joseph Sisco, one of Rogers' assistants. His first meeting with Golda, at her home, was a disaster. Sisco tried to break the ice by talking about his long friendship with a Jewish family, the Goodmans, who served him chicken soup when he was a boy. "Sisco was trying to trade Mrs. Goodman's chicken soup for a pullback from Suez," Golda told her colleagues contemptuously. She wasn't so easily bought.

The American attempts to broker a peace were already doomed by U.S. domestic politics and Cold War rivalries. Nixon was under mounting pressure to be more supportive of Israel from a senatorial who's who that included Ted Kennedy and Bob Dole, Jacob Javits, Scoop Jackson, Edward Brooke, Stuart Symington, and Abe Ribicoff. And in January 1972, Nixon finally admitted what had hardly been a secret, that the Soviet Union was sending arms to the Middle East and that the United States needed to help Israel maintain the military balance in the region. In fact, Arab arms procurement had increased 300 percent since 1969, with thirteen Arab countries spending $2.43 billion and Israel just $790 million, according to the Swedish International Peace Research Institute.

By the dawn of 1973, the Middle East was again at a stalemate. But Golda's military advisers assured her that she need have no concern. Golda "has better boundaries than King David or King Solomon," pronounced Rabin. Standing on Masada, Dayan proclaimed that Israel had circumstances "the likes of which our people has probably never witnessed in the past and certainly not since the modern return to Zion."

If the Egyptians attack, he said, "we'll step on them. I will crush them. Let them come."

*   *   *

Despite the failure of the major powers to bring Israel and Egypt together, there was no shortage of would-be mediators—individual crusaders and diplomatic wannabes—anxious to step up to the dusty plate of the Middle Eastern conflict.

Small nations anxious to make their mark, or to prove their independence from the world's superblocs, entered the fray, the first a group of African leaders—the Four Wise Men, they were called—representing the Organization of African Unity. Theirs was a curious, or at least an original, conceit, that nations without enormous military might, great wealth, or the clout those characteristics imply might succeed where the United Nations or the heavyweights failed.

Africa's self-styled "messengers of peace"—Senegal's poet-president Léopold Senghor, President Ahmadou Ahidjo of Cameroon, Chief of State Yakubu Gowon of Nigeria, and President Joseph Mobutu of Zaire—descended on Israel in mid-November 1971 to bridge the Arab-Israel gap "by means of a dialogue." It was a chaotic event, to say the least. Mobutu arrived an hour after the first three presidents, requiring two formal welcoming ceremonies at Lod Airport. The name of his country had changed from the Democratic Republic of the Congo to the Zaire Republic one week earlier, forcing the Israelis to revise all their programs and invitations. And the specially ordered national flags were delivered late because the flag-maker's wife had just given birth to Israel's first quintuplets.

No one in Israel had much confidence that the peace effort launched at the June summit of the OAU would bear much fruit. But Golda retained a special affection and respect for Africans, so, for the first time in Israeli history, she ordered her cabinet to eschew Israel's emblematic informality for black ties for the welcome dinner.

Senghor, the leader of the group, approached the mission with high seriousness of purpose, proclaiming that the four leaders had a special opportunity to succeed where so many others had failed because of the unique bond between the Africans, Arabs, and Jews as "a trilogy of suffering peoples." But during four meetings with Golda, the Africans offered no original proposal or approach. And their subsequent visit to Cairo was no more fruitful. When they returned to the OAU with divided recommendations, the notion of African mediation was summarily dropped.

President Nicolae Ceauşescu and Prime Minister Gheorghe Maurer of Romania, the only Eastern European nation to maintain diplomatic ties with Israel after the 1967 war, were the next to throw their hats into the ring as go-betweens. In April 1972, shortly after a trip by Ceauşescu to Egypt, they invited Golda to Bucharest. The announcement of her trip, the first time an Israeli prime minister had visited Eastern Europe, touched off a frenzy of speculation about a possible peace initiative. As was her habit, Golda stayed mum. "I don't know whether I'll bring a peace treaty home in my bag," was all she would say.

During her three-day visit, Golda spent eight hours with Ceaușescu and another four with Prime Minister Maurer discussing how to bridge Israeli and Egyptian differences. Or at least that was the official story. But Maurer and Ceaușescu also quietly conveyed a message to Golda, allegedly from Sadat, offering a meeting between representatives of the two countries. Golda embraced the proposal enthusiastically and breathlessly awaited word that the Romanians had made the necessary arrangements. Ten days passed, then two weeks, but Golda never heard anything about the meeting again.

According to the Egyptian foreign minister, Mahmoud Riad, the entire affair was cooked up by the Romanians despite Egypt's expressions of disinterest in such a meeting. In his memoirs, Riad recounted numerous such failed diplomatic endeavors, by Ihsan Sabri, then foreign minister of Turkey, by the Netherlands' foreign minister, Joseph Luns, then secretary-general of NATO, all of which were similarly hushed up.

In January 1973, Pope Paul VI offered his services as a mediator during his first meeting with an Israeli prime minister. But the encounter was too tense to lead to anything productive, and its aftermath negated any possibility that the Vatican would be seen as a neutral arbiter. Minutes after Golda left the Vatican, the papal spokesman told the press that the pope had agreed to Golda's request to see him only because "he considers it his duty not to let sleep any opportunity to act in favor of peace, in defense of all religious interest, particularly the weakest and most defenseless, and most of all the Palestinian refugees."

The most infamous private peace effort was the Nahum Goldmann affair, which solidified Golda's reputation for inflexibility. Goldmann was a peculiar character even in the cosmos of Zionist odd ducks. An ebullient Lithuanian who'd been a leading force in German Jewry before the Second World War, he went on to become one of the major figures of Diaspora Zionism and the president of the World Jewish Congress. By the late 1960s, Goldmann was living in a gracious apartment on the

Avenue Montaigne in Paris, an aging swell pontificating condescendingly about a country where he'd never really lived, calling Israelis a "people who are impossible to like."

For Israelis, Goldmann was both an annoying and useful maverick. He negotiated the reparations treaties with West Germany and was a firm supporter of Zionist causes across the globe. But he had the irritating habit of acting as if he were an official of the Israeli government.

In the spring of 1970, Goldmann informed Golda that he'd been invited to meet with Nasser on the condition that she consent to the visit. "Nasser, all of a sudden, is so worried about the Israeli government that no Jews and no Israeli citizens should go see him without the permission of the Israeli government?" Golda replied, horrified at the prospect of a man she didn't trust talking with Nasser.

"How do you imagine the meeting will unfold?" she asked, and Goldmann quickly assured her that he didn't expect to negotiate with Nasser but would serve only as "a sounding board." Golda continued to press him for details. "You're going to sit there and Nasser will speak and then you'll say thank you very much. I've sounded you out?"

After a long discussion, Golda indicated that she needed to discuss the matter with her cabinet, but Goldmann objected, worried that Begin's people would undermine any chance that Golda might agree. Nonetheless, Golda raised the issue with her ministers, who dismissed the idea as absurd.

Two days later, Golda informed Goldmann of that decision, and within twelve hours, demonstrators were outside her office chanting, "Golda, if you can't make peace, GO!" and "We want a Goldmann-Nasser meeting." Seeing the invitation as a major breakthrough, a wide segment of the Israeli press castigated Golda mercilessly. University professors and intellectuals organized against her and a group of high school students wrote her a letter refusing to carry out their military service "under the circumstances."

Golda assured the public that she had repeatedly sought meetings with Egyptian officials, although she refused to offer details of any of the secret negotiations. But almost breathless at the prospect of an end to more than two decades of war, Israel's peace movement wasn't in the mood to believe her.

If they'd known the full details of the Goldmann affair, they might have cut Golda some slack. Although Goldmann maintained that Egyptian foreign minister Mahmoud Riad was involved in arranging his visit, Riad denied that account. According to him, Yugoslavia's Marshal Josip Tito had raised the possibility of a meeting between Nasser and Goldmann, but Riad had rejected the idea out of hand.

Despite the Goldmann fiasco, freelance peace negotiators, with their strange blend of vanity and idealism, continued to hound Golda, and she never barred her door. The most dogged was Marek Halter, a Jewish painter and political activist living in Paris.

Born in the Warsaw Ghetto, Halter had first met Golda shortly after the 1967 war, and three years later he approached her with a plan to open a dialogue between Israel and Egypt using a more appropriate intermediary than Goldmann, Lyova Eliav, secretary-general of the Labor Party and well-known dove. Some Egyptian friends had shown interest in the idea, as had Eliav. But Eliav wasn't sanguine about securing Golda's agreement.

Golda, however, surprised them. "Eliav isn't Goldmann," she said. "He's one of us, he's part of Israel. But I hope you realize how hard it is for me to start us on an adventure like that without knowing where it may take us."

Thinking to appeal to Golda's concern for her reputation, Halter reminded her how many people held her personally responsible for Goldmann's failed trip. But as indifferent as ever to popular sentiment, Golda seemed worried only about how serious the Egyptians were about talking. "Personally, I have nothing against it," she concluded. "Anything that may help to promote peace is important. . . . But I can't make a commitment all by myself. We have a National Unity government. It will fall if your plan succeeds. I don't mind that, but I wouldn't want it to fall for nothing."

Give me twenty-four hours, Golda asked, promising to talk to other members of her cabinet. A day later, she gave Halter permission to proceed. "I ask only one thing of you," she said. "Report to me directly on what happens. I hope you succeed. For us, for the Arabs, for all of us."

Elated, Halter returned to Paris and called his Egyptian friends, who promised to discuss the possibility of such a meeting with Nasser. Then Halter flew to Cairo himself. "Do you seriously believe it's possible to talk about peace with Israel?" Nasser asked. The two men discussed the tangle of forces at work, then Nasser ended the meeting cryptically. "You're an artist. I'm glad to have met you. Come back to Cairo."

A week later, Halter's Egyptian contacts called to set a date for Eliav's visit. Then came another phone call asking for a postponement. After several weeks of silence, a new date was set, but the meeting was then put off once more. It's impossible to know whether it would ultimately have occurred because, within days, Nasser was dead of a heart attack.

Despite that disappointment, Halter tried again in 1971 and secured Golda's agreement to enlist Pierre Mendes-France, one of France's leading socialists, as an intermediary. But Mendes-France fell ill and was unable to travel. She cooperated with Halter once more when he approached her in May 1972 about a pullback from the Suez Canal. Halter's Israeli contacts had told him that Golda would not be receptive. But Golda was never quite so inflexible as she was reputed to be. "I'm in favor of withdrawing a few kilometers to let the Egyptians reopen the canal," she told him. But allowing Sadat to move his army across the canal was "unthinkable."

Then she softened. "If they agree to negotiate—which is what we've always asked them to do—we'll discuss it with them then," Golda said. "But we don't have to state our intentions beforehand."

Halter, however, was looking for more, for some conciliatory promise that he could dangle in front of the Egyptians. "If you want them to agree to that meeting at least, they'll have to have something to gain from it," he argued.

"They have everything to gain from it!" Golda shot back.

"Israel has even more," Halter responded. "For the Arabs, what's at stake is the territories they lost in 1967. For Israel, it's peace."

"The Arabs don't need peace?" Golda asked.

Finally, she relented. "I agree to a withdrawal from the canal. I also agree to letting in as many Egyptian civilians as it will take to clear the canal. But when it comes to the army, even just a symbolic force, I'm on my guard. But if the Egyptians were to accept direct contacts, then . . ."

Free to sound out the Egyptians, Halter returned to Paris and reported back to Golda, in Yiddish via the Israeli diplomatic pouch, that Egypt's new foreign minister seemed interested. A secret meeting would be set up in London. But that peace prospect fell victim to terrorism.

CHAPTER FIFTEEN

When peace comes we will perhaps in time be able to forgive the Arabs for killing our sons, but it will be harder for us to forgive them for having forced us to kill their sons.

By 9 A.M. on September 5, 1972, Golda was locked in the cabinet room on the ground floor of the Knesset, frantically consulting by phone with German chancellor Willy Brandt. At dawn, eight terrorists from Black September, an offshoot of Yasir Arafat's Al-Fatah faction of the Palestine Liberation Organization, had stormed the apartments of the Israeli delegation to the XXth Olympiad. Within minutes, two Israelis were dead and nine others roped together on chairs and beds, the bullet-ridden corpse of one of their colleagues at their feet. If 234 Arab prisoners locked in Israeli jails and two leaders of the German Red Army Faction imprisoned in Frankfurt weren't released, the Israeli athletes would be murdered, the kidnappers announced.

For three years, terrorist raids had been mushrooming in Israel and beyond. During Golda's first years in office, fedayeen based in Lebanon and Syria had repeatedly snuck across the border to bomb Israeli apartment buildings, set off grenades in markets, fire bazookas at school buses, and plant bombs in public places.

As Israel's domestic security became savvy in thwarting their assaults, the terrorists had moved into Europe, using Western European capitals as the bases for new operations. They attempted to assassinate Ben-Gurion while he was on a trip to Denmark, bombed a Jewish-owned department store in London, planted explosives in the Israeli government tourism office in Copenhagen, and tossed a hand grenade into the headquarters of Zim Shipping in London. During a raid on the El Al office in Athens, they murdered a Greek child.

Every day brought a new report—attacks on embassies in Bonn and The Hague, the bombing of a Swissair plane bound for Tel Aviv, letter bombings on three continents, the murder of the Israeli consul in Istanbul, and the killing of the wife of an Israeli diplomat at the embassy in Paraguay.

By 1972, Palestinian terrorists had established a worldwide infrastructure of training camps and offices. The defeat of Syria, Jordan, and Egypt in the Six-Day War had seriously eroded the credibility of Arab state leadership and opened the way for Arafat, a long advocate of open guerrilla warfare, to seize control over the PLO, which the Arab League had founded in 1964. But Arafat's militancy was too tame for George Habash and the Popular Front for the Liberation of Palestine, a Marxist-Leninist cadre within the PLO sworn to eradicate the "Zionist entity" through military struggle.

In the months before Munich, their assaults had become increasingly spectacular. Four terrorists hijacked a Sabena plane out of Brussels and landed it in Tel Aviv to force the release of 117 fedayeen—only to be apprehended by the Israelis. A three-man hit squad from the Japanese Red Army, supported by a breakaway faction of the PFLP, opened fire with machine guns in the waiting lounge at Tel Aviv's Lod Airport, massacring twenty-six travelers, including twelve Puerto Rican pilgrims. And they had already begun plotting the assassination of Golda herself, although the missile strike order by Abu Yusuf, Arafat's chief deputy, against the airplane carrying her to Rome was thwarted by Israeli intelligence just moments before landing.

In the face of the onslaught, Golda was helpless. Holed up in "Fatah-land," the region along the southwest slopes of Mount Hermon in southern Lebanon, they'd taken over villages and set up a network of supply lines. When Golda sent planes to attack those bases, the United Nations howled, condemning Israel for its attacks while staying silent about the provocation. And the Europeans, lukewarm about ferreting out the terrorists in their midst, were giving in to their blackmail by releasing convicted hijackers and bombers. A tradition of terrorism was taking root, and the international community had no strategy for dealing with it—indeed, was encouraging it by caving in to terrorist blackmail. And every attempt to force the United Nations to face the new reality was thwarted by the Arab and Soviet blocs.

"This thing has spread like an infectious disease the world over," Golda told the Knesset. "Can it be possible that governments . . . should acquiesce in this state of affairs on the presumption that they will emerge unscathed? Is it possible that there is no power in the entire world, in dozens of countries in Europe, North America, and elsewhere . . . to put an end to this?"

The Mossad urged Golda to order agents into Europe to assassinate the leaders of the terrorist cells, but Golda resisted their pleas to take the war to the enemy. "Golda believed that the European states would wake up in the light of the terrorist offensive on their soil," said Zvi Zamir, head of the Mossad.

Then came Munich. Golda wanted to send the Sayeret Matkal, the commando unit that had freed the passengers of the Sabena plane hijacked five months earlier, to Germany. But the German authorities refused and instead opened negotiations, offering the Black September terrorists a virtually unlimited amount of money for the release of the athletes or high-ranking Germans as substitutes for them. When they refused, Chancellor Brandt contacted Arab leaders, but none was willing to intervene. Anwar Sadat wouldn't even come to the phone.

Finally, pretending to give in to terrorist demands for safe passage to Cairo, the Germans planned a rescue mission at the airport using a team of snipers—little more than weekend competitive shooters, in fact—and a group of armed German police hidden on the Boeing 727 the terrorists would think was their ride to freedom. But the police unit on the airplane mysteriously abandoned its mission, armored cars were caught in traffic, and a bloody firefight broke out between outgunned German snipers and the Palestinians, ending in the murder of all the Israelis.

Golda watched in agony as the world ignored the lesson of Munich. The five dead terrorists were delivered to Libya, where they received heroes' funerals. The three surviving terrorists were imprisoned by the Germans but were quickly released after a new set of hijackers seized a Lufthansa passenger jet.

Kurt Waldheim, the secretary-general of the United Nations, placed the question of terrorism on the agenda of the General Assembly, the first time a secretary-general had ever added a substantive agenda item. But a number of Arab and African countries took exception to the initiative he hoped to launch, so his antiterrorism proposal was shunted off to the legal committee for study, where they languished and died.

To friends, Golda spoke of the horror of watching Jews being blindfolded and massacred again on German soil while the rest of the world played volleyball, and of the agonizing image of Jews being carried home in coffins while the Olympic bands played on.

"Whoever looks on passively when Jews are being murdered and Israeli planes hijacked . . . is dreadfully mistaken if they think the matter ends there," Golda warned quietly at the memorial service for Israel's dead. "For if an Israeli plane is hijacked then no plane in the world is safe anywhere, and if Arab assassins are allowed to think that Jewish blood can be spilled with impunity, then the lives of people everywhere are in jeopardy.

"Just as we overcame them on the field of battle, just as we tracked them down in Israel, on the West Bank, in Gaza and over the border, so we will find them wherever they may be. . . . Let each of them know that the State of Israel and the people of Israel have no intention of letting terrorism prevail."

So on September 15, she approved the operation the Mossad had long advocated, an operation to hunt down and assassinate leading terrorists, requiring that each mission receive her explicit approval. "For Golda . . . to consciously plan [and] kill people, there was a hesitation," said Aharon Yariv, who Golda appointed as special adviser on counterterrorism after Munich. "But there was no other way to deal with the matter."

Publicly, she laid the blame squarely on the shoulders of the Europeans. "Thanks to your inertia and your acquiescence, terrorism will increase and you too will have to pay the price," she told an interview from the Italian weekly magazine *L'Europeo* shortly after the massacre. "Up until today there has been too much tolerance on your part, a tolerance which, allow me to say, has its roots in an anti-Semitism still not extinguished."

*   *   *

The Europeans were loath to accept any responsibility for the rising violence in their own cities and the terror in their airports and flung Golda's accusations back in her face. "It's hard to understand how the Jewish people, which should be merciful, behave so fiercely," Pope Paul VI lectured her, reflecting growing European sentiment.

But Golda was as unwilling to consider her complicity as the Europeans were to acknowledge theirs. "Your Holiness, do you know what my own very earliest memory is?" she retorted. "It is waiting for a pogrom. . . . Let me assure you that my people know all about real harshness and also that we learned all about real mercy when we were being led to the gas chambers of the Nazis."

Golda's rejection of the accusation that Israel was acting unjustly toward the Palestinians by denying them a homeland was not a pose for international consumption. In her view of history, the Arabs were a single people, like the Jews, and the notion of a Palestinian people made no more sense than the idea of a Russian or Moroccan Jewish people. When she arrived in Tel Aviv, after all, the Arabs themselves thought of Palestine as a fragment of Syria, and as late as 1956, Arab leaders still defined it that way. Arafat himself fed Golda's pan-Arabist view. "Palestine is only a small drop in the great Arab ocean," he said. "Our nation is the Arab nation."

As Golda was quick to point out to her detractors, historically Palestine had extended from the Mediterranean Sea across the Jordan River to the Iraqi border until the British hacked off a piece of it to create a kingdom for King Abdullah. When the United Nations divided the rest of Palestine between the Jews and the Arabs, the Arabs didn't create a Palestinian state; the Arab area was absorbed into Jordan, and the only complaint was that it didn't include the Jewish region as well.

"For nineteen years, from 1949 to 1967, the Palestinians were on the West Bank, which was administered by Jordan from the Eastern Bank, which was Jordan, and on the Gaza Strip, which was administered by Egypt," Golda protested. "I would like to understand why there was no Palestinian entity at that time, why there was no Palestinian people? Why did they not set up a state of their own? The vast majority of Palestinians were on the Eastern and Western Banks. In Jordan, they became Jordanian citizens, were elected to parliament, and served as prime ministers and foreign ministers. If they didn't like the monarchy, why didn't they establish a republic? If they did not like the name 'Jordan,' why didn't they call it Palestine?"

Again and again, no matter where she traveled, Golda returned to that same theme. "Why did the West Bank Palestinians accept . . . annexation [by Jordan] . . . without an uprising of explosives and mines?" she asked. "Why did they awaken as a Palestinian entity only after the Six-Day War?" The answer, she was convinced, was that the entire fracas about Palestinian nationalism was a political ploy, yet another club to bash Israel.

Heavily invested in her self-image as a progressive person, Golda was stung by international derision of her position on the Palestinians. "There is no such thing as Palestinians," she was widely quoted as saying. What she really said made a more subtle distinction, "There is no Palestinian people. There are Palestinian refugees. When was there an independent Palestinian people with a Palestinian state? It was not as though there was a Palestinian people and we came and threw them out and took their country away from them. They did not exist."

Golda was angry and frustrated by what she saw as a double standard. Why are Israelis condemned as imperialists for holding on to the West Bank, the Golan, and the Gaza when no one raised any criticism of Jordan or Egypt, who held the same land for much longer? she asked. "Where was the holy principle against the acquisition of territory by force when they seized the Old City? . . . How was the Old City . . . acquired by Abdullah? By serenading?"

The Palestinian leadership's goal of creating an Arab state comprising both the West Bank and Israel itself—and their use of terrorism to achieve it—made it even more difficult for Golda to muster up much sympathy for the Palestinians. "We don't want peace," Arafat declared repeatedly. "We want victory. Peace for us means Israel's destruction and nothing else."

His sympathizers might not have taken Arafat literally, but Golda, the most literal of women, did. "When Arafat speaks for the Palestinians, he doesn't say he wants to have a small Palestinian state on the West Bank," she explained. "He says that Israel is Palestine. . . . Arafat has made only one great concession. He said he would permit those Jews who were in the country prior to 1917 to remain. . . . So we others should have to break up and again become refugees in the world."

For Golda, the Palestinian problem wasn't a matter of statehood or what to call the Arabs of the West Bank but of the refugees languishing in miserable camps in Gaza and on the West Bank. While still foreign minister, she'd pleaded at the United Nations for a comprehensive strategy for the resettlement of the half million Arabs who had fled Israel at its founding, offering to compensate them for the property they'd left behind. But the fate of the refugees had become hopelessly entangled in wider politics, and she found no takers. When asked in 1956 why Iraq didn't bring in refugees to deal with its manpower shortage, an Iraq cabinet minister said candidly, "Oh, we couldn't do that. That would solve the refugee problem."

Reminding the delegates that Israel had absorbed 500,000 Jewish refugees from Arab countries, "practically the same number as that of Arabs who left the area which is Israel," Golda railed annually against such manipulation of the refugees for political ends. "How can you reconcile the outcry over the fate of the refugees living on international charity with the fierce opposition to any plan of constructive development, of resettlement, and of integration designed to rehabilitate these unfortunate people?" she asked.

"The Arab leaders exploit the Arab refugees in their countries for the purposes of their political war against us."

By the time she became prime minister, however, Golda's views were hopelessly out of synch with the reality that years of struggle against Israel and the nationalist fervor sweeping the Third World had awakened a new self-identity among the Arabs on the West Bank and Gaza.

No single voice was better placed to convey that message to Golda than that of Arye—Lyova—Eliav, one of Golda's fair-haired boys. For someone of Golda's ilk, Eliav had all the right stuff. He'd fought in the Haganah and in the Jewish Brigade during World War II, and after the war, he smuggled 2,000 Holocaust survivors past British blockades on an old banana boat. In the early years of the state, he'd been Eshkol's right-hand man in the nightmare of resettling tens of thousands of immigrants, rescued Jews from Egypt, and worked undercover in the Soviet Union.

After the 1967 war, Eliav had told Eshkol that he wanted to resign in order to conduct a one-man survey of the West Bank and Gaza. "I am not a Middle East expert," he explained. "I have so far only gotten to know the Arabs through the barrel of a gun." But he sensed that Israel's future would be decided in the occupied—what the Israelis called the administered—territories.

"There's a people there, a national movement with its own institutions, its own heroes," he reported back to Eshkol after visiting refugee camps and villages, interviewing mayors and meeting with residents, naively assuming that the prime minister would be pleased to have an accurate firsthand report. Let me set up a new national authority to build housing and industry for the refugees and "show what Israel can do," he proposed.

Eshkol shunted him off on Golda. "What Palestinian people?" Golda asked churlishly. "What are you talking about? Lyova, charity begins at home."

When no one else in the government showed much more interest, Eliav began to write. "The Palestinian people today have all the attributes of nationhood," he argued in a three-part series published in *Davar*, the Labor Party paper. "They have national consciousness. . . . They have a Palestinian history of decades, marked by struggles and wars. They have a Diaspora with a strong affinity to their birthplace. They have national awareness of a common disaster, common victims, sufferings, and heroes. . . . The Arab Palestinian nation is perhaps the nation with the most obvious signs of identity and the strongest national unity, among the Arab nations.

"The problem of our ties with the Palestinian Arabs now takes precedence in the complex of our ties with the Arab world."

The articles fell into a black hole of governmental myopia, so Eliav had them printed up in a pamphlet called "New Targets for Israel" and mailed to the entire party leadership. Unlike the other critics of emerging government policy, he was an insider and assumed that his ideas would have currency with his comrades.

Still, his ideas were ignored until he threw his hat in the ring to become the new secretary-general of the party. By then, Golda had become prime minister and she happily supported Eliav's rise. Lyova and his wife, one of the Holocaust survivors he snuck into Palestine, were close to Golda. When she needed a secret emissary to the king of Morocco or to the Kurdish rebels in Iraq, he was the person she sent.

"I was a man in his forties who'd done all the right things," Eliav explained. "She thought I might be the right person to become a prime minister."

But once Eliav became a party heavyweight, Marlin Levin of *Time* magazine noticed what Golda had ignored, that he disagreed with the government's position on the Palestinians, and decided that the rise of a dissident to party leadership might be a good story.

The day that the issue of *Time* carrying Levin's article, "The Lion's Roar," arrived in Israel, Golda invited her protégé to her home. While she made tea, Eliav couldn't help noticing a copy of *Time* on her kitchen table, open to Levin's article, with several sentences underlined. "The first thing we have to do is to recognize that the Palestinian Arabs exist as an infant nation," Levin had quoted him as saying. "We have to recognize them. The sooner we do it, the better it will be for us, for them, for eventual peace."

Over tea and cake, Golda confronted her protégé. "I assume you will deny several sentences in it?" she said. No, Eliav responded. Those are my views. "Why didn't I know?" Golda asked. Eliav shook his head. "I really couldn't say," he answered, reminding her that he'd laid out the same arguments in *Davar* and in a pamphlet he'd sent her.

Eliav understood that the overheated rhetoric of Arafat and Palestinian attacks on civilians pricked at all of Golda's worst fears of an Arab pogrom. But he hoped that he'd be able to persuade her to think about the Palestinians in a new way. Instead, Golda asked him to call a meeting of the party Central Committee. "I intend to say to them that they have a young and clever secretary-general but that they also have an old and foolish prime minister, and that they'll have to decide," she told him. "I'll ask if they want to stay with the old stupid prime minister or the young, clever secretary-general."

Stunned by the reaction, Eliav didn't argue, but he did suggest that forcing the issue could divide the party, knowing full well that the party was Golda's last redoubt. "Let us agree to disagree and work together as long as we can," he offered. Dragging on a cigarette for a few moments, Golda concurred.

"That was the end of the love affair in her kitchen," Eliav recalls, chuckling. "That day I turned from white to black." Within weeks, he was hearing the rumor that Golda was worried about his mental health.

*   *   *

No matter Golda's personal views on the Palestinians, Israel still needed a policy for dealing with the occupied territories, what writer Gershom Gorenberg called Israel's "accidental empire." At the end of the 1967 war, Israel had awoken master of a new domain, twenty-six thousand square miles of territory—three times the size of Israel itself—with a population of 1.1 million Arabs, and no plan for what to do with it or them.

As was his wont, Eshkol had dithered, passionately, one minute swept up in the romance of presiding as master of Gaza, site of denouement of the ancient Samson and Delilah tale, the next panicking at the realization that Gaza was "a rose with a lot of thorns," as he put it, 280,000 Arabs primed by years of overcrowding and misery to jab Israel in the side. Eshkol vacillated similarly over every other acre of the new territories except the Sinai, which everyone expected to trade for peace with Egypt, and East Jerusalem, which less than 10 percent of the Israeli population was willing to give up.

For almost two years before Golda came into office, the entire Israeli government was caught in a labyrinth of "yes but," "no way," and "not now" on the dilemma of the territories. Mapam feared that annexation of any land would block the path to peace; Menachem Begin's minions ordained Greater Israel to be a single nation, indivisible; religious Zionists saw Messianic design in the new Jewish hegemony over the Tomb of the Patriarchs in Hebron; Ahdut HaAvoda, also on the left, split between judiciousness and an ancient nationalism; and the military couldn't decide precisely what would best serve Israel's strategic interests.

The squabble was endless, ugly, often bitter. The day after the end of the war, Dayan announced on American television that Israel would keep both the Gaza Strip and the West Bank, although Eban was telling the world that Israel would annex neither. A dozen generals waxed lyrical about the virtues of strategic depth while the justice minister spoke out in horror against Israel's becoming a colonial power in an age of decolonization. And when she took office, Golda told everyone to shut up since there was no need to decide anything until the Arabs were ready to talk.

Although she had been a reluctant supporter of partition, neither the West Bank nor Gaza found a place on Golda's personal map of Israel. Her Judaism was too secular and too light on ancient history to entangle her in Old Testament nostalgia, and her nature was too practical to blind her to the daunting demographics. Who could think about Rachel's Tomb when retaining it meant keeping another 1.1 million Arabs inside Israel's borders? Preserving Israel as a Jewish state, not a biblical one, was her only priority.

"I want a Jewish state with a decisive Jewish majority which cannot change overnight," she pronounced shortly after becoming prime minister. "If the Arabs agree to sit down with us around a negotiating table, many of us would be prepared to give back these territories." Three years later, she still had not changed her view. "I think we should be prepared to return the populated areas as the price of peace—and at the same time avoid gaining tens of thousands of Arabs as part of our population."

So in her early years as prime minister, Golda resisted the annexationists, even in her own ranks. When Aranne, her minister of education, and Teddy Kolleck, the mayor of Jerusalem, suggested that the Israelis maintain the Jordanian curriculum in the city's schools, she supported them, even allowing Jordanian inspectors to cross the border into the city so that accreditation could be maintained. In June 1969, when the military government of the West Bank requisitioned 300 acres of land on which twenty-five Arab families farmed, Golda intervened at the request of the mayors of Hebron, Bethlehem, Beit Jala, and Beit Sahar to stop the expropriation.

For similar reasons, she threw her weight against allowing Arab residents of the West Bank and Gaza to work inside Israel, although 50,000 were already doing so. "It is better for the Arabs of Judea and Samaria to work close to home and we should help them do so," she said. "Besides, the Zionist ideal has been to accept happily the challenge of manual labor—to build this land of ours. What is happening now is that too much of that building is being done for us by Arab labor.

"All the Jews will be getting PhDs, and the Arabs will be doing the dirty work."

But Israel was a raucous democracy, and Golda's personal sentiments could not easily hold sway, although she had the firm—and politically weighty—support of the other two members of the troika that had deposed Ben-Gurion, Sapir and Aranne, both committed doves. If she tried to give up too much, her National Unity government would fall apart. But she was also aware that if she opted for Begin's maximalist position, she would lose both the Americans and many of her closest colleagues on the left.

To make matters worse, scores of Israelis at both ends of the political spectrum were boxing her in by creating new realities on the ground. By the time she became prime minister, settlers had already established themselves in the Etzion Bloc, the key area between Jerusalem and Hebron. Small kibbutzim had been established on the Golan Heights and the ultra-Orthodox had reestablished a Jewish presence in Hebron itself.

The spectacle of Israelis taking the law into their own hands to constrain government action was the sort of open challenge Golda could never abide. More than once, she ordered the military to physically remove Jews from would-be settlements on the West Bank, but she knew that every move she took was politically dangerous. So while the fruits of victory had become overripe, Golda, like Eshkol before her, opted for blind expediency and postponed a decision that was, in fact, too pressing for delay.

Nonetheless, Yigal Allon, her deputy prime minister, devised a comprehensive plan for the West Bank to guide the government in its immediate planning. To provide Israel with maximum security and the smallest possible increase in its Arab population, he proposed that Israel annex a strip of land six to nine miles wide along the Jordan River, most of the sparsely populated Judean desert along the Dead Sea, and a wide swath of land around Greater Jerusalem. The heart of the West Bank—its great Arab cities and most densely populated towns—could be returned to Jordan or become the basis for an autonomous state. A mild supporter of some version of Allon's design, in the absence of a peace partner, Golda refused to set off the internal political explosion she knew would ignite the minute the government adopted it, or any other plan.

But Golda had to face the conundrum of what to do with the West Bank and its residents while she waited for the Arabs to agree to negotiations. Should the government incorporate them fully into Israel? Isolate them entirely? Who would provide water and education? To whom should the Arab residents pay taxes? Would their relatives from Egypt and Jordan be allowed to visit them? Should Israel build industry there to provide work for the Palestinians?

The actual governance of the territories fell to the Ministry of Defense, which left Dayan to make such daily decisions, and Dayan was the most ardent annexationist in the government. "We should see our presence in the territories as permanent," he'd advocated in a secret memo written to the cabinet in October 1968. To consolidate that hold, Israel should build towns along the mountain ridge near Hebron, Ramallah, Nablus, and Jenin to "dismember the territorial contiguity" of the West Bank.

In his imagination, those towns would provide West Bank Arabs with jobs, and the Israeli government would modernize local hospitals and roads, education systems and power grids. While the Israeli military would provide security, the civilian administration would remain Jordanian, except in the area of Israeli settlements. West Bank Arabs would thus continue to vote for representation in the Jordanian parliament and travel on Jordanian passports.

Dayan's chimera of two peoples living in the same territory but swearing allegiance to different countries, only one of which had ultimate control, wasn't a classic political formula, he acknowledged. But turning the West Bank into the mistress of two nations would lay the groundwork for a firmer future peace by binding Arabs and Israelis together economically and creating a network of relationships. Knowing Israelis and prospering under their benign rule, he believed, would inevitably blunt years of enmity.

That sort of "benevolent colonialism," as Sapir called it, horrified most of Dayan's cabinet colleagues. Nonetheless, as the sort of emir of the West Bank, he gradually began putting his policies in place. He abolished travel restrictions to allow Arab residents of the West Bank and Gaza to travel throughout Israel and the territories without permits. Then he established an Open Bridges scheme so that people and goods could move freely between Israel and Jordan. Family members from Jordan, Egypt, Syria—or anywhere else, in fact—could visit relatives on the West Bank, Gazans could send their children to study in Cairo, and political figures in both areas could travel without restriction. Finally, despite enormous opposition from men like Sapir, who objected to using Arab laborers as Israel's "hewers of wood" and "drawers of water," residents of the occupied territories could cross the old Green Line to work in Tel Aviv and Jerusalem.

The opening of Israeli markets to Palestinian employment and to exports of their fruit, vegetables, and fish pushed up the real gross national product in the occupied territories by an average of almost 30 percent per year. Private investment skyrocketed, as did jobs and new businesses. In 1971 alone, more than 106,000 Arab visitors crossed Dayan's open bridges, up from 16,000 in 1968.

Cockily ignoring the concomitant rise in terrorism, Dayan blithely assured his colleagues that his program was working and urged the party to abrogate its "oral doctrine," its internal policy that held that Israel would annex Jerusalem, the Golan Heights, and Sharm el-Sheikh but not the West Bank, and rewrite its platform, which said nothing about future borders, to reflect his pipe dream.

But Dayan wasn't done. Next he lobbied the government relentlessly to accelerate the pace of Jewish settlement in the territories and to begin building a deepwater port at Yamit, in the northeast corner of the Sinai Peninsula. On both those issues, he was in the clear minority. "Why do we need a deepwater port in a military buffer zone?" Sapir asked caustically.

Ever the political operative, Golda took no position, trying to hold the Labor Party together against the emerging alliance between Begin's coalition, several centrist groups, and a large number of Land of Israel supporters, all more amenable to Dayan's approach than was Labor. If Dayan threw his lot in with the right, she knew that Labor would wind up as the new opposition fighting full annexation.

Placating Dayan proved nearly impossible because he kept upping the ante. In the midst of the Grand Debate, a series of party discussions designed to build a consensus about the territories, he added a new demand: that Israelis be given the freedom to buy land on the West Bank. When Eban and Sapir objected that such a policy would inevitably undermine prospects for peace, he sneered and proclaimed that he would not remain in the party if it retained a dovish cast.

Golda tried to resist him, arguing that allowing Israelis to purchase West Bank land would inevitably provoke real estate speculation that would hurt the local population. On the floor of the Knesset, Dayan's allies in the opposition fought back viciously, comparing her to the British, whose White Papers barred Jews from purchasing land outside given regions. When she still refused to give in, they spread the rumor that she was caving in to pressure from the United States.

The Grand Debate tore at the fabric of Labor during 180 hours of meetings, involving eighty speakers, and it ended with a whimper at a party meeting in April 1973. For six months, Golda had exercised iron control to prevent any change in the party platform, and thus a split in the party itself. But as the final evening of debate wore on, it became clear that none of the major combatants was mollified. Eban and Sapir repeated their laments against creeping annexationism while Dayan asked huffily, "Does Zionism end at the Green Line?"

Unable to contain her frustration, Golda cut off discussion. Since the debate had solved nothing, she asked Yisrael Galili, a close adviser who was notorious for producing the type of vague language that obscures all differences, to draft a platform that would be victory "not for the hawks or the doves, but for the party." Galili tried to work in secret, but the arguments within the party leaked to the press, especially when Dayan added yet more demands: that the new party Knesset list include fewer doves; that Jewish development around Jerusalem be stepped up; that three dozen new settlements be planted on the Golan Heights, the West Bank, and in Gaza; that businessmen be given tax credits for investment in the territories; and that the government commit itself to a huge amount of new funding—more than $300 million—for urban settlements. He shrugged off the accusation that he was making annexation inevitable. But his schemes represented a revolutionary change in government policy that, to that point, had limited most settlements to security areas.

As late as June 1973, Golda still hadn't budged in her opposition to Dayan's policies. "It is unthinkable that there should be under our rule a population composed partly of citizens of Israel and partly of non-citizens," she declared. "I don't want a binational state and I don't want to be obliged to count apprehensively every day how many Jewish babies have been born."

In July, however, smelling the danger that the shifting political winds posed to her party, she gave in to the pressure and declared that the territories "strengthened the foundations of the state." On the surface, the Galili document, which Golda wound up drafting, looked like a compromise: Dayan got less than he was looking for. The port at Yamit would only be studied, and while private land purchases were approved in principle, the government retained the right to regulate them. But no one was fooled. Dayan had won. The new guidelines, the underpinning for the creeping annexationism Labor had long rejected, were a dramatic departure from its old moderation, a telling testament to the power of a dogged minority to change the course of a nation.

The official acceptance of the plan by the party secretariat was a chilling moment in the history of Labor, and in Golda's political life. The deal had been sealed, but Eliav could not resist speaking. "This document has been brought to us by flailing the lash of time and the scourge of panic and haste," he shouted. "This document chastises with scorpions what I understand to be the values of the Labor movement.

"There are many in this hall and in this land, throughout this movement, in town and country, whose souls weep in silence because of this document. I feel myself the emissary of this public. I will be the voice of this ideological Jewry of silence."

When he finished, Golda rose slowly, her fury at his apostasy barely contained. "I have lived through fifty years of political activity . . . and we never before had a comrade who set himself up as a messiah," she roared. "Lyova Eliav, who is in truth a pioneer and volunteer, has taken upon himself to be the voice of the Jewry of silence in this land, the Jewry that is afraid to speak its mind and weeps in silence. What kind of picture is drawn here? Whip, lash, fear, silence. What is all this? Are we speaking of the Jewry of Russia or Syria?"

The document passed unanimously, 78–0, many of the party's doves having stayed away. Galili hailed the agreement as a rejection of maximalism that would hamper peace while Allon bit his tongue and professed that the platform did not represent "crawling annexation." And Golda vowed disingenuously that the new watchdog committee set up to supervise private land purchases would guard against speculation or the takeover by Jews of Arab sections of the West Bank.

Golda knew full well that Dayan had limned a fantasy. The territories had to be a bargaining chip, not Israel's future. But she was never one to throw herself on a sword over principles that weren't core to her personal beliefs. Lacking a realistic peace partner with whom she might make a bold gesture, she bowed to the inevitable, hoping that she could find a way to hold back Dayan.

*   *   *

On March 14, 1972, standing before a gathering of Palestinian and Jordanian notables in the royal palace in Amman, King Hussein announced a proposal of his own for the future of the West Bank, a federated state of both the East Bank, the current Transjordan, and the West Bank. Each region would be run semiautonomously by elected people's councils while the central state would be ruled by the king, a cabinet, and parliament. Other "liberated" territory, by which everyone assumed he meant the Gaza Strip, would be welcome to join.

Hussein's plan was a version of what he and Golda had been working toward since their first meeting in Paris in the fall of 1965, when Golda was Israel's sixty-seven-year-old foreign minister and Hussein a twenty-nine-year-old king who'd already been on the throne for twelve years. Hussein had begun meeting with the Israelis two years earlier, encounters generally arranged by his London physician, Dr. Emmanuel Herbert, with Chaim Herzog, the director general of Levi Eshkol's office.

Early on, the young king had won Israel's support in his relationship with the U.S. government, which wasn't sure whether it should continue to prop up an anachronistic, moderate monarch in an era of Nasserian radicalism. Golda had repeatedly urged Washington to maintain its alliance with Hussein—even to sell him arms. "We can't have Nasser sitting in Jerusalem," she said, raising the specter of war on the West Bank should he be ousted.

Israel and Jordan also had to deal with the ordinary problems of neighbors, and Golda entered the picture when Hussein sought to resolve the most vexatious of them, the sharing of water from the Jordan River. Always a target of suspicion for his pro-Western inclinations, Hussein didn't have enough clout to untangle the water drama, which was hopelessly intertwined with Syrian attempts to choke off Israeli's growth. But breaking the ice, he said, was ultimately more important.

"We talked about our dreams for our children and grandchildren, to live in an era of peace in the region," Hussein told historian Avi Shlaim years later. "And . . . she suggested that maybe a day will come when we could put aside all the armaments and create a monument in Jerusalem which would signify peace between us and where our young people could see what a futile struggle it had been and what a heavy burden it had been on both sides."

The Israeli-Jordanian contacts even survived Jordan's entry into the

1967 war despite Israeli disappointment that Hussein had ignored Eshkol's pleas that he not join the hostilities. All the Israeli leaders understood that the king was caught on the horns of an impossible dilemma, trapped between Israel, his own Palestinian population, and pressure from the other Arab heads of state. So not long after the war, Hussein and Abba Eban had held a spate of meetings to discuss how to implement UN Security Council Resolution 242 and then to thrash out the conflict between Israel's war against the fedayeen based in Jordan and the king's anger at Israel's regular incursions into his territory to stop them. In 1968, Eban and Allon presented him with the Allon plan. "I was offered the return of something like 90-plus percent of the territory, 98 percent even, excluding Jerusalem," the king told Shlaim. "But I couldn't accept. As far as I was concerned, it had to be every single inch."

When Golda became prime minister, Hussein tried to arrange another meeting, part of his attempt to regain the land that he'd lost in the war. Golda resisted, sending Allon or Eban to impart the bad news that Israel would not quickly surrender the West Bank. But on September 6, 1970, Golda received a vivid reminder of how closely Israel's fate was intertwined with Jordan's when the PFLP launched its boldest attack, hijacking three airplanes and landing them at an old British Royal Air Force base in Jordan called Dawson's Field promptly renamed Revolution Airport. The terrorists offered to free all 310 hostages unharmed in return for the release of all fedayeen prisoners in Swiss, German, American, and Israeli jails.

The standoff on Jordanian soil was yet another humiliation to King Hussein, whose power was openly threatened by the strongholds—mini-states, in fact—in Amman and northern Jordan established by the PLO and the PFLP. Guerrillas raced through the streets of the capital with loaded weapons, registered their own vehicles, and set up a parallel system of customs checkpoints and taxation. Regular fighting had broken out between Jordanian security forces and the Palestinian groups, which had 50,000 armed fighters. Caught between the terrorists, who had substantial support among the half of his population that was Palestinian, and his own East Bank Jordanians and Bedouins, disgusted at the growing anarchy, Hussein could do little more than beg. But time and again, he negotiated with guerrilla leaders only to watch them break their agreements.

The summer before the triple hijacking, after declaring that before Palestine could be liberated, they would have to liberate Amman, the PFLP attempted to assassinate the king. Civil war had broken out and the Iraqis, with almost 12,000 troops in Jordan, threatened to "take all necessary measures" to protect their fedayeen brothers.

Predictably, then, Hussein's attempts to mediate an end to the hostage crisis at Dawson's Field were rebuffed by the PFLP, and the king seethed at the challenge to his authority. Although most of the 310 hostages were quickly transferred to Amman and released, the flight crews and Jewish passengers, a total of fifty-six people, were kept as hostages and, on September 12, the empty planes were blown up on his soil in front of the world media.

Fed up, Hussein surrounded Amman with loyal troops, replaced his civilian government with military officers, and ordered the guerrillas to remove their weapons from Amman. When they refused, he imposed martial law and sent his army into the city and the northern guerrilla strongholds. The fedayeen, however, proved more determined than Hussein had anticipated, and on the third day of the fighting, Syrian tanks crossed the border to reinforce them. If the Syrians committed themselves to a full-scale attack, Hussein knew he couldn't survive.

That night, Golda was in the midst of a forty-minute speech at a United Jewish Appeal–Israel Bonds dinner at the New York Hilton when Ambassador Rabin was called off the dais. "King Hussein has approached us, describing the situation of his forces, and asked us to transmit his request that your air force attack the Syrians in northern Jordan," Henry Kissinger told him. "I need an immediate reply."

The U.S. government was in a panic, unsure how to keep Hussein in power without inviting a Soviet response. Israel, they knew, could not stand for an overthrow of Hussein. But it was unclear how far Nixon would go in supporting Israeli military action on Hussein's behalf given his desire for rapprochement with the Soviets.

With an audience of 3,000 applauding Golda in the background, Rabin turned the tables on Kissinger. "I'm surprised to hear the United States passing on messages of this kind like some sort of mailman," he said. "I will not even submit the request to Mrs. Meir before I know what your government thinks. Are you recommending that we respond to the Jordanian request?"

Unprepared, Kissinger promised to call Rabin back. Thirty minutes later, after consulting Nixon, who'd been bowling in the White House lanes, he reported, "The request is approved and supported by the United States government." Still, Rabin wasn't satisfied.

"Do you advise Israel to do it?" Rabin asked.

"Yes, subject to your own considerations," replied Kissinger.

An hour later, Rabin transmitted Golda's answer: At first light, an Israeli reconnaissance plane would fly over the area and, if necessary, act. But Golda wanted a firm commitment that the United States would protect Israel from Soviet or Egyptian reprisals, replace any military equipment lost in the action, and turn over eighteen promised Phantoms earlier than scheduled.

The Israelis did more than send up a reconnaissance flight. They flew over the Syrian armored brigades, virtually buzzing them, to make their presence felt, and sent ground forces to the border in broad daylight. Rabin informed Kissinger that the IDF believed that air strikes might not be sufficient and that Israel might have to send in troops on the ground. Nixon approved the plan but Kissinger couldn't bring himself to broach the topic to King Hussein, knowing that being saved by Israel would destroy whatever credibility he had left in the Arab world.

The combination of American protests to the Soviets, the Jordanian military resistance, and Israel's threatening moves, however, cooled Syrian ardor for battle, buying Hussein a respite in his civil war with the Palestinian resistance. But that moment left both Golda and the Jordanian king with a keen awareness of their mutual dependence and the commonality of their enemy.

Several months later, then, Golda invited Hussein to visit her in Tel Aviv, the first of at least seven more get-togethers. The king flew his own helicopter to a rendezvous point on the Israeli side of the Dead Sea and was picked up by an Israeli helicopter, whose pilot surrendered the stick to allow the king to make a pass over his beloved Jerusalem. No agreement was struck that day. Hussein had already rejected the Allon plan, and Golda was not inclined to accept either of his ideas for the future of the West Bank—an independent state or a federated Jordan—without some change in the pre-1967 border.

While Hussein knew that a "separate peace" was not yet realistic for him politically, he continued meeting with the Israelis—in an air-conditioned limousine in the desert, on a missile boat in the Gulf of Eilat, on Coral Island near Israeli-occupied Sinai, and in the Midrasha, a concrete fortress north of Tel Aviv run by Israel's intelligence agency. He and Israeli officials hammered out joint strategies to fight terrorism. They cooperated on irrigation systems, pest control, and bridge building. And in a meeting on the Red Sea in early 1970, Hussein and Dayan negotiated the withdrawal of Israeli troops from seventy-five square kilometers of the As-Safi area in the Arava in exchange for Hussein's promise to station a unit of his own army in the area to prevent terrorists from firing onto Israel's Dead Sea Works.

Over time Golda and Hussein established a telex "hot line." When the king heard that Golda's sister Sheyna had died, he graciously conveyed his condolences. And the ever courteous and gallant Hussein always brought gifts—gold pens embossed with the symbol of the Hashemite crown for Eban and a German-made G-3 assault rifle for Allon. On Golda's seventy-fifth birthday, he sent her an exquisite strand of perfectly matched pearls, the stones of paradise, according to the Koran.

Gradually, Israel and Jordan had settled into a de facto peace. Anything more was impossible, and not only because Golda and Hussein could find no way to overcome the hurdles of distrust and pride. The pro-Western king recalled all too keenly what had happened to his grandfather Abdullah after he moved toward an agreement with the Jews. In

1951, at the age of sixteen, Hussein had entered Al-Aqsa Mosque with Abdullah and watched him murdered by a Palestinian terrorist, a fate he escaped only because of a medal pinned to his chest that deflected the bullet aimed for his heart.

*   *   *

Given that history, Hussein's 1972 proposal for a federated state was a calculated risk, dependent on the support of West Bank residents, and he didn't get it. Too little, too late, Palestinians said. "Jordan should have done this back in 1948," Mohammed Abu Shilbaya, an influential writer for the East Jerusalem newspaper *Al Quds*, told the *New York Times*. "Now everything is damaged and ruined."

The PLO rejected the federated state "categorically and conclusively," and the West Bank suddenly became rife with rumors that Hussein and Golda had hatched it during one of their regular late-night phone conversations. Hussein is an "agent of Zionism and imperialism who will share the same fate as Wasfi Tal," said Arafat, referring to the Jordanian premier who'd been assassinated four months earlier by a group of Palestinians.

Golda had long assumed that some version of Hussein's proposal would be the West Bank's future since a third country in the old British Mandate of Palestine seemed utterly unrealistic to her. "It's not a viable state by any means," she reminded everyone who would listen, pointing to its small size and lack of access to the sea. "The only purpose that this state can serve is to be a bridgehead against Israel."

Reuniting the peoples of the two banks of the Jordan, on the other hand, seemed entirely logical. "In Jordan . . . the majority of the people are Palestinian," she explained. "What is the difference between somebody who lives in Nablus and went over to Jordan and the one who remained in Nablus?"

Yet, like Hussein, Golda knew that the international political climate was not yet ripe for such discussions. Speaking to the Knesset about the king's proposal, then, she lashed out at Hussein angrily. "The king is treating as his own property territories which are not his and are not under his control. He crowns himself king of Jerusalem and envisions himself as the ruler of larger territories than were under his control prior to the rout of June 1967. In all this detailed plan the term peace is not even mentioned."

But whether on the floor of the Knesset, where the rhetoric against Hussein was overheated in the extreme, or in speeches and press conferences, Golda carefully—studiously, given how meticulously she parsed her words—never rejected his federation plan.

CHAPTER SIXTEEN

I'm not cynical at all.
I've just lost my illusions.

O n May 3, 1973, Israel celebrated Golda's seventy-fifth birthday with all the hoopla she'd disingenuously asked the nation to forgo. Kibbutz Revivim put on a version of *This Is Your Life* for her, complete with skits, songs, and secret guests. A group of high school students presented her with seventy-six pink roses and her cabinet ministers gave her a painting by Mordechai Levanon, one of Israel's most famous artists. German chancellor Willy Brandt cabled her warm congratulations, and King Hussein sent along that perfectly matched set of pearls.

Falling just three days before Israel's twenty-fifth Independence Day, it was the perfect moment for Golda to announce her retirement. The "caretaker" prime minister, who'd been in office for four years, was at the peak of her popularity. Polls indicated that 73 percent of the Israeli population was anxious for her to remain at the helm.

Year after year, she'd been voted the most admired woman in America and Britain, beating out indigenous queens and heroines. After her verbal run-in with the pope, Italians denounced the pontiff for his rudeness to Israel's grande dame. Her political clout was so formidable that when she announced that she would attend the Socialist International meetings in Paris in January 1973, the French practically begged her to stay away, fearing that her mere presence would spotlight the Gaullist arms embargo against Israel and sway votes away from Prime Minister Georges Pompidou in the March elections.

And while senior officials at the U.S. State and Defense Departments complained bitterly that she was impossibly intransigent, Richard Nixon loved her, as he later confessed quite publicly, an odd admission for an American president she lectured relentlessly, odder still for a man as reticent as Nixon.

Although Golda would never feel quite as secure with anyone else guiding the country, she knew it was time to leave. She wasn't merely tired; she was sick. During her first six months as prime minister, she suffered a few migraines, although the ones that felled her seemed to occur on Saturdays. But gradually Lou Kaddar began receiving more phone calls from Golda, her voice barely recognizable, asking that she send over a doctor to give her a shot. Years of kidney stones and gallbladder attacks, phlebitis, and the white leatherette cigarette case she always kept close at hand were catching up with her, and her lymphoma had recurred, forcing her to set aside time every Monday for treatment. Still, Golda brushed aside all concerns for her health. When an aide suggested that she take a vacation, she asked, "Why do I look tired?" No, he said, but I am. "So you take a vacation," she insisted.

Those closest to her couldn't imagine how she kept going. "The woman radiated force," said Eli Mizrachi. "She radiated authority and strength and conviction. . . . Golda was undoubtedly an example of superiority of the spirit, of the mind over the body. Otherwise, there's no explanation for how she carried on."

Yet after declaring in 1971 that she would retire, Golda coyly—and that occasional and incongruent coyness was one of her preeminent qualities—refused to commit herself. When the *Jerusalem Post,* Israel's

English-language daily, asked her who she wanted to succeed her, she quipped, "What? So you're sacking me?"

But by October 1972, she seemed to have made up her mind. "I am old," she told a reporter. "I am feeble. . . . My heart is functioning, but I am not able to continue at this insane pace forever. Lou only knows how many times I have said to myself, 'To hell with it all, to hell with everyone. I did my part. Now others should do theirs.' . . . Many do not believe that I will leave. But they should start to believe. I'll even give you a date: October 1973."

With no viable candidate save Dayan, whom Golda had begun calling Dr. Jekyll and Mr. Hyde, however, her party colleagues begged her to remain in office. Although she waffled, ever the party loyalist and a sucker for that sort of flattery, she finally bowed to the inevitable a month after her birthday. When she announced that decision, the party Central Committee broke out in cheers, if not from open enthusiasm then from sheer relief that a split had been averted yet again.

Public affection for the anachronistic old lady still longing for the days of self-sacrifice was a strange quirk in a fast-paced, hypercompetitive consumer society. But despite soaring inflation and ugly social divisions, the Israeli economy was booming, the GNP growing between 8 and 14 percent per year, and private consumption rising steadily. Israel's foreign currency reserves, long precariously low, had hit $1.2 billion, three times what they had been four years earlier.

Peace negotiations through Jarring and the Americans had been nearly forgotten when Rogers was replaced by Henry Kissinger, who proved more amenable to Golda's manipulation if not her reasoning. Her annual American shopping trip in March had been a reassuring triumph. The House Foreign Affairs Committee applauded her entrance into their chamber, and she'd been greeted with a standing ovation at the National Press Club.

She and Nixon had collided briefly over a bill introduced by Senator Henry Jackson to deny Most Favored Nation trade status to the Soviets if Moscow didn't lift the ransom tax imposed on Jews trying to leave the

Soviet Union, a bill strongly opposed by the White House. "Don't let the Jewish leadership here put pressure on Congress," Kissinger had exhorted her sternly amid press reports that the administration would punish Israel if the bill were enacted. But Russian Jewish immigrants were trying to persuade Golda to do the opposite, to support the bill openly.

"I cannot tell the American Jews not to concern themselves with their brethren in the Soviet Union," she told the White House, appeasing them by informing the Russian Jews that she would not interfere in the internal affairs of the United States with any public gesture on either side.

So to the delight of the Israelis, Nixon filled all of her orders for military equipment. In addition to the 42 F-4 Phantom fighter-bombers and 80 A-4 Skyhawk attack aircraft he'd promised her in December 1971, he topped off Israel's arsenal with another four squadrons of combat jets— 24 Phantoms and another two dozen Skyhawks—and pledged that the United States would assist Israel in building Super Mirages, advanced jet fighters of its own design.

From time to time, a ripple of sobriety tempered the saucy aplomb of the country that existed with "one foot in war, one foot in peace, the body suspended in bedlam," as Amos Elon, Israel's leading journalist, described it. But having held off massive Arab armies three times, Israelis were blithely confident.

"What does the Israeli army need in order to occupy Damascus, Moscow, or Vladivostok?" began one common joke. The punch line: "To receive an order."

In another popular bit of mordant humor, Dayan and David "Dado" Elazar, the chief of staff of the IDF, were drinking coffee, bored with the absence of military action. "There is nothing to do," Dayan said. "How about invading another Arab country," Dado suggested.

"What would we do in the afternoon?" Dayan asked.

With security concerns requiring less of her attention, Golda imagined devoting her next term—surely her last—to domestic issues, especially to the problems of poverty. "I want to get things moving on the inside of the country," she told the Teachers' Union as she outlined a series of new initiatives drawn up by her Committee on Disadvantaged Youth. The final budget of her term was the first in Israel's history to devote more money to domestic problems than to defense. Her legacy, she was certain, would be one of social justice.

*   *   *

The landing of the Bell 206 helicopter at the Mossad safe house in Herzliya, just north of Tel Aviv, on September 25, 1973, should have burst Golda's bubble. After all, it wasn't every day that King Hussein snuck across the border for an emergency meeting. And the news he brought had an ominous ring: a sensitive Syrian source had informed the Jordanians that Syria was positioning for war.

Bellicose and entrenched in their negation of Israel's existence, the Syrians had long kept three divisions of troops opposite the Golan Heights. Until August, their presence had not been particularly troublesome since Israel ruled the skies. But that month, the Soviets installed SAM-6 batteries that turned routine flights into hazardous missions.

In the middle of September, four Israeli Phantoms braving the missile grid on a photoreconnaissance operation touched off a dogfight that downed twelve Syrian planes. In the wake of that battle, the Syrians moved their artillery closer to the border and boosted their armor until they had 800 tanks facing 77 Israeli ones. That movement had worried the general leading Israel's Northern Command, but no one on the General Staff of the IDF or in the Ministry of Defense believed that the Syrians would dare strike at Israel alone.

Golda's first question to Hussein, then, was automatic. "Without Egypt?" she asked.

Hussein paused and replied quietly, "I think they would cooperate."

It was midnight by the time the king departed and Golda had a chance to call Dayan to dissect Hussein's warning. Lou Kaddar, who'd served coffee and tea at the meeting, didn't understand the need for discussion. "My feeling . . . was that, although King Hussein didn't know all the details, he knew enough to understand that war would break out," she later explained. Zusia Kniazer, the head of the Jordan desk of Israel's military intelligence group, who had listened on closed-circuit television, agreed. "War!" he told the head of the Northern Command after Hussein departed. "Syria and Egypt are poised to attack Israel."

But after consulting with the other intelligence officers who'd heard the conversation, Dayan showed no alarm. The Jordanian king's relationship with Egypt wasn't good enough for him to be in the loop, he told Golda. Don't worry. We're watching the Syrian buildup carefully. There's no cause for panic.

Reassured, the following morning Golda flew to Strasbourg to deliver a speech to the twenty-fifth consultative assembly of the Council of Europe. Shortly after she arrived, two Palestinians seized three Soviet Jews and an Austrian customs official on board the *Chopin Express*, a special train that transported Soviet Jewish emigrants to Vienna, where they were processed in a transit camp for immigration to Israel. The kidnappers demanded that Austrian chancellor Bruno Kreisky shut down that camp as the price for release of the hostages. Declaring that he would not permit Austria to be turned into a "secondary theater of the Middle East conflict," Kreisky agreed.

Kreisky's decision outraged Golda, all the more because the Austrian chancellor was Jewish. She scrapped her planned speech to the council for a two-and-a-half-hour jeremiad against terrorists and their appeasers. "There are those who threaten with a gun and those who try to defend themselves so that the gun does not go off," she declared. "To say 'a plague on both your houses' is the greatest injustice. If there is a family of nations, everyone has the right to move around the world. These terrorists, at the point of a gun, have raised the question whether anyone should be allowed to use Austria's soil for transit."

While Golda was lobbying her European colleagues, Galili called to report that the security situation on the borders was looking dire and that she should return home promptly. But Golda was too angry with Kreisky to pay him much heed. Despite a council resolution proclaiming that no government should be bound by a promise extorted by violence, Kreisky, who later bragged that he was "the only politician in Europe that Golda Meir can't blackmail," refused to reconsider his decision. So rather than rush home, Golda flew to Vienna, hoping to change the Austrian chancellor's mind.

The signs that had provoked Galili's phone call weren't the movements on the Syrian border. Israeli intelligence had discovered that a division of Egyptian troops was heading toward the Suez Canal and that 120,000 Egyptian reservists had been called up. Still Dayan and the leaders of Israel's military intelligence scoffed at the danger, concluding that Egypt was engaging in routine fall military exercises. They similarly dismissed a report from the U.S. Central Intelligence Agency that elite Egyptian commando units were being deployed to new bases. Even the sudden silence of Egyptian radio traffic did nothing to dent Dayan's confidence enough for him to notify Golda or bring Allon, the acting prime minister, into the loop.

Those in charge—particularly Dayan and Eli Zeira, the head of Military Intelligence—were blinded not only by deep contempt for Arab military ability but by Israel's long-held belief that the Syrians would never risk war without the Egyptians and that the Egyptians would not take on Israel without long-range bombers, planes they had not yet acquired. The first assumption was little more than an exercise in logic, but the second was based on solid information from Israel's best and most reliable Egyptian intelligence source, code-named In-Law, who was, in fact, Nasser's son-in-law and a confidant of President Anwar Sadat.

The ability of Dayan and Chief of Staff Dado Elazar to read the tea leaves was further clouded by Zeira's habit of sharing only his own conclusions with his superiors without mentioning contradictory data or the disagreement of his subordinates. His self-assurance about his ability to anticipate Sadat's calculations had been reinforced five months earlier, when Dado grew alarmed about Egyptian troop movements. Golda had mobilized Israel's reserves despite Zeira's opinion that nothing was afoot. That alert and call-up had cost Israel an estimated $35 million. No one, least of all Dado, who'd been made to feel like the boy who cried wolf, wanted to repeat the same mistake.

The May call-up offered a glimpse into the enormous difficulty Israel faced in divining Arab intentions. Most countries rely on "tactical warnings," watching an enemy's troop movements and paying close attention to their threats. But if Israel, which maintained a standing army of only 135,000, had called up its reservists every time the Egyptians or Syrians let loose a barrage of artillery fire, moved tanks close to the border, or threatened invasion, the country would have been in a perpetual state of alert. Dayan, then, had developed his own approach, "strategic warning," which depended on analyzing political thinking and objectives—and on the Arabs' doing nothing to try to deceive him.

In planning the recapture of the Sinai, Sadat had counted on that sort of Israeli arrogance, what General Ahmad Ismail Ali, the Egyptian defense minister, called Israel's "wanton conceit."

Despite Dayan's conviction that nothing alarming was happening, he briefed Golda and her inner cabinet during her first afternoon back from her fruitless trip to Austria. Oddly, however, he seemed less interested in bringing Golda up-to-date than on seeking her counsel, the first in a series of strangely passive interactions between the brash general and the elderly prime minister.

Hearing the new reports, Golda worried. Might the Egyptians be poised to stage a diversion to allow the Syrians to attack? she asked. "Is there anything, any weapons system that we don't have that we could ask for right away from the United States?" Shouldn't we consider sending up more reinforcements? Her military leaders were nearly unanimous that nothing threatening was afoot. Still, Golda was troubled. "There's a contradiction between the signs on the ground and what the experts are saying," she told Galili.

Unbeknownst to Golda, the Israeli troops on the front lines shared her concern. In the north, another Syrian armored brigade had taken up a position along the border, and in the Sinai, the Israeli soldiers could see a steady stream of convoys coming in.

Privy to none of that information because of Zeira's hubris, Golda didn't grow seriously anxious until the following afternoon, when military intelligence intercepted a KGB radio transmission about the evacuation of the families of Soviet troops and experts from both Egypt and Syria. For Golda, news of the arrival of enormous Antonov 22s to ferry the Russians home was the first serious alarm bell. But finding no sympathy among her generals, she was left to worry alone.

By the morning, the morning of Yom Kippur eve, however, Dado was growing apprehensive. Reconnaissance photographs taken the previous afternoon along the canal showed not just a growing mass of armor and artillery but bridge-laying and water-crossing equipment. To hedge his bets, he ordered a C alert, just shy of Israel's highest, shifted more armored units to the Golan Heights, canceled all leaves for the holiday, and instructed his logistics team to prepare for a general call-up. The head of the air force had already discreetly called up his reserves.

Maybe we should keep the army radio broadcasting during Yom Kippur in case we need to call up reserves? Dado suggested to Dayan. Exasperated, Dayan dismissed the notion, worried about provoking national panic.

Dado had been Golda's personal choice for chief of staff, and Dayan had agreed to his appointment only reluctantly. The tension between the two men was palpable at the best of times, and the eve of war was not one of those moments. The two generals couldn't agree on what they were facing, a limited Egyptian raid backing up a Syrian artillery barrage or a full-scale offensive. So they trudged over to Golda's Tel Aviv office, where she routinely worked on Thursday and Fridays.

Dayan clung to the evaluation given by Zeira, who remained convinced that there was a "low probability" of war. And while Dado disagreed, he believed that the measures he'd taken would be sufficient to hold off any possible hostile Arab attack until the arrival of the reserves, standard doctrine in the IDF.

Golda, however, was less sanguine. "There is something," she said. Why would the Syrians deploy so massively if they planned nothing more than an artillery barrage? she asked. And Arab press reports of alleged Israeli troops massing on Arab borders sounded ominously like the equally fallacious stories from 1967. "Maybe this should tell us something," she mused.

Haunted by the evacuation of the families, she called an emergency cabinet meeting, a mini-cabinet meeting, really, since Eban was in the United States and most of the other ministers were en route home for Yom Kippur, which would begin at sundown. With no inkling of the contents of the full reports from the field, which Zeira was not disclosing, and Zeira still predicting that war was unlikely, the nine cabinet members had to decide whether to call up the reserves.

Dayan opposed any mobilization unless the Arabs made a hostile move. Galili, who'd commanded the Haganah, agreed, concerned that a mass call-up could alarm the Arabs and precipitate the very scenario they were attempting to avoid. Dado was divided, anxious to avoid a repetition of the embarrassment of May, yet apprehensive at the signs of a full-scale invasion.

The ministers concluded that they should delay mobilization until the picture became clearer, although the cabinet gave Golda the authority to call up the reserves on Yom Kippur if she saw fit. That afternoon, she instructed Simcha Dinitz, her ambassador to Washington—in Israel after the death of his father—to return to America. "Get the first flight out," she told him. And filled with foreboding, she canceled her holiday trip to Kibbutz Revivim so that she could stay close to her office.

As Israel turned still and silent for Yom Kippur, fewer than 450 Israeli infantrymen backed by 300 tanks and 50 artillery pieces were strung out along a chain of outposts known as the Bar-Lev Line on the eastern shore of the Suez Canal, facing more than 100,000 Egyptian soldiers with 1,350 tanks and 2,000 artillery pieces deployed on the west bank. On the Golan front, the Syrians had three infantry divisions, a total of 45,000 troops with 540 tanks, behind a line of armored brigades with another 960 tanks and 942 artillery pieces poised to attack Israel's two infantry regiments and 177 tanks.

Two-thirds of Israel's army, its reserves, were on vacation or at home for the holiday, their stripped-down tanks, ammunition, and guns scattered in bases and ammunition depots across the country.

That night, Golda sat down for dinner with her son and daughter-in-law, but she didn't talk or eat much, and then left early. Uneasy, she clung to the message her friend Haim Bar-Lev, the former chief of staff and her current minister of commerce, had left when he'd stopped by a day or two earlier. "Golda, I will sleep well tonight."

But Golda tossed and turned in the stifling heat. Then, at 3:30 A.M., her telephone rang with news from the Egyptian source the In-Law: Egypt and Syria would attack at sundown.

*   *   *

By 8 A.M., Golda was in her office, caught in an open fight between her chief of staff and her minister of defense. Since mobilizing Israel's reserves in thirteen hours was impossible, Dado wanted to launch a preemptive air strike. Dayan, still uncertain that war was imminent, rejected that option lest the Americans punish Israel for any precipitous move. He was even refusing Dado's request for full mobilization of Israel's reserves, more than 200,000 troops, declaring them superfluous.

To defend two fronts he'd need 50,000 to 60,000 troops immediately, Dado insisted. "And what about tomorrow?" he asked.

Dayan wouldn't budge. "Tomorrow, tomorrow, tomorrow," he said. You mobilize for the counteroffensive "only after the first shot."

Again, they'd taken their argument to Golda, a woman who acknowledged that she didn't know how many troops were in a division. Golda agreed with Dayan about the dangers of a preemptive strike. "None of us knows now what the future has in store for us, but there is always a possibility that we'll need someone's help and if we'll strike first, no one is going to help," she concluded. "I wish I could say yes because I know the meaning of it, but, with a heavy heart, I say no."

But she sided with Dado about the mobilization of the reserves. "If war breaks out, better to be in proper shape to deal with it, even if the world gets angry at us," she decided. Petulant as ever, Dayan didn't even try to conceal his pique. "If you want to accept the chief of staff's proposal, I will not prostrate myself on the road," he snapped.

Dayan was livid. "What will you do with all those reservists if no war?" he asked Dado after they left Golda's office. "You'll have one hundred thousand men hanging around." Dado was shocked that Dayan still doubted what was coming. "They won't hang around," he responded. "They'll go down to the front."

By the time Golda finished with her generals, U.S. ambassador Kenneth Keating was waiting to deliver a messenger from Kissinger, who'd spent the early morning on the phone with Nixon, who was at his home in Key Biscayne, with U.S. intelligence officials, the Egyptian ambassador, the Israeli embassy, the Soviet ambassador, and the UN secretary-general. Despite those wide-ranging consultations, Kissinger wasn't sure what to think. The Egyptian ambassador charged that the Israeli navy had attacked Egypt, which didn't make much sense since the Israelis had never been known for their reliance on their navy. And he was equally skeptical about intelligence estimates hinting that Israel was about to launch a preemptive strike. Golda's coalition partners from the religious bloc would not have agreed to such a violation of the holiest of Jewish days, he reasoned.

Nonetheless, he instructed Keating to advise Golda against a preemptive strike. "I will not open fire," she told Keating. "Moreover, Israel is not mobilizing fully to prevent such an act being interpreted as provocation." Keating didn't try hard to hide his skepticism, but Golda was adamant. "This time it has to be crystal clear who began the war, so we won't have to go around the world convincing people our cause is just." But she minced no words in conveying her expectation that the United States would reward Israel with much-needed weaponry in return for disregarding the advice of her chief of staff.

By noon, Israel's cabinet ministers had been pulled out of their synagogues, kibbutzim, and homes to face a pale and disheveled prime minister delivering the worst possible news. Dayan was upbeat about the situation in the south, predicting that Israeli troops would handily destroy the Egyptians if they attempted to cross the canal. But he was anxious about the Golan, which was close to Israeli settlements. Just as he was starting to air that concern, Golda's military aide knocked on the door: the invasion had begun.

Within minutes, air raid sirens abruptly shattered the quiescence of Yom Kippur afternoon. That whine might have terrified the political leaders of a country that had not fought three wars with great success in twenty-five years. But few at the table doubted the IDF's ability to dispose of the threat with dispatch. Only the troops on the ground understood that Israel was facing an entirely new type of warfare, and entirely new enemy.

For those in the south, the war began when 222 Egyptian jets flying over the canal struck at command posts, radar stations, air defense installations, gun emplacements, and antiaircraft missile batteries. Then, 2,000 Egyptian guns rained more than 10,000 shells on Israeli positions in the first minute. Before the Israelis could lift their heads out of their bunkers, Egyptian tanks moved to the top of ramps along the canal and began firing, demolishing all the lookout towers in Israel's widely spaced string of primitive forts. Fifteen minutes later, in almost perfect coordination, 4,000 Egyptian infantrymen and commandos slid down ramps to the water and loaded into 720 rubber and wooden dinghies.

As the astonished Israelis watched the perfectly executed maneuvers, they waited for their tanks to pulverize the Egyptians troops before they could blast through Israel's sand embankments and build bridges to bring their tanks across the canal. But the Egyptian infantrymen unleashed not only shoulder-fire rocket-propelled grenade launchers, but weapons Israelis had never faced: Sagger antitank missiles, which could be guided directly to their targets, turning foot soldiers into tank-killing machines. By 10:30, the Egyptians were moving heavy tanks across the waterway on pontoon bridges. Every long-held Israeli battlefield doctrine was quashed before the war was twenty-four hours old.

On the Golan, things were no better. The Syrians wanted to regain the heights in a single day—before Israel could move in its reserves—and they planned to do so by overwhelming the Israelis with sheer numbers. Led by engineering tanks with bulldozers, they moved in assault tanks, followed by hundreds of gun tanks and armored personnel carriers, all moving under a SAM-2 and SAM-3 missile umbrella that paralyzed Israel's air force. By late afternoon, Israeli's armored battalions were running out of ammunition and fuel, and Syria's tanks were still coming.

Goliath had become David, but Golda showed no sign of panic when she appeared on television that evening. "We have no doubt that we shall be victorious," she said, her affect grave but firm. "We are also convinced, however, that this renewal of Egyptian-Syrian aggression is an act of madness. . . . Citizens of Israel, ordeal by battle has been forced upon us again. . . . We are fully confident that the IDF has the spirit and the power to overwhelm the enemy."

Later, Dayan offered a more sobering estimate, although he added, his voice almost cracking, "The people of Tel Aviv will be able to sleep well tonight."

At a 10 P.M. cabinet meeting, Dado laid out his strategy: The next day, with the arrival of 450 reservists, they would stop the Syrians. Then they would turn their attention to the Sinai and mount an offensive two days later. But at one point, in the early hours of the morning, he admitted that he was unsure of how to proceed because the Israeli military simply didn't know how to operate on the defense. "We know how to do it from the books, but we've never actually done it," he said wistfully.

*   *   *

If the people of Tel Aviv had known how dire the situation really was, few would have slept at all. By the following morning, chaos reigned among Israeli troops in the south, and the central command structure was close to collapse. But with miles of desert between the canal and Israeli population centers, Dado couldn't concentrate on the Egyptian front. It was in the north, where the distance between the Syrian troops and Israeli settlements was less than ten miles, that the danger was pressing.

When Israeli pilots flew over the Golan front at dawn, they saw unending lines of tanks all pointed south, toward the Galilee. They tried to launch waves of ground support sorties, coming in low and fast to force the Syrians to run for cover. But Syrian troops brought out highly mobile SAM-6s, missiles the West never believed the Soviets would ship overseas. Suddenly, the sky was painted with streaks of black smoke from exploding Israeli aircraft.

Faced with a weapon more accurate and sophisticated than anything the Americans had developed, by late afternoon the Israeli forces were falling back in confusion, the troops near panic. Israel had already lost thirty-five planes and 10 percent of its combat force, and it had no armor between the invaders and the Jordan River bridges leading into the Galilee.

A master of self-control, Dado exuded a calm he did not feel. He and his staff had prepared Israel for the wrong war. He needed to scrap every assumption, every plan the Israeli military had ever made, and figure out a way to hold off destruction until his reserves were in place.

Dayan didn't believe it could be done. We can't stop them, he told the commander of the northern front. We must abandon the Golan, blow the bridges over the Jordan River, and hold the Syrians off from the other side. More than any other figure, Dayan had long assured the nation that Israel's borders were secure after the 1967 war. He'd been wrong, wrong about Israel's borders, wrong about Arab fighting ability, wrong about Egyptian intent, and his own fallibility threw him into a panic.

By the time he returned to Tel Aviv, Dado had countermanded his order to blow the Jordan River bridges. Dayan believed that the chief of staff was being unrealistic, but he had no energy for fighting. He needed to move and suddenly flew south. There, the situation was equally alarming. The Egyptian infantry held the water line along the canal, and Israel's allegedly invincible strongholds, the Bar-Lev Line, had become traps for the men defending them. If the Egyptians managed to take the Mitla and Gidi passes, two of the three routes over the mountains at the northern edge of the Sinai, all Israel had to keep them from streaming into Tel Aviv was two depleted divisions.

When Dayan returned to the Pit, the nickname for Israel's military headquarters, the staff looked to him expectantly for a daringly bold plan for victory. Instead, their old hero had transmogrified into Jeremiah, the Old Testament prophet of doom.

"The destruction of the Third Temple is at hand," he prophesized. The First Temple, Solomon's, had been razed by the Babylonians in 586 BCE, and the Second, Herod's, by the Romans in AD 70. The only road open to them, he admonished, was to pull back and fight to the last man, as had the Jews at Masada.

Leaving the entire war room rattled and aghast, Dayan raced to Golda's office and confessed, "Golda, I was wrong about everything. We are headed to catastrophe." He offered to resign. Golda wouldn't hear of it. The last thing a shaken Israel needed at that moment was the disappearance of its mythic hero.

When Dayan left, Golda wandered out into the hall. Concerned, Lou followed and found her "dissolving. She was wearing an outfit I detested, a sort of grayish khaki," Lou later recalled. "She was almost the same color."

Golda broke down in tears. "Dayan is talking about surrender," she whispered. "You must arrange for me, tonight . . . you must go to the house of friends. . . . He's a doctor. I'll make them give you some pills so that I can kill myself so I won't have to fall into the hands of the Arabs."

Golda lingered in the hall and regained her composure. Then she took charge. All of the characteristics her critics had long considered to be weaknesses—her inflexibility, her overbearing nature, her iron grip—became her greatest assets as she faced war without the man she'd depended on in security matters. She became not only the national morale booster and weapons procurer, but also Israel's "generalissimo," wrote Zeev Schiff, defense editor of the Israeli daily *Ha'aretz*. "It was strange to see a warrior of seven campaigns and brilliant past chief of staff of the IDF [Dayan] bringing clearly operational subjects to a Jewish grandmother for decision."

Galili, who had been in Golda's office when Dayan arrived, ran to the

Pit to bring back Dado. The chief of staff did not soft-pedal the trouble they were facing, but neither did he preach biblical apocalypse. Withdrawal to the Sinai passes, he warned, would be absurdly costly in lost facilities. "Today we hit bottom," he said. "Tomorrow I predict that we'll be able to get our chins above water."

Dado asked Golda to draft Bar-Lev, the former chief of staff, to fly north to look for alternatives for regaining the initiative. By 10 P.M., Bar-Lev was back in Golda's office after a lightning visit to the front. Serious but manageable, he characterized the situation before laying out his ideas for redeployment and a counterattack once a fresh division of reserves arrived.

"The great Moshe Dayan!" Golda cried out, in relief. "One day like this, one day like that!"

That night, when the cabinet ministers arrived for their second meeting of the day, they found Golda poring over maps of troop deployments. The Syrians had been stopped, for the moment, although at the loss of a full brigade. Dado reported that he was going to throw everything he had into keeping them contained while the southern command launched a limited defensive counterstrike.

Dayan exploded without restraint, describing the awesome size of the enemy forces, the quality of their equipment, and the striking caliber of their troops. "The Third Commonwealth might be destroyed" if they didn't take drastic measures, he warned, echoing his terror of the moment. The Israeli forces must withdraw into the Sinai and establish a secondary line out of range of Egyptian missiles, he proclaimed.

The ministers were uncomfortable with the concept of abandoning in order to defend. If Israel seemed to be in that much danger, other Arab nations might be emboldened to join the fray, placing them in far greater danger. And if the war ended with a loss of Israeli ground, they would be sitting ducks.

Rejecting Dayan's pessimism, they endorsed Dado's strategy. The only question remaining was whether the United States would keep Israel resupplied to continue the fight, and Golda already sensed Kissinger's hesitation.

She did not misread him. Despite his avowal of hearty support for Israel, Kissinger smelled opportunity.

*   *   *

With its overconfidence tempered and its reserves pouring into position, Israel began to turn the tide on Day 3, although it was hard to discern the outlines of that turnaround through the waves of destruction and the rising casualty figures. With the southern part of the Golan front collapsing, Israel still faced the real prospect of a Syrian invasion of the Hula Valley, where Golda had lived on Kibbutz Merhavia. And in the Sinai, Dado's limited counteroffensive had turned into a disaster when his orders not to try to dislodge the Egyptian bridgeheads or cross the canal were ignored by both the commander of the southern forces and Ariel Sharon, who was sure that he could mop up the entire front in two days if only central command would unleash him. Their overconfidence cost Israel 70 of its remaining 170 tanks and left Egyptian troops several kilometers deeper into the Sinai.

Meanwhile, Dayan was spreading alarm. "We do not now have the strength to throw the Egyptians back across the canal," he told Israeli newspaper editors in an off-the-record interview, a preview of what he expected to tell the nation that night on television.

Dayan's voice trembled as he talked about Egypt's "unlimited equipment. It's fantastic. It's terrible to fight against such things. . . . The question is: What will come next." No doubt they'll launch an attack with maximum strength, he predicted, and Israel will have to retreat.

Dayan believed that he was providing Israelis with an honest assessment of their situation, but Gershom Schocken, the editor of *Ha'aretz*, counseled against sharing that view with the nation. "If you say on television tonight what you have told us, that will be like an earthquake for the consciousness of the Israeli nation, the Jewish people, and the Arab nations," he warned him. Another editor called Golda, who replaced Dayan on television with the former chief of military intelligence, who offered a more sanitized version of the truth.

"[Dayan] was beaten and defeated," said the editor of *Ma'ariv*. "It's lucky for us that Golda wouldn't let him appear on television. He would have announced our surrender."

Some sources contend that despite Golda's outward calm, she also panicked that day by authorizing the arming of Israeli nuclear weapons at Dayan's request. The German magazine *Stern* reported that she ordered thirteen small nuclear devices taken out of secret bunkers in the Negev and loaded onto modified Phantoms and Kfir aircraft, while an Arab publication disclosed that she allowed Israel's Jericho missiles to be armed with nuclear warheads. Both those accounts, however, are open to serious question since Israel had not yet begun to produce Kfir planes and the guidance systems of the Jericho missiles were so unstable as to be useless for such delivery.

The American writer Seymour Hersh alleged that Israel maintained three nuclear-capable battalions and threatened to use 203mm nuclear artillery shells to blackmail the United States into airlifting weapons to Israel. But the source he cited for that information, an Israeli scientist, vigorously denied Hersh's published account, and Hersh provided no evidence at all for his blackmail assertion.

Assuming that Israel had developed nuclear weapons, Kissinger did worry that Golda would authorize a nuclear strike should the country seem in danger of being overrun. But the evidence that she went so far as to issue that order is insufficient to the widespread assumption, popularized in a dozen publications and on Broadway, that she did so. Nor would it have made much sense. If her goal had been to blackmail the United States, the threat alone—better still, a rumored threat—would have been sufficient. And despite the ominous news, Tel Aviv was not in imminent danger. While making provisions for her own suicide in the event of an Arab seizure of Israel was entirely consistent with Golda's character, making that decision for the rest of the country—and lofting nuclear-enhanced artillery shells at enemy troops would have subjected Israel itself to devastating radiation—was simply beyond even her arrogance.

In any event, by the late afternoon of the third day of the war, any real threat to Israel's survival had begun to recede. And by the following afternoon, Israeli troops had begun pushing the Syrians back across the 1967 armistice line and the Israeli air force started bombing deep inside Syria, inflicting heavy damage not only on the morale of the Syrians, but on the general staff and air force headquarters in Damascus, oil refineries, a radar station, and power plants.

In the Sinai, the bungled counterattack designed by Dado had been costly in scarce matériel and human life, but it had halted the Egyptian advance, and Bar-Lev had taken over command of the southern forces to bring some discipline to their endeavors. They still faced a costly and demoralizing haul, but the Egyptians had gone as far as they would advance.

For Golda, the new battlefront became Washington as she struggled to secure military supplies to replace Israel's dwindling stocks. The Israel Defense Forces had lost two-thirds of their tanks in the Sinai, and Bar-Lev would need heavy weapons and ammunition for a counterattack against Egypt. With the Europeans embargoing weapons to Israel and the British refusing to sell her spare parts for Centurion tanks, Golda had nowhere to turn other than to the United States.

Assuming that Israel would again defeat the Arabs handily, Kissinger had decided to play a dangerous game, hoping to avoid too grand an Israeli victory that would humiliate the Arabs into intransigence and invite more Soviet intervention in the region. His only weapon was to withhold the supplies Golda was seeking and he would get caught, he assumed, only if the Israel lobby in the United States began to scream or if the Soviets started a significant resupply of the Arabs. The key to containing the former was Dinitz, the Israeli ambassador, who could sound the alarm to mobilize the Jewish lobby. The Soviets, he assumed, were as committed to containing the war as he was.

Initially, Kissinger didn't face much difficulty because Golda's requests were minimal: small arms, ammunition, bomb racks, and Sidewinder missiles, none of which would change the course of the war, all of which Israel could pick up secretly in unmarked El Al planes. He facilitated the sale of those weapons so "they will have something to lose afterwards," he told Alexander Haig.

But on October 9, the fourth day of the war, Kissinger's grand design began to unravel. The CIA confirmed that Israel wouldn't win an easy victory and was running dangerously low on supplies. To make things worse, the Soviets mounted an airlift of armaments to Syria and Egypt, and Washington discovered that they'd already launched a sealift, a clear indication that Kissinger's estimation of the Soviets had been seriously misguided.

Still, Kissinger resisted Golda's entreaties and it fell to Dinitz to pressure him. Golda's favorite son, Dinitz was no match for the American secretary of state, who stroked Dinitz's ego with open access to his office and invitations to his home. When Kissinger promised swift attention to Israel's plight, Dinitz assumed that Kissinger was being forthright. Only later did he discover that the State Department hadn't issued the necessary export licenses or that the total number of planes promised by Kissinger was wildly exaggerated. Kissinger kept mollifying him by complaining that Secretary of Defense James Schlesinger was tying up the resupply in bureaucracy—and Dinitz never caught on to the game.

When Golda realized that the promised weapons were entangled in some sort of Washington political play, she asked Dinitz to arrange an immediate secret meeting for her with Nixon. Kissinger was horrified at that prospect, which would undercut his careful calculations. Nixon didn't agree to her request, but he did order Kissinger to begin supplying Israel with replacement weapons immediately, another potential hazard to his strategy. The language of Nixon's order, however, was vague and included no timetable for delivery, and Kissinger was a master at exploiting such loopholes. So the delays continued as Kissinger kept up the pretense that he was waging a "one-man fight" against the Pentagon bureaucrats.

Golda grew frantic. In the middle of one night, she woke Dinitz up twice, instructing him to contact Kissinger immediately. When Dinitz reminded her of the hour, she said, "Never mind the time. Wake him up and tell him what I'm requesting. We need the help today because tomorrow may be too late. And tell him I said he can sleep as much as he wants after the war."

By Thursday the eleventh, Golda was calling members of Congress, George Meany of the AFL-CIO, and Jewish leaders. "I would not have come to you if I did not think the situation would improve in the next few days," she cabled Nixon. "You know the reasons why we took no preemptive action. . . . If I had given the chief of staff authority to preempt, as he had recommended, some hours before the attacks began, there is no doubt that our situation would now be different."

Still Kissinger delayed the weapons while hiding behind his vociferous complaints about the Pentagon—although his supporters argue that he was, in fact, innocent of such charges. In their accounts, all the delays were the fault of elements within the Pentagon, especially William Clements, deputy secretary of defense, a Texan with close ties to the oil industry, who did everything possible to subvert Nixon's order.

But former Defense Department officials counter that Kissinger mandated an unattainable level of secrecy in the delivery of the equipment, knowing full well that such stringent requirements would cause long delays. And it strains the imagination to believe that a deputy undersecretary of whatever would defy a presidential order, or that Kissinger would not have involved Nixon personally if he had.

*   *   *

For the first forty hours of the war, Israel fought on the northern front on Arab terms, understrength units facing massive forces. But as the reserves arrived and the air forces began blunting the Syrian drive, the Israelis inexorably pushed the Syrians back, kilometer by kilometer. On the fifth day, they trapped the last two Syrian brigades holding out on the Golan front. Tank crews began chalking "On to Damascus" on the sides of their Centurions.

Still, Dayan remained mired in pessimism, exhorting that Israel should dig in along the old truce line in the Golan and prepare a fallback line in the Sinai. Meanwhile, they should build up the IDF, mobilize the elderly and teenage boys, and distribute antitank weapons to the entire populace to prepare for an enemy advance on Haifa and Tel Aviv.

"The only way to win is to fight," countered Dado, convinced that Israel needed a decisive victory to deprive the Arabs both of land and of any sense of triumph. He wanted to strike a crippling blow against the Syrians by moving his troops toward Damascus, getting close enough to force Syria to beg for a cease-fire. Once they did, he could shift armor south and launch an offensive to break what was turning into a dangerous deadlock with the Egyptians

As usual, it fell to Golda to make the decision, and Golda was worried about speed. When the change in the tide of the war sunk in, she knew that the UN Security Council would pass a cease-fire resolution that would not be friendly. They always did, and few UN members had so much as denounced the Arabs for attacking or acknowledged Israel's restraint in avoiding a preemptive strike this time around. So she took another drag on her cigarette and said, "If it is within our power to deal a crushing blow to the Syrians and force them to plead for a cease-fire, that will be a tremendous achievement. On to Damascus."

While Israeli troops in the south chafed at the bit for a chance to turn the tables on the Egyptians, Golda continued her battle against Kissinger since no offensive was possible until her military was resupplied. None of her pleas or complaints seemed to break the logjam. Only when Nixon realized how massive the Soviet airlift had become did he ram open the pipeline. On Saturday the thirteenth, he ordered an immediate, massive military resupply operation that was hampered only by the refusal of all the European countries but Portugal to permit U.S. aircraft flying weapons to Israel to refuel on their soil.

Even then, Kissinger tried to keep the supplies to a minimum by suggesting that they use only three planes for the airlift. "Look, Henry," Nixon responded, "we're going to get just as much blame for sending three as if we send thirty or a hundred, or whatever we've got, so send everything that flies."

It was an awesome sight, wave after wave of gargantuan C-5A Galaxy transport planes rumbling into Tel Aviv. Thousands of Israelis gathered on the beach to watch the aircraft that Golda called "prehistoric flying monsters" ferrying smart bombs, missiles, and ammunition. When Golda heard the news, she wept.

The Russian airlift, begun five days earlier, had already delivered 3,000 tons of war supplies to Arab countries.

With its supply stream guaranteed, Israel was ready for its final push in the Sinai, and the Egyptians provided the perfect opportunity when they attempted an advance toward the Mitla Pass. What began with a ninety-minute artillery barrage became one of history's fiercest tank battles, on the order of General Bernard Montgomery's clash with the Afrika Corps at El Alamein.

That evening, Bar-Lev called Golda to report, "We are back to being ourselves and they are back to being themselves." She gave the order for Israeli troops to cross the canal the next night.

By the following morning, in the middle of Sukkoth, the holiday celebrating the Jewish passage through the Sinai desert after their escape from captivity in Egypt, the Israelis were behind Egyptians lines, racing toward the roads to Cairo and destroying the ground-to-air missiles that had turned Egyptian skies into a death trap for the Israeli air force.

That same afternoon, Alexei Kosygin, the Soviet premier, arrived in Cairo to urge Sadat to agree to a cease-fire. He refused. Smelling victory, so did Golda.

But after rejecting all of Kissinger's early suggestions for a cease-fire, the Soviets had concluded that the price of the war was escalating too quickly and invited Kissinger to Moscow to develop a joint cease-fire strategy. There was little that Golda feared more than a peace imposed by outside powers. "They compromise, Israel loses," she always said. "Sadat must, I think, be given time to enjoy his defeat and not to immediately, by political manipulations, turn it into a victory. . . . For God's sake, he started a war, our people are killed, his in the many thousands are killed, and he has been defeated."

No matter Golda's sentiments, Kissinger was not about to lose an opportunity to woo the Russians.

As the IDF raced to retake their old position on Mount Hermon from the Syrians and solidify their lines in Sinai, Golda bit her nails, waiting to see what Kissinger and the Soviets would cook up. Dinitz checked in with the White House regularly only to be told that there was no word from Kissinger.

Then, on Sunday evening—Israel time—the American ambassador brought Golda two messages, one from Kissinger laying out the cease-fire document, the other from Nixon urging her to announce Israel's immediate acceptance. Almost simultaneously, Dinitz was informed that the UN Security Council would vote on the truce terms six hours later, that it would begin in less than twenty-four hours, and that not a single word could be changed.

By then, Israeli troops controlled a swath of Egypt twenty-five miles long and twenty miles deep. While some Egyptian forces were still entrenched on the Sinai side of the canal, the Israelis had almost completed the encirclement of the Egyptian Third Army.

Golda was livid that the United States was trying to impose an immutable piece of text on her without any consultation. Why twenty-four hours rather than thirty-six or forty-eight hours, all Israel needed to finish what Egypt and Syria had started? Kissinger, she concluded, was trying to snatch victory from her.

Deciding whether to accept the terms was a painful dilemma. Israel had already suffered twenty-five hundred casualties and could get no news about dozens of prisoners of war captured by the Egyptians and Syrians. Begin was breathing down her neck in violent opposition to any truce, and she would face him in elections postponed from October 31 to the end of the year. And military leaders were intent on finishing off the Egyptians by demolishing their armies to recoup their honor.

But the Soviet-American cease-fire, which Sadat had readily accepted, called for direct negotiations between Egypt and Israel to establish a just and durable peace. Since no Arab country had ever before agreed to direct negotiations, it was a staggering diplomatic victory.

That night, the cabinet debated for five hours and, at three A.M., authorized Golda to send Nixon a message of acceptance on one condition, that Kissinger come to Israel to discuss the terms fully.

On the morning of the twenty-second, Kissinger, who'd just won the Nobel Peace Prize, arrived from Moscow, dreading his meeting with Golda. He'd never negotiated directly with the woman whose relentless ability to pick apart every word and dig in her heels was legendary in State Department circles.

"I presume she is wild with anger at me," he remarked to Eban, who picked him up at the airport.

Golda's greeting was even chillier than Kissinger expected. Why didn't you contact us during negotiations? she quizzed. When Kissinger earnestly explained that the Soviets had jammed the communications equipment in his plane and from the embassy, she guffawed. "How did you stay in touch with your president, then?"

Although Kissinger had pronounced the cease-fire to be a package deal, Golda demanded changes, most important an immediate exchange of prisoners. Unwilling to fiddle with the document, Kissinger told Golda that Leonid Brezhnev, the Soviet head of state, had promised that the exchange of prisoners would take place immediately after the cease-fire. That was precisely the wrong thing to say to Golda, who'd learned from William Rogers the danger of verbal commitments.

What else did you agree to that's not written down? she asked, badgering him relentlessly. The truce called for "negotiations toward peace," the kind of vague language Kissinger favored. What kind of negotiations? When? In what forum? Nixon had mentioned direct negotiations. Why wasn't that language included? If Golda was going to be robbed of victory, she was determined to understand the price.

Golda suspected that Kissinger was seeking to buy an end to Arab oil embargo threats with Israeli currency, so she was not an easy sell, and Kissinger understood why after Dado and the other military commanders described how close Israel was to smashing the Egyptian armies.

"Well, in Vietnam, the cease-fire didn't go into effect at the exact time that we agreed upon," Kissinger told them. Hearing that statement as tacit permission, after Kissinger left, the cabinet decreed, "If the Egyptians fail to live up to the cease-fire, the IDF will repel the enemy at the gate."

The truce was scheduled to begin at 6:53 P.M., and by the time Kissinger returned to Washington, the Russians were complaining that Israel had violated its terms by tightening its noose around Egypt's Third Army. "I won't stand for the destruction of the Third Army," Kissinger screamed at Dinitz. When Dinitz protested that Israel had not initiated any action, Kissinger responded with flat disbelief.

Golda usually communicated with Kissinger in writing, but after hearing from Dinitz, she picked up the telephone. "You can say anything you want about us and do anything you want," she said, barely controlling herself, "but we are not liars."

It turned out that a commander in the Egyptian Third Army had tried to break out of his encirclement. In response, Israel did precisely what Golda had promised she would order when she agreed to the cease-fire. Kissinger, however, wasn't interested in the facts. They were not "relevant," he told the Israelis after American intelligence confirmed Israel's account.

The situation grew tenser when, moments before a second cease-fire could begin, Israel announced that its troops had surrounded the Third Army, sealing it off from the rest of Egypt. Stop them even if you have to send American troops, Sadat fumed. To reinforce that message, the Soviets put seven divisions of their own forces on the alert and threatened to send them in.

Israel had stopped, but Kissinger was in a panic, facing the collapse of both détente and a Middle East peace. To preserve his bargaining power, he demanded that Israel open a supply corridor to the Third Army. "You want the Third Army?" Kissinger yelled at Dinitz. "We don't go to a third world war for you."

On the night of October 25, Golda's cabinet deliberated until four A.M. about their response to what had become an ultimatum. Finally, Golda calmed down enough to cut through the rhetoric, including her own. Israeli troops were on the verge of capturing a 40,000-man army and forcing Sadat to admit his defeat, but the price of that victory would be painfully high. Do we stand pat and risk both Soviet intervention and American wrath? Golda asked. Which is more important, capturing the entire Third Army or good relations with the United States?

"There is only one country to which we can turn and sometimes we have to give in to it—even when we know we shouldn't," Golda lectured the ministers who were balking. "But it is the only real friend we have and a very powerful one. We don't have to say yes to everything. But there is nothing to be ashamed of when a small country like Israel, in this situation, has to give in sometimes to the United States."

CHAPTER SEVENTEEN

Whoever wants to be prime minister
deserves what he gets.

Motti Ashkenazi took up a solitary vigil across the road from Golda's office two days after his discharge from reserve duty in the Sinai. It was raining, one of those damp and gray February days that keeps Jerusalemites off the streets and leaves drivers defogging their windshields every three minutes. But the placard the slight thirty-three-year-old carried was too poignantly provocative to be ignored:

GRANDMA, YOUR DEFENSE MINISTER IS A FAILURE AND 3,000 OF YOUR GRANDCHILDREN ARE DEAD.

For Ashkenazi, a doctoral student in philosophy at the Hebrew University, the demonstration wasn't so much a political act as a cri de coeur he'd been fantasizing since the afternoon when a deluge of mortar shells and rockets had turned the small outpost he'd commanded into an inferno. His reserve unit, an odd collection of ill-trained immigrants and older men, had held out for five days after the Egyptian invasion. When they were rescued, Ashkenazi began dreaming of asking the question that was on the minds of most Israelis: How did this happen?

His protest didn't remain solitary for long. Other reservists arrived in ones and two, and then a mother joined them. One afternoon, Golda looked out of the window and spotted a middle-aged man holding a sign reading, MY SON DIDN'T FALL IN BATTLE. HE WAS MURDERED AND THE MURDERERS SIT IN THE DEFENSE MINISTRY.

After the catastrophic first days of the war, Israel had won a stunning victory, but the price was too high for celebration. More than 2,600 Israelis had died in less than three weeks—per capita, three times America's losses during ten years of fighting in Vietnam. Although Israeli troops were dug in twenty-six miles from Damascus and sixty-five miles from Cairo, the myth of the Israeli superman had been shattered. Their aura of invincibility gone, Israelis felt naked and exposed.

Golda was barely hanging on, haunted by her knowledge that she had let Israel down. "It doesn't matter what logic dictated," she explained. "It matters only that I, who was so accustomed to making decisions . . . failed to make that one decision. It isn't a question of feeling guilty. I, too, can rationalize and tell myself that in the face of such total certainty on the part of our military intelligence . . . it would have been unreasonable of me to have insisted on a call-up. But I know that I should have done so, and I shall live with that terrible knowledge for the rest of my life."

To one journalist, she added candidly, "I will never be the same Golda I was prior to the war. Yes, I smile and laugh and listen to music, I tell stories, I listen to stories. . . . But in my heart, it is not the same Golda, and until eternity I will never be."

But faced with the marathon challenges of grabbing diplomatic victory out of the ambiguous end to the war, preparing for national elections, resolving the economic crisis caused by a conflict that cost more than Israel's annual budget, and calming the increasingly ugly mood of a country coming to grips with a new self-image, Golda couldn't indulge her own frailties.

Winning the peace, Golda knew, would not be easy. Israel was a near pariah, deserted even by her oldest friends in Africa. Terrified of an oil embargo, the Europeans wouldn't so much as evince sympathy over the two-front invasion. And the Soviet Union seemed committed to unending Arab adventurism. No matter how much the admission damaged Israelis' haughty sense of independence, Golda knew that the country had become perhaps dangerously dependent on the United States. And the United States had its own priorities, which diverged from Israel's in an age of oil embargoes and détente.

Within a week of the second cease-fire, then, Golda packed her suitcase with baggy suits and sensible shoes, grabbed her trademark black handbag, and flew to Washington, stopping en route only for a visit to the troops along the canal. Three weeks earlier, she'd visited the northern front, and in the midst of a scene of destruction and carnage had come upon a group of soldiers half asleep as they worked on a crippled tank. "Golda, is this all worth it?" one of the young men, gray with dust from three straight days of battle, had asked her, looking his prime minister straight in the eye.

Clutching her handbag, she spoke softly despite the din of war. "If it is just for us, for three million Jews in this country, then your question is legitimate," she answered. "But if it is for the thirteen or fourteen million Jews, for the whole Jewish people, for the very existence of the Jewish people in history, it has to be worthwhile."

This time, as she steeled herself for her trip to Washington, the troops were more encouraging. "Golda, be strong," shouted soldiers along the canal, awed by the sight of the old woman who'd flown ten hours on helicopters to visit them. Having survived on coffee and cigarettes for almost three weeks, Golda was beyond exhaustion. "I don't even have strength to be strong," she responded.

Golda wasn't in Washington for an hour before she began feeling worse. Opening the newspaper, she saw a huge photograph of Nixon embracing Ismail Fahmy, an envoy from Sadat, who'd preceded her to America.

Golda was well aware of what Kissinger hoped to achieve in their negotiations: an Israeli withdrawal that would free Sadat's Third Army. "It's ridiculous," she fumed at her their first meeting. "They start a war and lose. And they want [victory] handed to them."

But Kissinger held a devastating club to use against her, the return of Israel's prisoners of war. Egypt had acknowledged holding 231 Israelis captive and Syria held several dozen more about whom they would release no information. The televised images of the captives, gaunt and shackled, had paralyzed the entire nation, and Golda had been besieged by their families, to whom she could offer no concrete words of hope.

Kissinger didn't hesitate to wield that weapon and maintained that the Arabs would not release the prisoners—or even provide their names— until Israel agreed to return to the first cease-fire lines, where Israel had been before the IDF surrounded the Egyptian army.

"I won't do it," vowed Golda. "I won't go to my people and tell them I accept this plan. . . . I faced a woman the other day. She had lived through Hitler and came here with one son. . . . She is dying of cancer. . . . She appeals to me, 'Release my son. . . .' What can I say to her?"

Golda suggested an entirely different strategy. Rather than talk about Israeli withdrawal, Israel and Egypt should exchange prisoners and then straighten out their hopelessly entangled cease-fire lines by moving all the Egyptian forces back to the west side of the canal, where they'd started, and all the Israeli troops to the Sinai side. In the process, the Third Army would be freed from its encirclement.

Kissinger saw right through her. Since Golda's approach would force Sadat to give up the pretense of victory he'd gained when his troops crossed the canal, the secretary of state refused to discuss it.

One hour into their first bargaining session, they were at an impasse.

Like most Western diplomats, Kissinger paid little attention to the public declarations of Arab leaders, assuming that they were posturing for domestic and pan-Arab audiences. He accepted, then, Sadat's protestations that he wanted peace, considering the only serious stumbling blocks to be Israel's reluctance to return land captured during the 1967 war.

Having listened to Arab public statements for fifty years, Golda believed that Arab rhetoric was sincere, that they considered Israel an illegitimate entity forced on the Middle East by Europeans wracked with guilt over the Holocaust. She wasn't unaware of the shift away from the old "drive the Jews into the sea" oratory. But knowing that President Habib Bourguiba of Tunisia had suggested to Arab leaders that a more reasonable public tone would force Israel onto the defensive, she saw the change as a tactic, not a retreat from Arab designs to destroy the Jewish state.

"There can be no greater mistake in assessing the current situation in the Middle East than to assume that the conflict continues because of a specific political Arab grievance—the plight of the Arab refugees, the Israeli presence on the West Bank or in the Sinai, the reunification of Jerusalem," she asserted. "The heart of the problem is what caused the Six-Day War. . . . Simply put, the root issue is the Arab attitude to Israel's very existence. . . .

"They don't want us here. That's what it's about. It isn't true that they don't want us in Nablus or Jenin. They don't want us, period."

Golda hadn't flown to America to fight with Kissinger but to see Nixon, who she instinctively trusted. And the president greeted her with the reassuring warmth of an old friend. "Now we have something else in common," he told her, falling into the banter that characterized their relationship. "We both have Jewish foreign secretaries. Yes, Golda acknowledged, "but mine speaks English without an accent."

While Nixon was a great admirer of Golda, whose moxie had won his heart, he had decided that the time had come to end the conflict in the Middle East, which was interfering with his plans for détente. "We have to squeeze the Israelis when this is over," he'd told Kissinger the day the first wave of the American weapons landed in Tel Aviv. "And the Russians have got to know it."

Sensing that change in the president's thinking, Golda nonetheless hoped to maneuver Nixon into backing away from what was, essentially, Kissinger's strategy. Nixon, however, was distracted by his sinking presidency. While Golda was battling Kissinger for the airlift of weapons, Nixon's vice president, Spiro Agnew, had resigned and pleaded no contest to criminal charges of tax evasion and money laundering. During Kissinger's cease-fire discussions in Moscow, Nixon had been enmeshed in firing Archibald Cox, the Watergate special prosecutor.

"We are aware of the enormous suffering that you have undergone," said Nixon, oozing empathy. "The problem is now to move on toward the goal. . . . The goal is not to have another war. You have already had four."

Gently, he reminded her that the Arab use of their oil weapon put Israel in a perilous position. "If this cease-fire breaks down and Europe and Japan freeze this winter, Israel will be in a hell of a spot." Then he issued a not-so-veiled threat. "I could leave you to the UN," he said.

Golda interrupted. "That court of high injustice?"

Golda tried desperately to convince Nixon that he was being unfair, but the subtext of the discussion was clear: if the United States is going to underwrite Israel's existence, you're going to have to pay by giving up territory and saving the Egyptian Third Army. All roads, then, led back to Kissinger.

The next night at 10 p.m., Kissinger met with Golda at Blair House. He preferred late-night sessions when his own energy was high, assuming he would have a natural advantage. He had not yet learned how little Golda slept or how thoroughly she, too, thrived on arguing at 2 a.m. Kissinger's short-term goal was to induce Golda to open a relief corridor to the Third Army and then to order a pullback of Israel's troops to the original armistice lines.

"Where is the October 22 line?" she asked sarcastically, knowing that no one was sure. Once Egypt provided the names of the Israeli prisoners of war, allowed the evacuation of the wounded, and permitted the Red Cross to visit the others, she would allow the UN to open a corridor. But the road itself would remain under Israeli control, she insisted, so that her army could ensure that the Egyptians didn't try to rearm.

Having already promised both Sadat and the Russians that Egypt would have a resupply corridor under its own control—a detail he didn't share with Golda—Kissinger could not back down.

Neither would Golda. The Third Army will not be released or resupplied until Israel's prisoners of war are returned home, she told him. We are not asking for a favor. This is Israel's right under international law. "To leave our prisoners there for a . . . length of time, that we can't live with."

When browbeating didn't work, Kissinger switched tactics. "The one thing the Arabs have achieved in this war, regardless of what they lost, is that they've globalized the problem," he warned her. "They have created the conviction that something must be done, which we've arrested only by my prestige, by my trip, by my maneuvers."

It was classic Kissinger: I, Henry Kissinger, am your only friend. I'm willing to help you, but you have to help me to do so. The strategy had won him a Nobel Peace Prize, announced just two weeks earlier, but Golda was immune to it.

Kissinger then tried reassuring her that the prisoner exchange would begin as soon as Egyptian troops were assured a regular supply of food and water, although he had barely raised the matter in his meeting with the Egyptian foreign minister the day before.

Still, Golda didn't yield. When he departed at 1 A.M., he left her with an ominous vision. "Think about our other problem, our nightmare, the Russians going in there."

As Kissinger emerged from the meeting, an Associated Press photographer captured an image of him with Golda that did not hide the tension. "Was it that bad?" a reporter asked him. Kissinger didn't bother to answer. The atmosphere verged "on the abrasive," the State Department blandly acknowledged.

The only levity breaking the strain was Kissinger's response after Golda shrugged and said, "Look, what do you want from me? I was born in the last century."

Kissinger smiled. "The nineteenth century is my specialty."

The next night they went at it again. "We didn't begin the war," Golda reminded him. While that decision had cost Israel dearly, it didn't count for very much with Kissinger.

"Madam Prime Minister . . . you didn't start the war, but you face a need for wise decisions that will protect the survival of Israel," he responded. "This is what you face. This is my honest judgment, as a friend."

Golda feared that Kissinger's advice was tantamount to an order. "You're saying we have to accept the judgment of the United States?" she asked. "I'll call the cabinet and tell them we have to accept the U.S. position or see the destruction of Israel."

"We all have to face the judgment of other nations," Kissinger replied.

But Golda had a response for everything. To Kissinger's lecture about the importance of maintaining the cease-fire, she responded with a complaint about the Arab blockade of the Bab el Mandeb Strait, which had cut Israel off from her supply of Iranian oil. After he broached the issue of a continuing supply of food and water to the Third Army, she turned the discussion to Arab violations of the Geneva Convention on prisoners of war. When he played his favorite tune, conjuring up an apocalyptic vision of the dire consequences of not making peace, she interrupted with a lecture of her own about all the dire consequences Jews had suffered in their long history.

Golda and Kissinger were oil and water. She never had much patience with intellectuals and she worried, too, that the American secretary of state was a bit of a self-hating Jew. While he'd been raised in an observant home, Kissinger had been sworn into office on a Saturday, thus forcing his Orthodox parents to walk to the White House, and used Nixon's personal copy of the King James version of the Bible rather than the Old Testament published in his hometown in Germany that his parents had brought with them.

Beyond the personal, how could she trust Kissinger knowing that his priority was to protect America's oil supply and détente with the Soviets rather than the security of Israel? At times, his arrogance overwhelmed her. "There's no greater expert on American policy than Kissinger," she once retorted after he'd issued yet another treatise on how Israel should conduct itself. "But on Israeli policy, I have my PhD."

Kissinger didn't trust Golda either—and she was frustrating his carefully laid plans. Never one to begin negotiations until he felt the field was ripe, Kissinger believed that the Middle East was perfectly configured for a major peace initiative. Sadat, he had decided, was serious about ending the conflict and willing to work through the United States. If Golda gave him some leverage, he could well emerge as the American middleman and sideline the Soviets.

But Golda remained adamant that she would allow no resupply of the Egyptian army until prisoners of war were exchanged. "You are not giving me anything to go to Cairo with," objected Kissinger, who was scheduled to fly to Egypt that week. "Do you want justice or the prisoners?" Stubborn but rarely impractical, Golda cared about the prisoners of war above any other issue and finally agreed to allow the opening of the corridor—but only if Egypt did not control it. Again, Kissinger could not bring himself to admit that he'd boxed himself in with a promise to Sadat.

All night, they fought, and it was never pleasant. "You know, all we have, really, is our spirit," she told Kissinger toward the end of the meeting. "What you are asking me to do is to go home and help destroy that spirit, and then no aid will be necessary at all."

Golda had planned to rest the next morning before her return to Israel. Instead, she got a taste of Kissinger's wrath. Alexander Haig, Nixon's chief of staff, called to say, "The president is furious." Then Nelson Rockefeller stopped by and pleaded, "Please don't be stubborn."

But Kissinger didn't fully appreciate that Golda was not a run-of-the-mill foreign leader. Earlier in the week, she'd had breakfast with fourteen senatorial heavyweights—including Jacob Javits, Abe Ribicoff, Hubert Humphrey, and Ted Kennedy—and had met with eighteen leading members of the House of Representatives. Her friends George Meany of the AFL-CIO and Arthur Goldberg, former Supreme Court justice and U.S. ambassador to the UN, were urging her to stand fast. And she hadn't even mobilized the American Jewish community.

"The Almighty placed massive oil deposits under Arab soil," an Israeli diplomat commented to a colleague at the U.S. State Department. "It is our good fortune that God placed five million Jews in America."

*   *   *

Despite Kissinger's warnings Golda's obstinacy would hold back progress toward a semblance of peace between Israel and Egypt, tellingly, perhaps, the Egyptians and the Israelis were already succeeding in their negotiations. In exchange for Israel's initial resupply of the Third Army, Sadat had agreed to hold direct talks about military disengagement with the Israelis at a military level. So in the bitter cold on Sunday, October 29, Major General Aharon Yariv, a veteran military intelligence officer, and General Muhammad al-Ghani al-Gamasy, the chief of operations of the Egyptian armed forces, saluted each other awkwardly under a hastily erected tent stretched between four Israel tanks at Kilometer 101 on the Cairo-Suez Road.

"For the first time in a quarter of a century there is direct, simple, personal contact between Israelis and Egyptians," Golda told the Knesset, unable to conceal her excitement. "They sat in tents together, hammered out details . . . and shook hands."

It quickly became clear that Sadat cared less about an Israeli withdrawal to the first armistice line than he did about a supply corridor to his trapped Third Army, a subtlety Kissinger had either missed or ignored. After his ebullient rhetoric during the first days of victory, Sadat had not admitted to his people that Israel had captured more than 500 square miles of Egyptian territory and 8,372 of its soldiers. "The story of the encirclement was used according to the methods of Goebbels," he announced at a press conference. "They have put me in a dilemma: Should I wipe out their force, which is squeezed between the two parts of the army? Or should I obey the cease-fire?"

The destruction of the Third Army, then, posed the risk to his presidency.

It took al-Gamasy and Yariv just two hours to settle on a resupply and prisoner exchange agreement, so they moved on to the more fundamental question of the disengagement of forces. After a bit of posturing by both sides, al-Gamasy suggested that Israel withdraw about thirty-five kilometers from the canal and be replaced by United Nations peacekeepers. Once that withdrawal began, he said, Egypt would lift its blockade of the Bab el Mandab Strait that choked Israeli shipping and thin out its forces on the west bank.

It was a startlingly generous offer, not quite good enough, perhaps, but it held out the possibility of a wide neutral zone along the canal that would prevent further conflict. Building on that momentum, during eighteen meetings the two men moved their countries toward peace. Before Kissinger even reached Egypt, Yariv and al-Gamasy were already discussing a phased disengagement and the reopening of the canal and were well on their way to the very agreement for which Kissinger would take credit a week later.

But Middle East peace was being staged as the Henry Kissinger show, so all attention was focused on his first meeting with Sadat, who had already decided to make the United States the center of the struggle for peace in the Middle East. Kissinger arrived with a six-point agreement in his briefcase providing for the scrupulous observation of the cease-fire; daily deliveries of food, water, and medicine to the encircled areas; immediate negotiations for Israel's return to the earlier armistice lines within the framework of separation of forces talks; the replacement of Israeli checkpoints along the Cairo-Suez road with UN personnel; and the exchange of prisoners.

Fully expecting that Sadat would oppose his proposal, Kissinger was shocked when the Egyptian president agreed to everything in less than two hours. "I wanted to show him that I am a very, very reasonable man—unlike Mrs. Meir," Sadat later told friends. "She will haggle over every point."

With Sadat's agreement in his pocket, Kissinger took off on a trip to the pyramids, sending Joe Sisco to Israel. I'll have an announcement as soon as he returns, Kissinger told the press, adding, "that is, if we ever see Joe again," reflecting his fear of what would happen in Jerusalem.

Kissinger had a fondness for the sort of vague language that allowed all sides to claim diplomatic victory. Having been burned more than once by such locutions, Golda worked through the document Sisco brought with a fine-tooth comb and demanded greater linguistic rigor. The agreement, after all, didn't spell out the number of checkpoints along the supply corridor or explicitly bar the transport of military supplies along it. The timetable for the release of the prisoners of war was hopelessly ambiguous, and there was no mention of Israel's formal control over the corridor, which Kissinger had already promised. Her cabinet and opposition leaders had more complaints, most important that the agreement didn't mention the lifting of the Egyptian blockade of the Bab el Mandeb Strait.

When Kissinger heard from Sisco about what he considered to be "quibbling," he fired off two messages, one to Golda warning that any changes would sabotage the entire accord, and another to Nixon, seeking help.

Kissinger saved his agreement by resorting to what would become his favorite tactic, American guarantees on points omitted from agreements. Golda loathed such guarantees. To a woman who'd embraced Zionism in large measure because she believed in Jewish self-reliance, trading it for American promises seemed a poor bargain since it left Washington free to interpret the guarantees and to decide when or if to take violations seriously. She remembered all too well what happened in 1970, when Egypt broke Rogers' temporary cease-fire agreement by moving SAM missiles, and the United States let the matter pass.

For Golda, Middle East peace had to hinge on Israeli-Arab relations, not Israeli-American relations or Arab-American relations. By imposing himself between the two sides, she believed, Kissinger would only delay a final resolution to the conflict.

Guarantees are diplomatic Band-Aids, she told Kissinger, a poor substitute for directly negotiated agreements.

They're only "icing on the cake," Kissinger countered.

If there was a cake, Golda said, Israel wouldn't need any icing. And what good was the icing without the cake?

Ultimately, Israel had no choice but to accept the terms. The pressure from Washington was relentless, and the families of Israel's prisoners of war were demonstrating outside the Knesset and the American embassy. WHERE IS MY FATHER, one placard asked. Another read, NIXON, YOU GAVE THE EGYPTIANS A CEASE-FIRE, NOW GIVE US BACK OUR SONS.

Golda understood that by caving in to the Americans, she had not so much signed an agreement with the Egyptians as turned herself over to Kissinger, and few others missed that distinction. "My dear General, what does the phrase 'disengagement agreement' mean?" Yariv asked al-Gamasy when the two men signed the document at Kilometer 101. Puzzled, the Egyptian responded, "It means to place the troops away from one another." Yariv shook his head. "No. . . . It is a Harvard expression and it is Kissinger who will work out the explanation for it, and you and I will not be able to do anything about it until Kissinger says what he means."

The agreement might have been a bitter pill for Golda to swallow, but it went down easier as Golda watched a DC-6 leased by the International Red Cross touch down at Tel Aviv's Lod Airport with the first planeload of Israeli prisoners of war. Among those released were nine soldiers who'd been held captive for years, after being grabbed during the War of Attrition. In their 20 x 20 cell, they'd taught themselves to knit and had brought Golda what they'd produced: an off-kilter Star of David that they told their jailers was a religious symbol rather than the flag of Israel.

The day after the disengagement agreement was signed, Golda interrupted the never-ending series of meetings for a personal errand, an errand of anger and conscience. For years, she had been the Socialist International's biggest star. When she arrived at their annual meetings, Pietro Nenni, the patriarch of Italian socialism, kissed and hugged her, and Willy Brandt pulled her aside for confidential discussions. One year earlier, she'd been elected deputy chairman of the international organization of Socialist, Social Democratic, and Labour parties.

But during the war, the Socialists in power had refused to sell her weapons or parts, or allow the U.S. planes ferrying in war matériel to refuel at their airfields. Golda understood why they had betrayed her. But she wanted to force them to speak the words out loud.

So she flew to London for a special meeting of the Socialist International called at her behest. Twenty-one Socialist political leaders, including nine prime ministers, gathered, as Golda demanded an explanation.

> *I just want to understand ... what socialism is really about today. Here you are, all of you. Not one inch of your territory was put at our disposal for refueling the planes that saved us from destruction. Now suppose Richard Nixon had said, "I am sorry, but since we have nowhere to refuel in Europe, we just can't do anything for you after all." What would all of you have done then? You know us and who we are. We are all old comrades, long-standing friends. .... Believe me, I am the last person to belittle the fact that we are only one tiny Jewish state and that there are over twenty Arab states with vast territories, endless oil, and billions of dollars. But what I want to know from you today is whether these things are decisive factors in Socialist thinking too?*

When Golda was finished, Bruno Pitterman, the Austrian who chaired the gathering, asked if anyone wanted to speak. Not a single hand was raised. Not a word was uttered. The silence turned deafeningly tense. In the back of the room—from somewhere behind the prime ministers of Austria, Belgium, Germany, Malta, Mauritius, the Netherlands, Norway, and Sweden—someone whispered not so quietly, "Of course they can't talk. Their throats are choked with oil."

*   *   *

Scheduled for October 31, the elections for the eighth Knesset had been postponed to December 31, and in the midst of her negotiations with Kissinger, Golda also fought for her political life. Although exhausted and nearly broken, she could not allow Begin and Likud to take over the country, certain they would destroy any possibility of wringing peace out of the conflict. Already hard-liners were rejecting every overture made at the Kilometer 101 talks, every proposal from Kissinger. Israel had won a magnificent victory, Begin declared, and shouldn't have to pay for another Arab defeat because of Western greed for oil.

Regaining her mandate tested all of Golda's political skills. Angry and confused by the massive losses and shocked at the surprise of the war, the Israeli public demanded an explanation, and, reluctantly, given her aversion to scrutiny of the operations of the government, Golda appointed a commission to inquire into the reasons for the intelligence and military failures before and during the early days of the war to be chaired by Shimon Agranat, chief justice of the Supreme Court. Until the commission issued its findings, she argued, the government should remain intact.

Her more immediate challenge was to hold the Labor Party together, and it wouldn't be easy. Having learned little from the war, Rafi still clung to all its old assumptions, that Israel had to remain on the Golan Heights and in the Jordan Valley and needed a deepwater port by Gaza, perhaps more than ever. Like Begin, Rafi dismissed Sadat out of hand, certain that the Egyptian leader's bargaining position had been hopelessly weakened by the siege of the Third Army.

But chastened by the body count and refusing, at last, to be held captive to Dayan, the doves argued that all of the old assumptions had become obsolete on Yom Kippur and that the party platform had to be redrawn. Allying herself with the moderate doves, Golda agreed.

On the surface, the new platform they wrote didn't seem drastically different from the old one. It still called for a strong military; defensible borders rather than those of 1967; a peace agreement with Jordan as the Palestinian state; and continued settlement in the occupied territories. But in the new iteration, those defensible borders were to be based on "territorial compromise," a revolutionary phrase. Settlements would be built "giving priority to national security considerations," a clear rejection of Dayan's maximalism. And the word "peace" appeared a record fourteen times, a triumph for those ready to take Israel in a new direction.

But the anger and anxiety unleashed by the Earthquake, as Israelis called the political and psychic tremor set off by the war, couldn't be quenched by linguistic tricks. For six years, Israelis had taken it as a given that the Arabs could not mount an effective attack against them and thus would eventually be forced to accept peace on Israel's terms. When that framework of belief collapsed, Israelis inside and outside the Labor Party needed someone to blame. The national target became Moshe Dayan.

Grieving relatives shouted "Murderer!" in front of his house, the old adoration turned to venom by the national angst. When Dayan entered one party meeting, he was greeted with a cacophony of boos. "Moshe, you created the feeling in Israel that we were not going to be attacked," said Michael Bar-Zohar, an Israeli writer and political figure close to Dayan, who counseled him to resign.

Golda did not join in that chorus. Her critics charged that she feared that Dayan's departure would become a slippery slope that would wash her away as well. But at that moment, axing Dayan might well have bolstered her authority.

Indeed, when the minister of justice screamed at Golda that she had to oust Dayan, she spat back, "Why don't you demand MY resignation?"

But that wasn't his point. "You are not responsible," he replied. "You are not the minister of defense."

That sentiment was widely shared by those who had watched Golda take over after Dayan's collapse. "She revealed the strength of a warrior without fear," Galili said. "She gave us a sense of hope and the opportunity to do what was necessary to turn things around."

Still, public confidence in Golda had been severely eroded. Shortly after the signing of the first disengagement agreement, a *Ha'aretz* poll found that only 45 percent of the voters supported her, a drop of 20 percent in six weeks. Golda might have been a political colossus, but in the new light of postwar Israel, she began to look like an aging dinosaur who ran the country like her mother had run her delicatessen in Milwaukee, rather than as a modern leader guiding a complex bureaucracy.

Those same doubts wracked the party as well, and Golda abruptly summoned the 615-member Central Committee to figure out, once and for all, "who is for whom and what is for what." Allon, spokesman for the moderates, opened the proceedings by calling for all the cabinet members but Golda to step down.

For hours, Labor's leaders slung mud at one another and issued passionate harangues against the new platform, against Dayan, and against how demoralizingly sad Golda appeared on television. The occasional speaker reminded members that Israeli troops were still in danger. But weeks of pent-up fear and anger, years of pent-up frustration, could not be easily contained. For too long, Labor had compromised itself into national leadership with fuzzy language that obscured the reality that it was less a coherent political party than a conglomeration of factions and special interests driven as much by ego as by a common ideology.

Toward evening, Dayan requested the floor. As if nothing had changed, he reasserted his commitment to building an Israeli port in the Gaza Strip and to Jewish land purchases on the West Bank. Ignoring the anger on the streets and in the hall, he laid down the gauntlet against party policies with which he disagreed. "I will turn in my party card and walk out," he declared, seemingly unaware that Labor was no longer inclined to bend to his desires.

Having declined to open the proceedings, Golda listened, stoically, to forty speeches. It was after midnight when she finally rose to the podium:

> I could offer excuses for myself and say that the information we had was not so clear . . . that it would have been illogical to be stubborn and demand a call-up of the entire military when military leaders were saying otherwise. . . . Maybe I should have listened to my own intuition. . . . It will be with me my entire life that I did not.
>
> Earlier, I said that someone has to take responsibility, and I put myself first. I received much criticism that I don't appear good on television, and this is not good for the morale. They say that I look sad. At my age, should I begin wearing make-up? I would do so if I thought it would help. But I am realistic about these things. Like everyone else, I am sad, even more so because I am the Prime Minister. I cannot say

that I didn't have the information. . . . It was a fatal mistake to rely on the information that I received. . . . I am not in the mood to make excuses.

After rambling for almost an hour, Golda issued her coup de grâce: If you don't want me to lead the party into the election, that's fine. Say the word. In fact, let's have a vote, a secret ballot so that no one can raise charges of intimidation.

Numbed by fifteen hours of debate and terrified, as always, of a split in the party, no other candidate was offered. It came down to a straight vote of confidence in Golda's leadership. Meekly, the delegates trudged downstairs to the voting booths. When the ballots were counted, 281 Labor leaders affirmed their support for Golda, with 33 against and 17 abstaining.

Normally, Israeli election campaigns were boisterous affairs, but the ongoing diplomatic tensions and the mood of the country quelled the usual fracas. Campaigning on television for the first time, Labor portrayed itself as the party with the experience and vision to achieve a postwar peace through reasonable compromise and attacked Likud as retrogrades bent on holding on to every inch of occupied lands. Likud, in turn, declared that Golda and Labor had forfeited their right to govern and raised the specter of a Munich-style capitulation should they be given a mandate to remain in power.

The danger they conjured up was all the more acute in some eyes because Golda had agreed to send Eban, a notorious dove, to a Middle East peace conference in Geneva. Using his Vietnam peace talks as a model, Kissinger wanted a meeting of all the warring parties not only to nail down final disengagement agreements but also to negotiate a comprehensive peace agreement. Golda had long dreamed of peace negotiations with the Arabs, although in her fantasy they involved only Israel and the leaders of the Arab states. Kissinger intended the gathering to be held under the auspices of the United Nations with the Palestinians in attendance.

Assuming from long experience that anything run by the UN would turn into an orgy of anti-Israel vitriol, Golda suggested that the United States and the Soviet Union convene the conference, and she rejected any PLO presence out of hand. Furthermore, while she happily agreed to sit with the Egyptians and the Jordanians, she refused to have any truck with the Syrians until they provided her with the names of prisoners of war held in Damascus and allowed the Red Cross to visit them.

After weeks of trading history lectures and apocalyptic scenarios, Kissinger cajoled Golda into accepting the UN mantle as symbolic, with the secretary-general disappearing after the opening ceremony. Syria became a nonissue when its president announced that he would not attend. And a compromise was struck that excluded the PLO and Yasir Arafat from the initial talks while leaving the door open for them to join at some future date.

The opposition railed against participation, certain that the deck was stacked against Israel. The United States, they were convinced, was conniving to protect its oil supply and the Soviet Union inextricably allied to hard-line Arab states. Even when Israel acted moderately—by not launching a preemptive strike, for example—the whole world still condemned Jerusalem, they trumpeted. What was the point, then, of moderate behavior?

Golda was skeptical about the chances that the meeting would yield anything substantive. "One must not expect that we will sit down around a table and have lunch together and after lunch we will sign a peace agreement," she cautioned. But she couldn't resist the opportunity to send Jews to sit at the negotiating table with Arabs for the first time in more than half a century.

The conference itself turned out to be one of those events that were of enormous importance symbolically but utterly irrelevant in practical terms. Eban arrived with a proposed peace treaty in his briefcase only to discover that the Egyptians wanted to put an empty table between him and their delegates. Then he was forced to listen to the Egyptian foreign minister denounce Israel for "exploitation and racist practices."

Finally, he was permitted to deliver one of his typically eloquent speeches, in which he invoked Abraham, "our common ancestor," and ended by quoting from the Koran in perfect Arabic, "If they incline to peace, then turn toward it and put your trust in God."

Then the conference adjourned for the day, reconvened the following for twenty minutes, and adjourned indefinitely.

Beyond Geneva, the domestic backdrop lent the entire election a nightmarish quality. Relatives of the prisoners of war trapped in Syria went on the rampage and smashed the windows of the Knesset building in their demand for stronger action to bring their men home. More than 150,000 reservists—15 percent of the labor force—were still on the borders. And the population still hadn't figured out what moral to derive from the war. Almost 85 percent believed that the Arabs remained committed to destroying them, but more than half nonetheless expressed a willingness to return at least some of Egypt's territory, seemingly in the hope that at least the Egyptians would change their minds and leave them in peace. A week before the election, one-quarter of the voters were still undecided, the proportional representation system leaving them with no attractive choice.

The final tally wasn't very different from what pollsters believed it would have been if there had been no war: Labor won 40 percent of the vote and Likud 28.9 percent, leaving Golda with a rebuke rather than a rebuff, tarnished but not decimated. It was unclear whether such sentiment would survive the peace process.

*   *   *

While Golda struggled to turn that shaky mandate into a majority government, she simultaneously moved Israel inexorably toward a final disengagement with Egypt, and Henry Kissinger gave new meaning to the word "shuttle," dashing back and forth between Cairo and Jerusalem with a dozen aides and his own personal news crew. His goal was an agreement with no political content—no mention of peace treaties or diplomatic relations. Golda, on the other hand, was determined to get something in exchange for Israeli withdrawal, and that something was what she'd been seeking for years: direct negotiations and a peace treaty. Golda was willing to play word games with Kissinger, to call their agreement a pledge of nonbelligerency, for example. She was willing to cede territory. But she was not willing to pull Israel's troops back without something resembling peace.

At its most basic, Kissinger's first shuttle was a gloriously baroque diplomatic extravaganza staged to force Golda to change her mind. Despite the Kissingerian face put on negotiations—the romance of a shuttle that turned the paunchy secretary of state into a diplomatic cross between Lawrence of Arabia and Superman—the outlines of the military disengagement had already been worked out by Golda and Sadat through Yariv and Gamasy at Kilometer 101. Largely ignored by an international media obsessed with Kissinger, the two men not only had hammered out the rules of the supply road to the Third Army but had begun exchanging maps of proposed Israeli withdrawals and buffer zones and dickering about the number of UN personnel needed and the thinning out of Egyptian forces.

But on November 28, Yariv abruptly cut off all discussion. Sadat announced that the agreements being crafted were "not to his liking, led nowhere, and were characterized by Israeli schemes and intrigues." The international media informed the world that a serious breakdown had occurred, raising the specter of new hostilities.

In reality, the breakdown had been orchestrated by Kissinger, who expected to be the star player in the peace drama. "For God's sake, stop the Yariv–al-Gamasy thing," he'd instructed Eban, miffed at the prospect of being upstaged.

That was strike two against Kissinger in Golda's book since the Kilometer 101 talks, Israelis and Egyptian negotiating directly rather than by proxy, was her vision of how peace would be forged. Peaceful relations weren't a matter of a piece of paper but a process of living and talking together, she often said. Kissinger's strategy of interposing himself as the intermediary—forcing both sides to go running to Mommy America— undermined that possibility.

Sadat, however, was obsessed with Kissinger and American mediation. "The United States holds most of the trump cards, since Israel entirely depends on it," he told to a reporter from *Le Monde*.

As he flitted back and forth between Israel and Egypt, Kissinger was forced to conduct two sets of negotiations with Israelis, first with her negotiating team of Allon, Eban, and Dayan, the second with Golda herself, who'd been felled by a nasty case of herpes.

"I guess I make her nervous," Kissinger, quipped, almost hopefully.

Even without the stress of dual bargaining sessions, for Kissinger, dealing with the Israelis was a form of torture. "All our sympathies for Israel's historic plight and affection for Golda was soon needed to endure the teeth-grinding, exhausting ordeal by exegesis that confronted us when we met with the Israel negotiating team," he said.

The Israeli negotiators had plenty of their own complaints about the diplomatic steamroller Kissinger had launched to flatten them, endless late-night meetings, about Germanic histrionics and carefully shaded duplicity. "It's not that he lied," said one diplomat. "He had a unique ability of explaining every situation in the manner most pleasing to the one who heard it."

When Kissinger conjured up the same apocalyptic images he'd painted for Golda in Washington, he overplayed his hand so egregiously and so repeatedly that the Israelis developed their own Kissingerian lexicon. A course described as "suicidal" meant that it would be difficult. A proposal branded as "impossible" was one that Sadat was unlikely to accept. Anything called "difficult" was clearly attainable. And "I'll see what I can get" meant that he had already secured that concession from the Egyptians.

The Israeli public found Kissinger even harder to take. While most told pollsters that they were willing to make concessions in return for life without the threat of conflagration, they felt poised on the edge of an abyss and blamed Kissinger for pushing them ever closer to the brink. IN EGYPT, IT IS KISSINGER, IN ISRAEL, IT IS KILLINGER, read signs that began cropping up at demonstrations. *Ma'ariv* ran a cartoon of an

Israeli soldier racing for an air raid shelter. The caption read, "Kissinger's coming."

With the negotiating team, Kissinger "played the role of an extremely talented teacher put in charge of a class of disturbed children," as one Israeli put it. But with Golda, he was more deferential, unable to resist her craggy face, seeing in it, he told others, centuries of Jewish suffering. From time to time, he showed her some of the hurt he seemed to feel at the public reaction to him. "When I reach Cairo, Sadat hugs and kisses me," he said. "But when I come here, everyone attacks me."

Not known for indulging self-pity, Golda quipped, "If I were an Egyptian, I would kiss you too."

Their encounters turned into a near-mythic clash of the Titans as Golda held out for some semblance of peace while Sadat offered nothing more than a continuing cease-fire. In discussing the use of deviousness in foreign policy, Kissinger once wrote, "I tended to share Metternich's view that in a negotiation the perfectly straightforward person was the most difficult to deal with." No one was more direct than Golda.

Yet Kissinger seemed to crave Golda's approval. "It struck me as strange that this university professor and secretary of state was unable to conceal such a furious affection," commented Syrian president Hafez Assad. Some of that affection stemmed from his appreciation of Golda's sardonic wit. He was particularly fond of recounting the probably mythic State Department story about Golda's first encounter with Robert Kennedy. America is becoming more tolerant, Kennedy allegedly told her. "Maybe in twenty years we'll even have a Jewish president." Golda paused for a moment then quipped, "We already have one. It doesn't help."

In later years, Kissinger admitted that Golda handled him superbly, playing the "benevolent aunt toward an especially favored nephew. So that even to admit the possibility of disagreement was a challenge to family hierarchy producing emotional outrage. It was usually calculated."

Usually the most cynical of skeptics, over time Golda mellowed and grew convinced that Kissinger would protect Israel, although not by allowing Israel to protect itself, as she would have preferred. Longing to wring peace out of the Yom Kippur disaster, she finally took a leap of faith. Defying all expectations, she put aside her demand for a peace treaty or pledge of nonbelligerency and accepted Sadat's offer of a continuing cease-fire.

When Kissinger returned from Egypt the following night at midnight, he brought Golda a clear sign that her faith had been justified. Sadat had agreed to a heavy thinning out of his troops on the west bank of the canal and sent Golda a personal note. "You must take my word seriously," he wrote. "When I made my initiative in 1971, I meant it. When I threatened war, I meant it. When I talk of peace now, I mean it."

Golda still wasn't sure about Sadat, but her reply indicated that she glimpsed promise.

"I am deeply conscious of the significance of the message received by the prime minister of Israel from the president of Egypt," she wrote. "It is indeed a source of great satisfaction to me and I sincerely hope that these contacts between us through Dr. Kissinger will continue and prove to be an important turning point in our relations. I, for my part, will do my best to establish trust and understanding between us."

*　　*　　*

Kissinger was indulging in a not atypical bit of self-aggrandizing humor when he joked about being the cause of Golda's painful bout of shingles. In truth, the stress of dealing with the secretary of state was mild compared to the torture of Israeli politics. Everyone in Israel was quarreling that winter. Reservists like Motti Ashkenazi attacked the government for throwing them into war ill prepared, the widows of the dead turned grief into blame, the families of the prisoners of war trapped in Syria marched in outraged certainty that their men were being abused, and workers walked out on strike over a new war tax. Juvenile crime had risen 37 percent and government economists were recommending a series of draconian measures—the devaluation of the Israeli pound, steep new taxes on luxury imports, and a reduction in subsidies of basic commodities like sugar, milk, eggs, and bread—that Golda knew would only add fuel to the political chaos.

"The state has gone through a devaluation," wrote Joel Marcus, a columnist for the daily *Ha'aretz*, "a devaluation in leadership ability, devaluation in spirit, in values, in morale, in faith."

It was Israel's greatest crisis of confidence since independence, and a disgruntled public took to the streets in a spontaneous series of demonstrations, sit-ins, and teach-ins. The movement had no name, no single ideology. Ultranationalists on the right chastised Golda for her close cooperation with Kissinger. Clamoring for greater flexibility in dealing with the Arabs, the left took out ads in newspapers calling for the entire government to resign. And the vast center simply exploded in angst.

Dayan remained the target. At a meeting of senior military officers, a colonel rose and called for his ouster. Then, during the parliamentary question period, a young Labor activist asked Dayan to account for his actions on the eve of the war.

Within five minutes, the young man was asked to withdraw his question. "Why are you protecting him?" questioner asked. The answer: Golda had threatened to resign if the question isn't taken off the table.

The more vigorously Golda protected her minister of defense, the more the venom against him spilled over onto her. Lines of young demonstrators carried placards around the Knesset: GOLDA, GO HOME. One day she emerged from her office and a woman yelled, "Murderer." Her approval rating fell to 21 percent.

"The streets will not tell me what to do," Golda resolutely informed her aides. With political instincts developed in an era when politics meant parties and party discipline was absolute, Golda had no idea how to cope with popular ferment or party divisiveness that couldn't be defused with backroom deals.

Throwing Dayan to the wolves might have quenched at least some of the popular desire for accountability. But Kissinger pressured Golda to keep him on the team, finding him more pliable than the rest of Golda's crew. And when Golda counted her Knesset votes, she knew she couldn't govern without Rafi, which would surely quit her government without Dayan. She might have made up for that loss by allying herself with a new party on the left, the Citizens' Rights Movement, which had bled off three of Labor's seats. But Golda was angry at the movement's leader, Shulamit Aloni, for defying party discipline. "Golda Meir had many great qualities, but these did not include tolerance of diversity," Eban said of Golda's refusal to embrace the CRM.

Trapped by outdated proclivities and the erosion of her authority, Golda stood by helplessly as Labor disintegrated, the doves threatening to bolt if Dayan remained in the cabinet, Rafi vowing to walk if Dayan didn't, the National Religious Party holding out for a more Orthodox definition of Judaism, and the right gaining strength every day. The longer the crisis endured, the more wary and angry the public became. Rallies of a hundred people became demonstrations of a thousand, and, tone-deaf, Golda couldn't see the writing on the wall.

"I don't think they use sensible methods," she said of the demonstrators. "They picket Labor Party headquarters night and day. They want us to do this and not otherwise. Fine, but I told them, 'Where do we confront each other? Where can I confront you? You give your opinion, I'll give mine and then there will be a democratic decision?' . . . Why should anyone, no matter what fine things he had done, be entitled to bring pressure on me to change my mind. That I don't accept."

But the protesters didn't want to join the party or dialogue with Golda. Their faith in the system of professional politicians making decisions behind closed doors eroded, they wanted change.

By early March, more than two months after the elections, Golda still hadn't formed a coalition behind a majority government, and party stalwarts suggested alternately that she call new elections or form a national emergency government embracing all of the nation's parties, as Dayan was advocating. Golda was dead set against either solution. The former held out the prospect of a victory by the right, and the latter—which she called a "national disaster government"—would paralyze all peace negotiations. Instead, she announced that she would form a minority government.

Even at that eleventh hour, however, she still hadn't figured out who would serve in her cabinet. Dayan abruptly decided that he would not serve "under existing conditions," which most people took to mean that he would pout on the sidelines until everyone apologized to him. "I have asked him, I am asking him, and I will ask him to serve in the new cabinet," Golda wearily told reporters. But she had already chosen his replacement, Yitzhak Rabin, the former chief of operations of the Palmach, chief of staff of the IDF, and current ambassador to the United States.

On the afternoon of March 3, she presented her new cabinet to a meeting of party leaders. With none of the fractures healed, speaker after speaker, from the left and the right, rose to gripe about Golda's refusal to appoint a National Unity government or call new elections, that Dayan should be ousted, or that Rabin didn't deserve to be defense minister.

When the speeches ended, Golda unleashed her exasperation. "Had I resigned last August or September, I would have been taking a wise step. I've sinned for the last forty-five years by allowing myself to paper things over. Under pressure from comrades and myself, I thought something needed to get done in this country and that the comrades thought that by putting Golda in charge, they'd overcome the internal conflicts. That's all over. The trick of Golda the Paperhanger doesn't help anymore."

The audience was riveted. Golda's venting had always been a fine show. But they were not prepared for the finale.

"Tonight, I shall hand my mandate back to the president because I have finished setting up the government," she announced. "I hope you will inform the president of the next candidate to lead the government." Then she picked up her purse and walked out, the party secretary-general running behind her yelling, "Golda, wait, Golda, come back."

That night, the usual delegations rushed to her residence, where she was watching the news of her resignation on television. The party Central Committee convened and begged her to remain, and party leaders urged the president not to accept her resignation. Assuming that this time Golda was serious, the media erupted in an orgy of speculation about new elections, a caretaker government, and the future of Labor.

Golda desperately wanted to be rid of the burden, of the feuding and nitpicking, but she had never run away from a fight and desperately wanted to bring the Israeli prisoners of war home. So after the full drama of repentance played out, she agreed to give coalition building one more try.

Less than a week later, she again presented her new government. At the last minute, Dayan conjured up a military intelligence report intimating that Syria was about to resume fighting. Ever the patriot, he humbly offered to remain in office so the country wouldn't be left in the lurch. No one was surprised, of course, although Likud asked why, if Israel was facing such serious danger, they had not been informed.

Golda had resolved the political crisis the only way she knew how, by wheeling and dealing, totally ignoring "the street," and it couldn't last. The public was too angry, and Labor too fractured, to maintain the discipline necessary to pull itself through.

Golda tried, attempting to calm the stormy national political waters by publishing an interim report from the Agranat Commission, which cleared her of any responsibility. She acted, they wrote, "properly and wisely." But they also exonerated Dayan, adhering legalistically to Israel's principle of governmental collective responsibility. By that standard, any government minister was blameless for actions taken that conformed to the advice of his principal advisers. With no such quasi-legal principle to shield them, then, Israel's military officers alone bore the full burden of blame.

To Israelis, the exoneration of Golda was understandable. What would an old lady know about defense matters? But while the commission concluded that the minister of defense was not the "super chief-of-staff," and thus insulated by the collective responsibility principle, the public knew better. Dayan wasn't just the minister of defense; he was Moshe Dayan, certified war hero. And if the men under his command—which is how Israelis thought of Dado and the others on whom the commission laid the full measure of responsibility—were going to be ousted, the least Dayan could do was to fall on his sword and follow them.

When he did not, the political squall gathered force into a gale. Rabin, who'd been shunted into the Ministry of Labor, recommended that the cabinet reject the commission report and send it back to be rewritten without blind consideration for ministerial responsibility. Polls showed that dissatisfaction with Golda was at 67 percent. Likud was demanding that the entire government resign. And Golda was caught in another turmoil for which she was uniquely unprepared.

This time, her consternation was complicated by a new bit of pique, over the growing national accusation that reliance on her kitchen cabinet might have imperiled Israel by keeping a full range of opinions off the table. It was an absurdly naive complaint given how informal her cabinet was and how obediently the ministers bowed to her will. But its weight reflected a deeper disquiet over the inbred political culture symbolized by her coffee klatches.

In early April, less than a month after Golda finally formed her government, Likud moved toward forcing another no-confidence vote and the Labor Party leadership met to consider whether the entire cabinet should resign to allow Golda to move Dayan out, or at least shift him to another position. "You realize that it won't end with Dayan," Yosef Almogi, the mayor of Haifa and a Rafi member, said to Golda, thinking that his caution would forestall Dayan's ouster. "They're really aiming at you."

"You're telling me?" Golda responded, her relief starting to surface.

All during the negotiations with Kissinger and the party, Golda had been rising at the crack of dawn to undergo 6 A.M. cobalt treatments at Hadassah Hospital, the most carefully guarded of her secrets. She was exhausted and her ambition, a strange blend of traditional careerism and idealism, had been sated. Remaining in office had become a debilitating duty rather than a need, and she was both frustrated and angry at being alternately pressured to remain at the helm and left without the support she needed to stay there.

On Thursday, April 10, dressed in a simple black coat, Golda rose before a meeting of the party leadership. "Five years is enough," she announced. "It is beyond my strength to continue carrying this burden. I don't belong to any circle or faction within the party. I have only a circle of one to consult, myself. And this time my decision is final, irrevocable. I beg of you not to try to persuade me to change my mind. . . . It will not help."

This time, as weary of the process as Golda was of office, the delegations arrived at her door more discreetly, although Sapir and Allon were so worried about finding a replacement that they stooped to asking Golda's children to persuade her to change her mind.

"I am very tired," she told her son, Menachem. "Please remember that I have really had enough."

*   *   *

The ending was anything but glorious. Golda couldn't simply pack up her office and slink off to Tel Aviv. Someone had to run Israel until the Labor Party could stop feuding long enough to choose a new leader and until that new leader could negotiate a coalition with the very parties that had stymied Golda for months.

Still a woman of fierce dislikes and long memory, she quietly threw herself into thwarting the candidacy of Shimon Peres, the one party figure she'd longest despised. To stop him, she turned to Yitzhak Rabin, the only leading name in Labor untarnished by the Yom Kippur War. Rabin was too much the technocrat for Golda's taste. But Eshkol had promoted him as a counterweight to Dayan, and he was a familiar figure in Washington. Most important, he was not Peres.

By a narrow margin, Rabin squeaked into office as the head of Labor, thus as prime minister, and immediately immersed himself in the nightmare of coalition building while Golda went back to the bargaining table with Kissinger to end the continuing roar of cannons across the Syrian lines. Since the Israeli-Egyptian agreement had been largely worked out at Kilometer 101, Kissinger's first shuttle had been brief, mostly a publicity stunt. But negotiations with the Syrians were an endurance test. Unlike Sadat, whose army had captured territory from Israel, President

Hafiz al-Assad could recoup his honor only with a victory at the bargaining table.

When Kissinger arrived in Israel to begin his Syrian shuttle on May 2, he made it clear that he expected Israel to return not only all the land captured in October, but also at least a symbolic slice of the Golan Heights, which Israel had held since 1967.

"Two wars in seven years, with the price we paid for it," Golda spit back. "Then Assad says he must get his territory back. I mean, that is chutzpah of the nth degree." His chutzpah was greater than Golda knew since Kissinger had shaved a considerable amount off the Syrian demand.

Kissinger chided Golda with a reminder that the Golan Heights had always belonged to Syria. "I didn't just get up one day in 1967 after all the shelling from the heights and decide to take Golan away from them," she reminded him angrily. "They say this is their territory. Eight hundred boys gave their lives for an attack the Syrians started. Assad lost the war, and now we have to pay for it because Assad says it's his territory."

The Israelis had no intention of holding on to their recent gains, which had brought them within twenty-five miles of Damascus, and Golda had already decided that Israel could afford to give up a slice of the Golan, the town of Kuneitra, the old provincial capital. An inveterate bargainer, however, she began by offering to split the town with the Syrians.

Even that concession provoked a spate of demonstrations. Prominent writers and university professors staged a hunger strike outside her home, declaring, "Our settlements on the Golan are the ramparts of Israel." Her office was flooded with petitions, with protests from Begin and some veteran kibbutzniks, all acting as if giving up some dusty fields on the northern border was tantamount to ceding Jerusalem.

Kissinger understood that Syria was intractable, so with little leverage over Syria, he vented all of his well-staged frustration on Golda, who he'd already discerned was more pliant than she seemed. "I am wandering around here like a rug merchant in order to bargain over one hundred to two hundred meters," he yelled. "Like a peddler in the market! I am trying to save you, and you think you are doing me a favor when you are kind enough to give me a few extra meters!"

Golda knew that Kissinger didn't scream at Assad like that. But there was a limit to how angry she was willing to get at him. Her penchant for never changing her mind about anyone was the stuff of legend in Israeli political circles. "She didn't forgive easily," said Ari Rath, former editor of the *Jerusalem Post*. "Once she had a grudge against someone, that person had to bear that cross for many years."

But Kissinger had become the rare exception. For decades, really since Morris refused to remain in a marriage over which he had no control, no man had stood up to her with such fervor. Kissinger frustrated her. And she knew that he mocked her to Assad, calling her "Miss Israel" and "the beauty of Jerusalem." But vulnerable and brokenhearted, Golda needed someone to help her end the agony, and Kissinger was her only option.

The day he brought her the list of Israeli prisoners of war in Damascus, she melted. After that, she still blew up every time he sought more concessions, but she cooled off more quickly. By the end of the process, when she'd already agreed to cede all the territory gained during the war, as well as the city of Kuneitra, and they were down to details about the number of Syrian troops who would be permitted in the limited-forces zone, Golda actually stopped haggling entirely. "Go get the best numbers you can," she told him, a leap of faith no one had ever imagined Golda was capable of making.

Nonetheless, the final agreement with Syria was almost derailed at the last minute when Palestinian terrorists snuck into the town of Ma'alot in northern Israel, captured four teachers and ninety schoolchildren, and threatened to blow them all up if Israel didn't release dozens of imprisoned guerrillas. When French and Romanian mediation attempts broke down, Israeli troops had little choice but to storm the building where the captives were held. They succeeded in killing three of the kidnappers, but, by then, twenty Israeli teenagers, mostly girls, were already dead.

Everyone knew that the terrorists received Assad's support. So thousands of Israelis, including a hefty swath of the government, encouraged Golda to break off all discussions with Damascus. Rejecting such intransigence, Golda suspended the talks for a single day. Terrorism, she said, was a symptom of peacelessness.

"It would be illogical to renounce the cure and allow the fever to pursue its deadly course," she announced. "We had all better get back to peacemaking."

CONCLUSION

You ought to thank us, your elders, that we've
left you still something to do.

By Thursday, November 17, 1977, the Israeli government was in a tizzy. Local flag manufacturers couldn't churn out Egyptian flags in time for Anwar Sadat's arrival. The military band had no sheet music to rehearse "By God of Old, Who Is My Weapon," the Egyptian national anthem. Protocol officers were unsure whether it was appropriate to offer a twenty-one-gun salute to the leader of a nation with which Israel was technically at war. And it still wasn't clear whether U.S. president Jimmy Carter, who'd been trying to prevent an Israeli-Egyptian rapprochement lest it undermine his comprehensive Middle East peace settlement, would scuttle the entire trip.

Golda watched the ruckus from New York, where she'd flown for the gala opening of *Golda* on Broadway, oozing skepticism. Despite her cordial exchange of letters with Sadat after the war, she remained heavily invested in the cynicism she'd brought to every hint of progress with the Egyptians, and nothing that had occurred in the four intervening years had dented her incredulity.

It had been a rough time for a woman who'd thrived on self-confidence and could no longer conjure it up after the Yom Kippur War. Rabin, who'd followed her as prime minister, had concluded the negotiations with the Egyptians that she had begun. But in May 1977, for the first time in Israel's history, the Labor Party that she'd help build and nurture went down in defeat, sweeping her old adversary Menachem Begin into power.

Until he withdrew from the National Unity government she'd inherited from Levi Eshkol over her decision to cooperate with William Rogers, Golda and Begin had overcome many of the enmities of the old *yishuv* days. Begin openly called her "a proud Jewess," and, behind his back, at least, Golda admitted that he was a gentleman. Still, the election of the man she'd long considered an irresponsible demagogue and whose assassination she'd advocated four decades earlier was a bitter pill to swallow. While she'd grown to admire Begin's integrity, she still abhorred his politics. Even closing in on the age of eighty, Golda still clung to her belief in socialism and in Labor Zionism as the soul of Israel.

Golda wasn't reeling merely from the political change in Israel. She remained Israel's most prolific overseas fund-raiser and a political heavyweight within Labor. But a woman who'd thrived on self-confidence, she no longer could conjure it up. Not only had she been rejected by a nation that blamed its long-beloved grandmother/prime minister for its lost innocence, but she was also battered by an eruption of criticism and scandalmongering from dozens of ministers, ex-ministers, and ministerial wannabes, backed up by much of the local press corps.

One former Knesset member breathlessly revealed that Golda had personally ordered him to slow down the immigration of the Iraqi Jews in the 1950s, an excruciating bit of libel against a woman who'd staved off those who had advocated that she do just that. Teddy Kolleck, the mayor of Jerusalem, accused her of putting her political party before the welfare of the nation.

"Doctrinaire and obsessed with the trappings of power, she believed that only she was right about any subject under discussion," wrote Chaim Herzog, who went on to become president of Israel.

The target of everyone's discontent, for the first several years after her retirement Golda had found relief from the barrage of negativity in the embrace of foreigners and in trips overseas, where she was still lionized.

"So you're tired," columnist Mary McGrory had written after Golda's resignation. "And believe me, you're entitled. But that's a reason to quit? We goyim had understood that Jewish mothers never resign. . . . But if you really mean it this time . . . had you thought of what you might do after? What I'm getting at is this: Have you considered coming back here? I'm not talking Minneapolis. I'm talking Washington, D.C."

William S. White, a nationally syndicated American columnist and Pulitzer Prize winner, had compared her to Winston Churchill, although he admitted that she is "the most earthy of commoners and he was the most aristocratic of aristocrats. But in the things that matter you don't lie down when your country is under attack, you don't go about begging for votes on your knees and you don't cry out for the chaplain and the medic. Golda and Churchill—these two formed a pair of aces indeed."

In London, she still played Albert Hall to an overflow crowd, had tea with Prime Minister Harold Wilson at 10 Downing Street, and was honored by members of Parliament. In Washington, Kissinger threw her a black-tie "family dinner," as he called it, with nine justices of the Supreme Court, a host of congressional heavyweights, and a smattering of Hollywood in attendance.

President Gerald Ford set aside time to chat with her at the White House. And when Jimmy Carter was running for the Democratic nomination for president, he asked to meet with her in New York, hoping that being seen with her would help him woo Jewish voters. (She agreed to the meeting. But suspicious of Carter from the first, she declined to make a public statement about the encounter or to permit photographs to be taken that might be used for his campaign.)

When she was at home, foreign dignitaries regularly stopped by to visit, as if she were a queen in exile. Walter Mondale came to call, and Princess Beatrix of the Netherlands dropped in on her seventy-ninth birthday. "I can say to this audience here gathered in the Knesset in Israel that no leader I have met, no president, no king, no prime minister, or any other leader has demonstrated in the meetings that I have had with that leader greater courage, greater intelligence, and greater stamina, greater determination, and greater dedication to her country than Prime Minister Meir," declared Richard Nixon when he flew in to meet with Prime Minister Rabin.

Even Sadat jumped on the Golda bandwagon, suggesting that Israel's new prime ministers paled by comparison. "I prefer dealing with a strong leader like Golda," he opined.

No matter where Golda traveled, the press besieged her, eager to snatch another pithy or sarcastic quote, and she was rarely shy about offering one. "I wouldn't want the West Bank even if were given to me as a present," she remarked on her eightieth birthday. When Moshe Dayan proposed that six new Jewish towns be built in the occupied territories, she told journalists, "I don't want another million Arabs who don't want us."

Her autobiography, ghost written by a publicist at the Weizmann Institute of Science, was a runaway best seller around the globe, producing a flood of adulatory fan mail. After years of refusing to authorize the production of any dramatic work based on her life, she finally succumbed to the pleas of the Theatre Guild of New York for permission to adapt her book to the stage—disappointing British songwriter Lionel Bart, of *Oliver* fame, who had hoped to turn her life into a musical. Instead, Bill Gibson, who wrote *The Miracle Worker*, a play that Golda admired, was tapped to write the script, Arthur Penn to direct it, and Anne Bancroft to star as Golda herself.

That agreement was yet another decision Golda came to regret. At the age of eighty, she'd begun slowing down and hadn't traveled abroad for more than a year. But with an Israel Bonds gala arranged for the play's opening at the Morosco Theatre on November 14, she invited a large group of family and friends to fly with her to New York, only to sit through the convoluted and confusing production in mounting horror. If I'd looked and sounded like Bancroft, I would never have been elected prime minister, she summarily informed the cast.

In the midst of that humiliation, Sadat announced to his parliament that he was willing to go "to the ends of the world" in his quest for peace. "Israel will be astonished when it hears me saying now, before you, that I am ready to go to their house, to the Knesset itself, and talk to them," he declared.

Golda had made similar offers on dozens of occasions and been thoroughly ignored. When Sadat suddenly stepped up to the plate, however, the world breathlessly lauded his audaciousness.

You must fly home immediately, Golda's friends urged her. But worried that she couldn't meet Sadat graciously, and hurt that she had not received a formal invitation from the government, she refused to pack her bags. "I'm not going to have people say, 'Look who's here!'" she exclaimed. "If I'm invited to meet Sadat, fine. If not, not."

The invitation finally came, and Golda flew through the night to arrive home in time to greet her old enemy. She landed in an Israel caught in the thrall of a global political drama that would culminate with the meeting between Begin, the old Jewish terrorist who'd finally become prime minister, and the Egyptian leader who'd sent his army across the Suez Canal. Taxi drivers had attached Egyptian flags to their antennas, and vendors sold T-shirts emblazoned with photographs of Sadat and Begin, reading, ALL YOU NEED IS LOVE.

"What's all the Messianic euphoria?" Golda asked cynically.

Golda arrived at the airport just as Sadat's plane was circling Tel Aviv in preparation for landing, thousands of Israelis standing in the streets and applauding. Emerging onto the red carpet laid across the tarmac and a fanfare of trumpets heralding his arrival, the Egyptian president made his way down the long receiving line, shaking the hands of the generals with whom he had long done battle.

When he reached Golda, the crowd froze. Before he left Egypt, Sadat's wife, Jihan, had warned him, "Mind your manners, especially with Golda Meir."

Sadat smiled, clearly tickled. Taking Golda's hand, he said softly, "Madam, for many, many years I wanted to meet you."

Golda returned his graciousness. "Mr. President, so have I waited a long time to meet you. Why didn't you come earlier?"

Sadat paused for a moment and replied, "The time was not yet ripe."

As Egypt and Israel moved toward the formal peace that had eluded her, Golda was haunted by that comment, angry and hurt that she was blamed for the continuing hostility when Sadat himself acknowledged that the season of peace had only just arrived. That it had come during the prime ministership of Begin, who had thrown up so many roadblocks to her peace efforts, stung so deeply that she never seemed to grasp that only a Begin could have been the architect of settlement with Egypt, just as only a Richard Nixon could have made the long journey to China. The announcement that he and Sadat had won the Nobel Peace Prize, then, was a devastating blow. "I don't know about the Nobel Prize, but they certainly deserve an Oscar," she remarked caustically, by then too tired and too demoralized to hide her chagrin.

By the time Begin and Sadat flew to Oslo for the prize ceremony, Golda had already been in and out of the hospital for almost four months with what doctors called a "viral illness." She could no longer see, and her liver was failing. On Friday afternoon, December 8, less than twelve hours before her two old adversaries were to receive perhaps the highest honor on the planet, Golda died, grabbing the headlines from them both.

* * *

Golda had a dying wish, a plan to clear her name that she'd pursued during the long months of Begin's rise as the great peacemaker, a grand gesture she tried to make even from her deathbed. Initially, she thought her old confidant Yisrael Galili would be her agent and that he would author an article that would vindicate her. Then she decided that calling a press conference to answer her critics directly would have more impact. Realizing that reason needed to be backed up by facts, she delayed inviting journalists in for that chat so that Eli Mizrachi, who'd worked in her office, could gather all the documents necessary to prove her case. But Begin dithered in approving her request for access to classified materials until Golda's health overtook her efforts.

It fell to Mordechai Gazit, director general of the prime minister's office, to disabuse Israelis of their tendency to confuse Golda's obduracy with Sadat's flexibility. A prominent dove, Gazit was an ironic agent of her defense. When Golda had first welcomed him into her office, she'd joked, "We do not always agree." Gazit interrupted her. "We never agree," he'd interjected.

But while Gazit churned out a meticulously researched and argued spate of pamphlets and articles about Golda's efforts as a peacemaker, no one in Israel was interested in facts. Dayan's disastrous prewar overconfidence and subsequent histrionics were conveniently rewritten out of the public mythos when he joined Begin's government as foreign minister, and Shimon Peres' adventurism, whether in French Guiana, France, or Dimona, largely forgotten. But Golda remained firmly engraved on the public consciousness as the standard-bearer of what Doron Rosenblum of Ha'aretz calls "the horrific Golda Syndrome." The symptoms he describes as: "arrogance toward and patronization of the Middle Eastern environment; an uncontrollable urge to be didactic; a blind spot that makes a Palestinian political presence completely invisible; and primarily endless self-righteousness, which sees everything in black and white—we are always right, the evil is entirely our enemy's, and everything is a justification for maintaining the status quo."

That simplistic portrait is a convenient and comfortable exercise in historical revisionism even if it doesn't quite rise to the level of historical accuracy. Fond of "what ifs" and "if onlys," Israelis conjure up imaginary scenarios of the peace that might have been—or the war that might not have—"if only" Golda hadn't been too stubborn or too blind to breach the wall of Arab hostility. But it rests on a few key moments—peace opportunities, they call them—and the belief that the Anwar Sadat who extended his hand to her in Jerusalem in 1977 was as ready for that historic gesture in 1972 or 1973.

By his own clear admission, however, all of Sadat's alleged peace gestures were contingent on the same precondition: that Israel commit itself in advance to yield every centimeter of land taken from Egypt, Syria, and Jordan in the 1967 war, including Jerusalem—and virtually no Israeli back then, and few today, was willing to concede all that territory.

"No Arab government offered to sign a partial agreement with us . . . without a commitment with a time table for full withdrawal neither she [Golda] nor the government nor any government 'til today was ready to make," explained Simcha Dinitz, former Israeli ambassador to Washington. "This is the basic mistake that all the supposed learned arbiters make."

Still, judging Sadat by his behavior in the late 1970s, the revisionists refuse to accept that he would not have budged from that absolute position before the Yom Kippur War. But every bit of evidence, including the testimony of all three of his foreign ministers and his wife, supports Sadat's own contention that the "time was not ripe."

"I do not agree with those among us and among you who assert today that Sadat tried to achieve a real peace before 1973," said his wife, Jihan. "Sadat needed one more war in order to win and enter into negotiations from a position of equality."

For most Israelis, however, Jihan Sadat's view is irrelevant, as is Sadat's clearest statement, that he was "convinced that she [Golda] could have done better and gone farther than Menahem Begin." Golda didn't know how to promote peace, they insist. Golda was an inflexible old lady incapable of compromise. All she offered the Arabs was an impossible choice between maintaining the status quo and humiliating capitulation.

Golda was rarely as intransigent as the mythologists have portrayed her, of course. She might have talked tough, but in the end, she was a pragmatist. So while she long railed against rapprochement with the Germans, she was the prime minister who accepted the appointment of Ambassador Pauls and invited Willy Brandt to Jerusalem, the first sitting German chancellor to receive such an invitation. Despite her dislike for Begin, she bowed to political necessity and included him in her National Unity government. And she was the architect of dozens of compromises that kept the Labor Party from splintering into dozens of fragments of ego and ideology.

In her quest for peace, she repeatedly met with King Hussein to forge a de facto peace. Time and again, she reached out to Nasser, and she agreed to every request from Marek Halter, the Romanians, and the Russians when they tried to bring her together with Sadat. As an opening to Sadat, she turned her back on the decades-old Israeli principle that they would never give up land for a partial peace. And by early 1973, she dropped her demand for direct negotiations with her enemies.

Golda was certainly not without her blind spots. Along with all the other early Zionists, she was cavalier about the seeds of anger and frustration that would spring up when they planted a Jewish homeland on someone else's soil, about the colliding nationalisms that inevitably poisoned the soil, and the difficulty of removing the toxins from such a harvest.

Her vision of the Palestinians was hopelessly clouded, although during her time in office, she shared that myopia with much of the planet. On the world stage, Israel's conflict was then seen as a struggle between Israel and the Arab states, between the Jews and the Egyptians, Syrians, Jordanian, and Iraqis. At the United Nations and at multilateral meetings, the Palestinians were still bit players.

Inevitably, perhaps, Golda measured the treatment of Israel against the backdrop of the history she had lived and witnessed. Scores of countries fought wars that created refugees, so why was Israel alone singled out as the oppressor when it didn't even fire the first shot? she asked. What about all the land taken from Jews in Arab countries and Germany? Shouldn't it be returned, or the descendants of the former owners be given a country of their own? So what if they had never constituted a nation before? Neither had the Palestinians.

But forging logic on the anvil of history rather than hammering it out by the flow of change trapped Golda in the past, in the terror to which she'd been raised, in nightmares and forebodings, and, ironically given her commitment to building a future in which Jews need no pity, in Jewish victimhood. Shortly after the 1967 war, on the floor of the Knesset, Shulamit Aloni argued that the Arabs on the West Bank considered the Israelis to be conquerors. Golda "stood up and said, 'How dare you say Jews are conquerors! We Jews are always the underdog,'" recalled Aloni. "She never went through a metamorphosis that today we are no longer a minority in a ghetto needing to build safeguards around us. . . . From her point of view, Jews were refugees, underdogs. . . . She lived with the fear of our being a minority and needing safeguards. The feeling of being an underdog that she brought from Russia and the Holocaust, this was her approach to everything."

Outsiders called it paranoia and cast Golda's as a Masada complex, such a powerful siege mentality that she saw Israelis on the brink of the impossible choice faced by the first-century Jewish rebels. "It is true we do have a Masada complex," she admitted, never ashamed of falling back on the lessons of the past. "We also have a pogrom complex. We have a Hitler complex."

She omitted perhaps her deepest complex, her Czech complex, reinforced daily by Arab rhetoric about driving Jews into the sea and Yasir Arafat's pronunciamentos that Arab Palestine must occupy the whole area from the Jordan to the Mediterranean Sea. "We are not Czechoslovakia and we do not want a Munich played on us," she declared, recalling the fatal meeting when Czechoslovakia was dismembered by European powers appeasing Hitler. "Before Czechoslovakia there was Spain. I remember as if it were yesterday a congress . . . which took place in London in March 1937. A representative of the Spanish Republic, after having in vain asked for aid, pointed his finger towards the Czech delegation, then at the Austrian delegation and said, 'It will soon be your turn and yours and yours.' It is exactly the same message that I am holding today at the disposal of the governments of Europe."

*   *   *

If Israelis were willing to admit to any tragedy in Golda's story, they would limn hers as a tale of Grecian proportions in which Golda's overarching hubris engulfed both her and them, to catastrophic effect. In that story line, a Golda deeply scarred by the powerlessness of Jews in Russia had all the right stuff—grit, doggedness, principle—to help dream a country into being and, as the shrewdest fish in a sea of political sharks, to domi-nate it. But with limited creativity and almost no tolerance for dissent, she glossed over deep social and ideological divides by force of personality. Her political repertoire was limited to bossing, charming, and cajoling, bereft of an ounce of finesse. Confusing party with state, she put the for-tunes of the Labor Party above those of the nation. And allergic to leaving anyone with so much as a smidgen of honor, she lacked the subtlety es-sential to a peacemaker, to devastating effect.

It's easy to point to the page on that script where the heroine began to look like an intractable old lady rather than a gutsy grandmother shield-ing her young. But that wasn't a moment in real time. The viewer sees it now because history, with its niggling tendency to edit reality and to shift moral values and perspective, unfolds in such perfect cinematic dissolve that it's all too simple to forget that we are no longer who we were or what really was.

Golda's life story doesn't turn on the essential elements by which Sophocles or Euripides turned their screws; she was as much poisoned by her environment as it was by her. The loyalist of party apparatchiks, she was trained to a political system that was anachronistic even before the establishment of the State, and she rose in that rigid hierarchy as its most persuasive international voice and its most prodigious fund-raiser. When she was already an old woman—too old, too thin-skinned, and too steeped in an outdated view of Israel, Jews, and Arabs—she was tapped to serve as prime minister, not because she was the most talented leader, the wisest,

or the most in tune with the Israeli zeitgeist, but because Labor was paralyzed by out-of-control egos, and Israel by crippling indecision.

Her people adored her for all the wrong reasons—for how safe her
towering strength made them feel and for the aplomb her edgy wit lent
them—rather than because they heard their own hopes and dreams reflected in her exhortations about socialism, equality, and self-sacrifice.
While she was celebrated across the planet as the first personification
of strong female political leadership, on the most pressing international
issue—the alarming rise of terrorism—she was cast aside as Cassandra
despite what history has shown to be her prescience. In her every attempt
to move Israel toward peace, she was hemmed in—by the great game between the United States and the Soviet Union and by Israel's political
landscape as much as by her own obduracy.

And despite the reality that her nation's political paralysis constrained
her from accomplishing much of what she longed to do, she was nonetheless forced to stay in office well beyond her time because there
seemed no other way for her to protect a nation at risk, from its neighbors, its refugees, its economic precariousness, and its own contentious
divisions.

A woman of greater wisdom might have resigned and let the younger
generation battle it out, no matter the cost. A leader of foresight might
have told her people everything they didn't want to hear, that the situation was not sustainable, that a dozen problems were woven into the national fabric, and that they were living on quicksand. A creative prime
minister might have devised new approaches to everything from ethnic
divisions to peacemaking. And an innovator might have burst the national bubble of arrogant self-confidence by explaining that the political
system was ossified or acknowledging that Israelis were not, in fact, the
new superheroes.

But the Israelis of the early 1970s weren't looking for wisdom, foresight, creativity, or innovation. They didn't want to be challenged; they
were thrilled to be led by a woman who amplified their smugness. Not-so-

silent coconspirators to the inflexibility and dogmatism they now deride, they lauded their prime minister, heaped her with the kind of approval ratings few world leaders could begin to hope for, and cajoled her to remain in power well beyond her day.

"Israeli society, for thirty years, a full generation, refused to look in the mirror," recalls Yaakov Hasdai, a researcher for the Agranat Commission and founder of Laor. "It looked for any explanation, hopes or dream that would allow them to sleep. . . . Golda was alone at the top. All around her were young Israelis, heroes, each taught to be Superman, and she thought they were."

When the Yom Kippur Earthquake rocked the national psyche, Israelis lashed out at Golda for not sharing their qualms or for allowing them to fester. She ceased to be the courageous leader who led the *yishuv* when all the male leaders were locked away, who faced down the British, raised the money to save the new nation from annihilation, created housing for thousands of new immigrants, drafted the first labor and social security laws, brought them under the protective wing of the United States, and kept faith with their spirit even when their bravest hero's belief in them flagged. She became Golda the Intransigent, who let them down.

After all, it was easier to rewrite history to blame everything on an old lady who was dead than to face inconvenient truths. Golda's fate, then, was decreed not by the gods, as in ancient Greece, but by historical circumstances, and the hubris of her tale is as much Israel's as hers.

*   *   *

In this age of the nonsectarian televised confessional, the tragedy of Golda Meir seems more personal than political. While she became the "somebody" her mother informed her that she never would be, Golda paid an enormous price for her success. Admired by millions at a distance and able to hold audiences spellbound with her American-accented Hebrew and her Yiddish-tinged English, she was such a lonely woman that her security guards took to calling her friends and acquaintances so she wouldn't languish without company during the long nights when sleep eluded her.

She died with a photograph of Morris still on the table at her bedside although he'd left her more than three decades earlier after finally admitting that he would never come first in her life. And having repeatedly become involved with men who put her on a similar footing, in her old age, she found little comfort. A woman of passion, she knew how to charm, but she couldn't give herself over to intimacy.

While her daughter, Sarah, staunchly defends Golda as prime minister, even two decades after her death she can never quite bring herself to defend her as a mother. "Would Milwaukee have really been such a bad place to live?" asks her son, Menachem, whose lack of enthusiasm for the country his mother built is sadly telling.

Golda herself wasn't without ambivalence toward her creation. She'd never imagined that a Jewish homeland would not become the world's beacon of freedom and justice, that, like any other country, it would need policemen and soldiers, an antipoverty bureaucracy, or lawyers to investigate government corruption. Or that despite her hectoring, her egalitarian utopia of idealist pioneers would turn into a dog-eat-dog capitalist society rife with consumerism and greed, and that she was too busy being a war president to devote much attention to turning that tide. She gave up any semblance of a personal life, then, for a country that turned its back on the very principles that drove her.

But if she'd been inclined to introspection, which she decidedly was not, Golda, child of a generation that had no patience to indulge self-pity, would never have cast her tale as a tragedy, or even as the story of an individual. Despite her political prowess, she didn't give up her life as an ordinary woman because she wanted to become a politician. She was a true believer. "She had faith when others wavered," Ben-Gurion once said. "She believed in the absolute justice of our cause when others doubted."

No matter how anyone else, now or then, measured her life, Golda had an unwavering standard for herself: "I can honestly say that I was never affected by the question of the success of an undertaking. If I felt it was the right thing to do, I was for it regardless of the possible outcome."

ACKNOWLEDGMENTS

Few clichés have a firmer basis in reality than the one that suggests that writing is a collaborate endeavor. Despite the image of the driven author locked away in the lonely grip of her muse, the truth is that neither writers nor their work thrive in a vacuum and that we are utterly dependent on an enormous cast of characters.

Golda came alive for me, flaw by flaw, triumph by triumph, thanks to the patience and generosity of scores of individuals in Israel and the United States who shared their memories. I am particularly grateful to Yigal Lossin, Raphael Rothstein, Ari Rath, Rinna Samuel, Nomi Zuckerman, Lyova Eliav, Meron Medzini, Yehuda Avner, and Aharon Yadlin. And I found my way to most of them thanks to the indefatigable Irit Pazner, without whom I could never have navigated my way through an endless list of pressing interviews.

For research assistance, I thank David Fachler, Chevy Weiss, Gennadii Kostyrchenko, and Andrew Blahnik, as well as the staffs of the National Archives of the United States; the Israel State Archives, especially Hagai Tsoref and Louise Fischer; the University of Wisconsin–Milwaukee; the library of the *Jerusalem Post*; the Central Zionist

Archives; the Oral History Division of the Institute for Contemporary Jewry at the Hebrew University; the Dorot Jewish Division of the New York Public Library; the Rasmuson Library at the University of Alaska–Fairbanks; the Stamford (New York) Village Library; and the Nelson Riddle Archives at the University of Arizona. And a special *spasiba bolshoii* to Artyom Zhdanov, who ensured that the photographs for this book were crisp and, as always, shielded me from mangling the Russian language.

I am grateful to my colleagues at the National University of Science and Technology in Bulawayo, who stayed on strike long enough in early 2006 for me to finish this book without pressure, and to Cleophas Muneri, Gibbs Dube, Vusa Maphosa, Hayes Mabweazara, Stenford Matenda, and Lawton Hikwa for welcoming me into the most collegial academic environment imaginable.

The Jewish community of Bulawayo swept me in with open arms, shared Golda stories with me over dinner at the Feigenbaums, kept me laughing at the Rubensteins, and grounded me at the Laskers. The friendship of Shelley, David, Benji, and Carly Lasker; Ida and Harry Shmeizer; Ivor, Pauline, and Zara Rubenstein; Ruth and Alan Feigenbaum; Elsa and Ray Roth; Brian Sher; Henry Sommer; Jason Roth; Elsie Furman; and Tanya Goldwasser kept me in touch with the best part of Golda.

Lisa Bankoff has now guided me through nine books with good grace, or at least as good as possible given the rising tensions, and Tina Wexler talked me down with the greatest of aplomb. Claire Wachtel provided the seed for this book. I am immensely grateful to them all.

Without my extended family, I couldn't stay grounded enough to function, not to mention to write. So thanks aren't enough. I've already overused the word "gratitude." And "appreciation" is an absurdly dry locution. Frank Bruni, Patrick Wright, Bruce Conroy, Cheral Coon, Jennifer Collier, and Michael Jennings: I'm relieved you don't need words since I don't have any.

For the seventh time, my husband, Dennis Gaboury, bore the brunt of the stresses of the bizarre nature of my work, steering us around Israel through the mazes of motorists trained in offensive driving; enduring repeated readings of a single paragraph; wending his way through piles of footnotes with me; and serving as the whipping post when the words wouldn't come but the deadline was approaching. Partners are the unsung heroes of publishing, and Dennis is most certainly mine.

Elinor Burkett
Bulawayo, Zimbabwe

NOTES

INTRODUCTION

3   "I remember so well": Richard M. Nixon, *RN: The Memories of Richard Nixon* (New York: Simon & Schuster, 1990).

4   "another state who is also a woman": September 25, 1969, Public Papers of Richard Nixon.

4   giggling about the flowers: Lou Kaddar interview, (56)88, Oral History Division, Institute for Contemporary Jewry, Hebrew University of Jerusalem (OHD).

4   At the National Press Club: Golda spoke at the National Press Club on September 26, 1969.

5   "We're not so fortunate": *New York Times*, September 28, 1969.

5   "You don't have to be Jewish": *New York Times*, September 30, 1969.

5   Golda responded with her coy: Ibid.

6   "such a sad face": Ibid.

10 Elie Wiesel: *New York Times*, October 7, 1969.

CHAPTER ONE

13 the bells of Kishinev: Andrei Shapiro, "A Singular Event in Jewish History: The Kishinev Pogrom," *Hagshama* (World Zionist Organization), 2003.

16 "Hebrew" immigrants: *New York Times*, September 13, 1912.

16 One afternoon . . . a drunken peasant: Kenneth Harris interview with Golda in *The Observer* (London), reprinted in *Washington Post*, January 17, 1971.

16 building mud castles: Peggy Mann, *Golda: The Life of Israel's Prime Minister* (New York: Coward, McCann & Geoghegan, 1971), pp. 7–8.

16 boarded up windows: Harris interview.

16 "That pogrom never materialized": Stephen Klaidman, "Golda Meir: She Lived for Israel," *Washington Post*, December 8, 1978.

18 It was a blessedly quiet Sabbath morning: Sheyna Korngold, *Zikhroynes* (Tel Aviv: Farlag Idpres, 1968), p. 42.

20 "Fasting is only for grown-ups": Peggy Mann interview with Clara Stern, Bridgeport, Connecticut, from Ralph G. Martin, *Golda: Golda Meir, the Romantic Years* (New York: Scribner, 1988).

CHAPTER TWO

24 "I could have stayed in Russia": Sheyna Korngold, *Zikhroynes* (Tel Aviv: Farlag Idpres, 1968), p. 62.

25 precisely how deep the American influence: Meron Medzini, *Ha-Yehudi-yah ha-geah: Goldah Meir ya-hazon Yisrael: biyografyah politit* (Tel Aviv: Edanim, 1990), chapter 2.

26 "She was an extraordinary person in every aspect": Golda Meir, *My Life* (New York: Putnam, 1975), p. 21.

26 American Young Sisters' Society: Ibid., pp. 38–39, and Marie Syrkin, *Way of Valor: A Biography of Golda: Myerson* (New York: Sharon Books, 1955), p. 31.

27 Christian boy threw a penny at one of Golda's friends: Interview with Regina Medzini, in Martin, *Golda: Golda Meir, the Romantic Years* (New York: Scribner, 1988), p. 35.

27 "to become a rebbetzin": Linda Maiman, "Golda's Milwaukee," *Milwaukee Journal*, October 31, 1976.

27 "I can tell you that Pa does not work yet": Martin, *Golda*, p. 32.

28 "But a very clever woman you'll never be": Ibid., p. 38.

28 "She was so beautiful and everyone spoiled her": Korngold, *Zikhroynes*, p. 72.

28 "such a horror of working in an office": Golda Meir interview on Panorama, BBC, August 9, 1971.

28 "how many times did you shout": Letter from Sheyna to Golda, from Denver, October 8, 1912, Israel State Archives.

29 "come to us": Sheyna to Golda, undated, Denver, Israel State Archives (ISA).

31 "The Land of Israel is acquired through labor": Golda Meir, *My Life*, p. 47.

32 "watching me like a hawk": Ibid.

33 "Oh, now I wish I could hear that music": Peggy Mann, *Golda: The Life of Israel's Prime Minister* (New York: Coward, McCann & Geoghegan, 1971), p. 45.

33 "he has a beautiful soul": Interview with Medzini, in Martin, *Golda*, p. 53.

34 "So who's this Mr. Somebody?": Mann, *Golda*, p. 47.

35 "I have repeatedly asked you not to contradict me": Quoted in Syrkin, *Way of Valor*, p. 35.

35 "oh how I would kiss you!": Morris to Golda, from Denver, January 22, 1915, ISA.

35 "I noticed this striking girl": Bill Marten interview with Isadore Tuchman, February 21, 1952, State Historical Society of Wisconsin, Madison.

35 Only twenty-three delegates, representing four chapters: http://www.louisville.edu/a-s/english/subcultures/Zionism.

36 "I don't know whether to be happy": Morris to Golda, August 1915, ISA.

37 "Yes, Gogole": Morris to Golda, October 5, 1915, ISA.

38 Yitzak Ben-Zvi and David Ben-Gurion were scheduled: Syrkin, *Way of Valor*, pp. 42–43; and Rinna Samuel interview with Golda, July 10, 1973.

38 "The truth is that I didn't have exact": Martin, *Golda*, p. 78.

39 "I don't want to shatter your dreams": quoted in Mann, *Golda*, p. 54.

39 "a national home for the Jewish people": For background on the Balfour Declaration, see, Walter Laqueur, *A History of Zionism* (New York: Schocken Book, 1972), pp. 181–98.

40 "one nation solemnly promised to a second": Arthur Koestler, "The Great Dilemma That Is Palestine," *New York Times Magazine*, September 1, 1946.

40 "My fellow brethren": Golda Meir, *My Life*, pp. 60–61, and Louis J. Swichkow, *Memoirs of a Milwaukee Labor Zionist* (Tel Aviv: The Diaspora Research Institute, 1975).

40 American Jewish Congress: Golda Meir, *My Life*, p. 68.

41 "her mouth was gold": Bill Marten interview with Isadore Tuchman, State Historical Society of Wisconsin.

41 "This is the life for me!": Author interview with Meron Medzini, December 15, 2004.

41 "Who leaves a new husband": Golda Meir, *My Life*, p. 67.

41 "At Poale Zion, whatever I was asked": Rinna Samuel interview with Meir, August 1974.

42 "Dearest, I arrived in Buffalo": Golda to Morris, December 5, 1919, ISA.

42 Golda got pregnant: Sheyna to Golda, April 7, 1918, ISA.

42 repeatedly urged Golda to slow down: Sheyna to Golda, quoted in Syrkin, *Way of Valor*, pp. 47–48.

CHAPTER THREE

45 Regina's fiancé announced: Radio interview with Golda Meir, 1969, reproduced in Marie Syrkin, ed., *Golda Meir Speaks Out* (London: Weidenfeld & Nicolson, 1973), p. 32.

46 SS *Pocahontas*: The story of the voyage is based on Meir's account in Golda Meir, *My Life* (New York: Putnam, 1975); Samuel interview with Meir, May 29, 1973; Sheyna Korngold, *Zikhroynes* (Tel Aviv: Farleg Idpres, 1968); and Ralph G. Martin, *Golda: Golda Meir, the Romantic Years* (New York: Scribners, 1988).

47 By the time they arrived in Tel Aviv: Barbara Mann, *A Place in History: Modernism, Tel Aviv, and the Creation of Jewish Urban Space* (Palo Alto, Calif.: Stanford Studies in Jewish History and Culture, 2006).

48 Kibbutz Merhavia: Description based on Golda Meir, "My First Days at Kibbutz Merhavia," *Jewish Affairs* (December 1970) (Johannesburg), and author interview with Meron Medzini, December 15, 2004.

48 "We don't know what will be": Martin, *Golda*, p. 111.

49 "a rebellion against the bourgeois norms": Lesley Hazelton, *Israeli Women: The Reality Behind the Myth* (New York: Simon & Schuster, 1977).

49 "When I returned to my room": Golda Meir, "My First Days at Kibbutz Merhavia."

50 "[Morris] was not able to tolerate": Interview with Golda Meir in Oriana Fallaci, *Interview with History* (Boston: Houghton Mifflin, 1976), p. 116.

51 "Why do you regard this work": Golda Meir, *My Life*, pp. 89–90; and Golda Meir, "My First Days at Kibbutz Merhavia."

52 "Golda sat and appeared a little regal": Syrkin, *Golda Meir Speaks Out*, p. 62.

53 "Why is giving food to people": Meir interview with Samuel, May 29, 1973.

53 "Ah, Palestine, Palestine, you beggarly little land": Quoted by Syrkin, *Golda Meir Speaks Out*, p. 63.

54 "It is a sad and shameful fact": Dafna Izraeli, "The Zionist Women's Move-

ment in Palestine, 1911–1927: A Sociological Analysis," *Signs* 7, no. 1 (autumn 1981), p. 107.

55 accompany a prominent British suffragist: Meir interview with Samuel, May 29, 1973.

55 "For him": Fallaci, *Interview with History*, p. 115.

55 "It lasted from the day we met": Ibid.

56 "To her affection for Morris": Syrkin, *Golda Meir Speaks Out*, p. 64.

56 "I was no nun": Meir interview with Samuel, July 16, 1973.

57 "They were practically starving": Quoted in Martin, *Golda*, p. 140.

57 "In Jerusalem I was a sort of prisoner": Golda Meir, *My Life*, pp. 100–102.

58 "They were interested in the services": Ibid., p. 109.

CHAPTER FOUR

61 "the inner struggles and the despairs": Her article appears in Rachel Katznelson Shazar, *Plough Woman: Records of the Pioneer Women of Palestine* (New York: Herzl Press, 1975).

63 "Morris was father and mother": Shapiro, quoted in Ralph G. Martin, *Golda: Golda Meir, the Romantic Years* (New York: Scribner, 1988), p. 151.

63 "But she wasn't always around.": Clara Stern, quoted in Martin, *Golda*, 150.

63 "I never had sympathy:" Meir interview with Rinna Samuel, July 16, 1973.

63 Histadrut was the heart: A good comprehensive history of Histadrut history appears in Michael Shalev, *Labour and Political Economy in Israel* (New York: Oxford University Press, 1992).

63 "dollar land": Mark A. Raider, "Emissaries in the Promised Land," *Judaism* vol. 49 (winter 2000), p. 359.

64 "The Palestinian workers have begun to dig": Michael Brown, "The American Element in the Rise of Golda Meir," *Jewish History* 6, nos. 1–2 (March 1992).

66 Imperial Labour Conference: Meir interview with Rinna Samuel, June 18, 1973.

66 Sarah and Menachem entertained themselves: Menachem Meir, *My Mother Golda Meir: A Son's Evocation of Life with Golda Meir* (New York: Arbor House, 1983), p. 28.

67 "felt it was her duty": Author interview with Menachem Meir, January 13, 2005.

67 Golda was felled by one of her regular migraines: Author interview with Sarah Rehabi, January 13, 2005.

67 "I know I will not bring the Messiah": Golda Meir, *My Life* (New York: Putnam, 1975), p. 118.

68 "easy to get": Passow quoted in Martin, *Golda*, p. 163.

68 "Men have always been good to me": Oriana Fallaci, *Interview with History* (Boston: Houghton Mifflin, 1976), p. 112.

68 her affair with Zalman Shazar: Detailed in Martin, *Golda*, pp. 152–56.

68 "He did not speak with his mouth": Meir interview with Rinna Samuel, July 16, 1973.

68 "We always avoid talking about serious things": Remez to Golda, June 14, 1930, quoted in Martin, *Golda*, pp. 171–72.

69 "Morris won't give me a divorce": Medzini, quoted in Martin, *Golda*, p. 197.

69 In Golda mythology: Meir interview with Rinna Samuel, July 24, 1973.

70 Even mainstream Zionist groups: Donna Robinson Divine, "American Jewish Women and the Zionist Enterprise," *Israel Studies* 11, no. 1 (spring 2006), p.205.

70 "Palestine today is a place where there are no gangsters": *Chicago Tribune*, February 20, 1929.

71 "Goldie brought us a waft": *The Pioneer Woman*, February 1934.

71 In one midwestern city, Golda attacked: Meir interview with Rinna Samuel, August 9, 1974.

72 "Golda was caught up in a running": Golda Meir, *My Life*, p. 141.

72 "When I first met Golda in Toronto": Steinglass, quoted in Martin, *Golda*, p. 192.

72 "Sometimes weeks would": Author interview with Sarah Rehabi, January 13, 2005.

72 One night Sarah asked Golda what she did: Dr. Jacob Katzman, quoted in Martin, *Golda*, p. 189.

73 "She certainly never should have had children": Medzini quoted by Martin, *Golda*, p. 142.

CHAPTER FIVE

76 "buy Jewish": The campaign is described well by in Marie Syrkin, *Way of Valor: A Biography of Golda Myerson* (New York: Sharon Books, 1955).

76   After each outbreak of Arab rioting: For excellent background on this dy-
     namic, see Tom Segev, *One Palestine, Complete: Jews and Arabs Under
     the British Mandate* (New York: Metropolitan Books, 2000). The docu-
     ments of the commissions are available online at the Avalon Project,
     http://www.yale.edu/lawweb/avalon/avalon.htm.

78   "Why not build a Jewish port": Golda Meir, *My Life* (New York: Putnam,
     1975), pp. 144–45.

78   "The sea is an organic": Syrkin, *Way of Valor*, p. 92.

79   "What I need is a wife": Menachem Meir, *My Mother Golda Meir: A
     Son's Evocation of Life with Golda Meir* (New York: Arbor House,
     1983), p. 36.

80   "She did not court her public": Syrkin, *Way of Valor*, p. 98.

80   Her most brutal fight in those years: Ibid., p. 93–95; and Golda Meir, *My
     Life*, pp. 143–44.

81   Next she faced down: Syrkin, *Way of Valor*, p. 97.

81   "There is a sadistic streak to her": Ibid., p. 96.

81   bragging about Histadrut efforts: Ralph G. Martin, *Golda: Golda Meir, the
     Romantic Years* (New York: Scribner, 1988), p. 225, based on interview
     with Yitzhak Eylam.

82   "In social questions she has a genius": Syrkin, *Woman of Valor*, p. 98.

82   she complained that Shamai: Martin, *Golda*, p. 244.

82   "God forbid if she didn't like you": Author interview with Nomi Zuckerman,
     December 15, 2004.

83   "I can still see his room": Ibid.

83   "I came to Palestine for one reason only": Ibid.

83   "What has Tel Aviv to offer": Morris Meyerson to Menachem, Abadan,
     Iran, 1945, reprinted in Menachem Meir, *My Mother Golda Meir*,
     p 19.

83   "She was incapable of gray": Quoted in Martin, *Golda*, p. 218.

83   "paper flying around": Author interview with Zuckerman, December 15,
     2004.

84   "Golda's mother was impossible": Ibid.

CHAPTER SIX

86   when the British sent Lord Earl Peel: Palestine Royal Commission Report
     Presented by the Secretary of State for the Colonies to Parliament by

Command of His Majesty, July 1937. His Majesty's Stationery Office, London, 1937.

87   "We are given an opportunity": Benny Morris, *Righteous Victims* (New York: Vintage, 2001).

87   "Some day my son will ask": Golda Meir, *My Life* (New York: Putnam, 1975), p. 157.

88   "l would agree that the Arabs leave": American Zionist Emergency Council, Minutes of Meeting of Executive Committee, January 14, 1946, pp. 1, 4, Central Zionist Archives, F39/385.

89   "Who else did we have?": Quoted in Ralph G. Martin, *Golda: Golda Meir, the Romantic Years* (New York: Scribner, 1988), p. 225.

89   From an office on Marc Aurel Street: For background on the smuggling, see Ehud Avriel, *Open the Gates!: A Personal Story of "Illegal" Immigration to Israel* (New York: Atheneum, 1975).

90   "Take our children away": Terry Morris, *Shalom, Golda* (New York: Hawthorn Books, 1971).

90   Evian-les-Bains: For background on the conference, see S. Adler-Rudel, "The Evian Conference and the Refugee Problem," in *Leo Baeck Institute Year Book*, vol. 13 (London: Horovitz, 1968).

91   "This will not be just another conference": Quoted in Peggy Mann, *Golda: The Life of Israel's Prime Minister* (New York: Coward, McCann & Geoghegan, 1971), p. 99.

92   the air of "a poker game": *New York Times*, July 6, 1938.

92   "We have very little bread": Robert Slater, *Golda: The Uncrowned Queen of Israel* (New York: Jonathan David, 1981), p. 93.

93   "There is one thing I want to see": Quoted in Peggy Mann, p. 104.

93   White Paper: The text of the White Paper is online at http://www.yale.edu/lawweb/avalon/mideast/brwh1939.htm.

94   "It is in the darkest hour": Peggy Mann, p. 110.

94   "It's inconceivable that we shall not": *D'var HaPoelet* (periodical of Histadrut), May 3, 1939.

94   Zionist Congress: Marie Syrkin, *Way of Valor: A Biography of Golda Myerson* (New York: Sharon Books, 1955), pp. 100–102.

95   "This time, we asked nothing": Menachem Meir, *My Mother Golda Meir: A Son's Evocation of Life with Golda Meir* (New York: Arbor House, 1983), p. 55.

95   "The Jews should act as though": Syrkin, *Woman of Valor*, p. 102.

95   "If war breaks out": Golda to Menachem and Sarah, August 21, 1939, Ge-
      neva, reprinted in Menachem Meir, *My Mother Golda Meir*, pp. 55–56.

95   It was an impossible dilemma: Syrkin, *Way of Valor*, p. 103.

95   "It was a very nice slogan": Meir, *My Life*, p. 165.

96   the steamer *Atrato*: Ehud Avriel, *Open the Gates!: A Personal Story of "Ille-
      gal" Immigration to Israel* (New York: Atheneum, 1975).

97   "There is no Zionism save": Mapai Executive Committee Meeting, No-
      vember 1943, quoted in Syrkin, *Way of Valor*, p. 114.

98   "They fight like lions": Golda Meir, *My Life*, p. 184.

98   "Our way in this country": "This Is Our Strength," *D'var HaPoelet*, May 5,
      1939.

99   "You are a nice, peaceful, law-abiding lady": Syrkin, *Way of Valor*, pp.
      106–13.

101   "Zionism and pessimism are not compatible": Ibid., p. 159.

101   "What can we do, first": Quoted in Martin, *Golda*, p. 239.

102   "It's bad enough that the rest of the world": Menachem Meir, *Golda Meir
      My Mother*, p. 63.

102   a Histadrut emissary reported: Ibid.

102   "I could see in his worried": Golda Meir, *My Life*.

103   "How can we, in Tel Aviv": Va'ad Ha'Poel meeting, April 10, 1943, quoted
      in Syrkin, *Woman of Valor*, pp. 114–16.

104   "I have sometimes wondered how": Golda Meir, *My Life*, p. 166.

CHAPTER SEVEN

105   "At this solemn hour": *Palestine Post*, May 10, 1945.

106   "Civilization has been appalled": Excerpts of Bevin's speech were printed
      by the *New York Times*, February 26, 1947.

107   "Jewish life is precious": Meron Medzini, *Ha-Yehudiyah ha-geah: Goldah
      Meir ve-hazon Yisrael* (Tel Aviv: Edanim, 1990), chapter 7.

108   "The American people": Released November 13, 1945. Dated August 31,
      1945, President Harry S. Truman, Public Papers of the Presidents, Harry
      S. Truman, 1945.

109   Etzel, which had seceded: For an accessible history of the Irgun, see the
      Irgun site, written by Yehuda Lapidot at http://www.etzel.org.il/english/
      Irgun.

109 "liquidating the possibility": Minutes of meeting of the political committee of Mapai, March 27, 1944.

109 "We have no alternative but to follow a new path": Medzini, *Ha-Yehudiyah*, chapter 7.

111 "What's wrong with workers from the Histadrut?" Ibid.

111 the Anglo-American Commission: The Commission report is online at http://www.yale.edu/lawweb/avalon/anglo/angtoc.htm.

112 spoke on behalf of the Histadrut: *The Pioneer Woman*, April 1946.

114 "cigarettes were permitted": Marie Syrkin, *Way of Valor: A Biography of Golda Myerson* (New York: Sharon Books, 1955), pp. 152–54.

116 "If Shertok [Sharett] is guilty": Quoted in ibid., p. 160.

117 "Maybe I wasn't important enough": Quoted by Abraham Rabinovich in *Newsday*, December 7, 1978.

117 "Kudos to a smart and energetic woman": *The Pioneer Woman*, March 1947.

117 "A lovely lady, a good speaker": Quoted in Ralph G. Martin, *Golda: Golda Meir, the Romantic Years* (New York: Scribner, 1988), p. 270.

118 setting up a *Judenrat*: Author interview with Chaim Hefer, December 14, 2004.

118 Shertok tried to calm: For examples of Sharett's frustration, see Ya'akov Sharett, ed., *Imprisoned with Paper and Pencil: The Letters of Moshe and Zipporah Sharett* [Hebrew] (Tel Aviv: Moshe Sharett Institute, 2000).

118 "What we want is complete independence": Quoted in Martin, *Golda*, p. 270.

119 I never lived through a pogrom: Minutes of Jewish Agency Executive Committee, Paris, August 2, 1946.

119 "We have hated death": Speech at funeral at Kibbutz Givat Hayim, quoted in Syrkin, *Way of Valor*, p. 137.

119 "If we don't do something active": Meir interview with Rinna Samuel, August 4, 1973.

120 Unless Bevin agreed to release: Medzini, *Ha-Yehudiyah*, chapter 7.

120 "Golda is inflexible": Sharett, *Imprisoned with Paper and Pencil*.

121 "For miles around, there was nothing": Quoted in Martin, *Golda*, p. 261.

121 "You and Ben-Gurion are destroying": Meir interview with Rinna Samuel, August 4, 1973.

121 "Look, you are afraid and I am afraid": Ibid.

121 "She had this gift of propaganda": Quoted in Martin, *Golda*, p. 278.

121 One afternoon, she fainted: Syrkin, *Way of Valor*, p. 136.

122   traveled to Basle: Ibid., pp. 165–68.

122   "what is our starting point?": Medzini, *Ha-Yehudiyah*, chapter 7.

123   "Why are we NOW pressing": Statement given in Yiddish, reprinted in English in Marie Syrkin, ed., *A Land of Our Own: An Oral Autobiography by Golda Meir* (New York: G. P. Putnam's Sons, 1973), pp. 58–65.

CHAPTER EIGHT

125   They made an odd pair: The most thorough account of Golda's meetings and the context in which they occurred is Avi Shlaim, *Collusion Across the Jordan: King Abdullah, the Zionist Movement, and the Partition of Palestine* (New York: Columbia University Press, 1988). See also Alec Seath Kirkbride, *From the Wings: Amman Memoirs 1947–1951* (London: Frank Cass, 1976). The *yishuv* view of the meetings was recorded most thoroughly by Zeev Sharef in his *Three Days* (London: W. H. Allen, 1962).

125   United Nations Special Committee on Palestine: For information on UNSCOP, see, Jorge Garcia-Granados, *The Birth of Israel: The Drama as I Saw It* (New York: Alfred A. Knopf, 1948).

125   "We don't anticipate spilled blood": Meron Medzini, *Ha-Yehudayah ha-geah: Goldah Meir ya-hazon Yiśra'el: Biyografyah polițit* (Tel Aviv: Edanim, 1990), chapter 7.

129   "The struggle against terrorism cannot be divorced": *New York Times*, February 11, 1947.

129   "We will not become a nation of informers": *New York Times*, February 23, 1947.

129   "Even if it means armed struggle": Menachem Meir, *My Mother Golda Meir: A Son's Evocation of Life with Golda Meir* (New York: Arbor House, 1983), p. 90.

129   "Terrorism is assisting Palestine's British": *New York Times*, March 30, 1947.

130   "the right measure of independence": Marie Syrkin, *Way of Valor: A Biography of Golda Myerson* (New York: Sharon Books, 1955), p. 162.

130   British detention centers in Cyprus: Ibid., pp. 171–76.

131   "We can hardly imagine a Jewish state": *New York Times*, September 2, 1947.

131   "The Messiah hasn't come": Quoted in Ralph G. Martin, *Golda: Golda Meir, the Romantic Years* (New York: Scribner, 1988), p. 288.

132   "For two thousand years": Ibid, p. 295.

132   "We will strangle": Larry Collins and Dominique Lapierre, *O Jerusalem* (New York: Simon & Schuster, 1972), p. 88.

133   "But how will I live if I'm blinded?": Ibid., p. 299.

133   On December 27: Syrkin, *Way of Valor*, p. 189.

133   the police station in Jewish Hedera: Ibid., p. 190.

134   "What you can do here, I cannot do": Peggy Mann, *Golda: The Life of Israel's Prime Minister* (New York: Coward, McCann & Geoghegan, 1971), p. 142.

134   "When she came here she stayed": Jeff Hodes interview with Henry Montor, October 14, 1975, Rome, Oral History Division Institute for Contemporary Jewry (OHD), 34(128), p. 54.

134   Arriving on a Friday afternoon: The best information on Golda's fund-raising trip was gleaned from interviews available at the OHD with Mathilda Brailove, Lou Boyar, Henry Montor, Joe Mazer, William Mazer, Gottlieb Hammer, Sam Feingold, Lee Horne, Ralph Wechsler, Leon Gerber, Sidney Lewine, Julius Livingston, Jeanne Daman, and Harry Beale.

135   Montor decided that Golda's only chance: 128(34), OHD, 54–56.

135   "The nations of the world": Her speech in Chicago was published as *A Report from Palestine* by the Council of Jewish Federations and Welfare Funds, General Assembly, Chicago, January 1948.

137   "Money is not any of your business": United Jewish Appeal interview with Golda Meir, January 19, 1971.

137   "marrying a Negro": Beale interview, OHD, p. 25.

137   "I was sure that they couldn't care less": Ibid. and Syrkin, *Way of Valor*, p. 210.

138   "we can't save the Jewish people": Montor interview, OHD, pp. 58–59.

138   "meet in Madison Square Garden": Jeff Hodes interview with Meir, quoted in Martin, *Golda*, p. 321.

138   "Someday, when history will be written": Golda Meir, *My Life* (New York: Putnam, 1975), p. 214.

140   to persuade them to remain: Mann, *Golda*, pp. 148–49.

144   "We need to go all the way": Medzini, *Ha-Yehudiyah*, chapter 8.

144   "Golda is too inflexible": Minutes of the meeting of Minhelet HaAm, May 12, 1948, ISA.

146   "I learned about the Declaration": *Jewish Newsweekly of Northern California*, June 5, 1998.

CHAPTER NINE

147   "need much larger sums": Menachem Meir, *My Mother Golda Meir* (New York: Arbor House, 1983), p. 121.

148   "We cannot go on without your help": Golda Meir, *My Life* (New York: Putnam, 1975), p. 235.

149   twisting her right leg: Menachem Meir, *Golda Meir My Mother*, p. 123; and *New York Times*, July 2, 1948.

149   Sharett barraged Golda: Marie Syrkin, *Way of Valor: A Biography of Golda Myerson* (New York: Sharon Books, 1955), pp. 234–35.

150   "APPOINTMENT OF SARAH": Peggy Mann, *Golda: The Life of Israel's Prime Minister* (New York: Coward, McCann & Geoghegan, 1971), p. 160.

150   schizophrenic world: For background on the Soviet position on partition, see Arnold Krammer, "Soviet Motives in the Partition of Palestine, 1947–48," *Journal of Palestine Studies* 2, no. 2 (winter 1973): 102–99.

151   "The Jews in the civilized world": "Critical Remarks on the National Question," written by Lenin in October–December 1913 and published the same year in the Bolshevik legal journal *Prosveshcheniye*, nos. 10, 11, and 12.

152   The Soviet Foreign Ministry had ordered: Gennadii Kostyrchenko, "Golda at the Metropol Hotel," *Russian Studies in History* 43, no. 4 (fall 2004), pp. 77–84.

153   "What were they thinking, sending an old": Lov Kaddar, interview, (56)88, OHD.

153   Male diplomats, she was told: Ibid.

154   "From the moment they arrived in Moscow:" Uri Bialin, "Top Hat, Tuxedo and Cannons: Israeli Foreign Policy from 1948 to 1956 as a Field of Study," *Israel Studies* 7, no. 1 (2002): 1–80.

154   "Golda had nothing to do": Kaddar, interview, (56)88, OHD.

155   Mordechai Namir . . . yanked the chain: Ibid.

156   "I would happily sacrifice my knowledge": Mordechai Namir, *Israeli Mission to Moscow* [Hebrew] (Tel Aviv: Am Oved, 1971) pp. 333–34.

156   A Jewish officer in the Red Army: Ibid., p. 336.

156   "[It is] as though the problem of Jewish immigration": Namir, *Israeli Mission to Moscow*, p. 60.

156   petitioned for exit permits: Ibid., p. 411.

156   One night a middle-aged Jew: Menachem Meir, *Golda Meir My Mother*, p. 132.

157 she urged him: Syrkin, *Way of Valor*, pp. 252–53.

157 she threw Friday: Kaddar, interview, (56)88, OHD.

157 submitted the first issue: Ibid.

158 Polina Zhemchuzhina: Namir, *Israeli Mission to Moscow*, pp. 83–84.

159 A savvy politician: Syrkin, *Way of Valor*, pp. 249–50.

159 "Let there be no mistake about it": *Pravda*, September 21, 1948.

159 Rosh Hashanah: Kaddar, interview, (56)88, OHD.

161 "The words shook the sun": Syrkin, *Way of Valor*, p. 252.

CHAPTER TEN

164 snapped Eliezer Kaplan: Dvora Hacohen, *Immigrants in Turmoil: Mass Immigration to Israel and Its Repercussions in the 1950s and After* (Syracuse, N.Y.: Syracuse University Press, 2003).

165 "A Jewish state that aims at a high": Address to workers' delegation, Tel Aviv, July 25, 1950, in Henry Christman, ed., *This Is Our Strength: Selected Papers of Golda Meir* (New York: Macmillan, 1952), pp. 50–51.

165 Golda faced her biggest hurdle: The debate over her plan was detailed in *Palestine Post*, August 9, 1949.

166 "We need cheap housing quickly": Marie Syrkin, *Way of Valor: A Biography of Golda Myerson* (New York: Sharon Books, 1955), p. 264.

166 "I went to our Parliament two weeks ago": Robert Slater, *Golda, the Uncrowned Queen of Israel* (Middle Village, N.Y.: Jonathan David, 1981), p. 98.

166 Amun-Israeli Housing Corporation: For background on Amun-Israeli, see Charles Abrams, "Israel Grapples with Its Housing Crisis," *Commentary* 11, no. 4 (April 1951), pp. 347–54.

166 "We pray for the day": Syrkin, *Way of Valor*, p. 287.

167 "If we have to choose": Address before the Executive Council of Mapai, Jerusalem, March 11, 1953.

168 made to send care packages: Syrkin, *Way of Valor*, pp. 287–88.

168 "any free ride": Golda's speech launching the bond drive was delivered on October 1950 in Washington, D.C. Pieces are reprinted in Menachem Meir, *My Mother Golda Meir: A Son's Evocation of Life with Golda Meir* (New York: Arbor House, 1983), pp. 142–43.

168 "You had to be a little crazy": Author's interview with Ralph Goldman, January 12, 2005.

168 "cables from good American friends": Quoted in Syrkin, *Way of Valor*, p. 288.

169   "Mapai does not promise a land": Ibid., p. 282.

170   "immigration will have to be": United Jewish Appeal Meeting, Chicago, June 1950, quoted in Syrkin, *Way of Valor*, p. 281.

170   In a new nation still teetering: The best analysis of the crisis was written by J. L. Teller, "The Israeli Voter Ponders the Moral Crisis," *Commentary* 11, no. 2 (February 1951).

171   "More immigrants or more shoes": Syrkin, *Way of Valor*, p. 282.

171   "I'm prepared to go not only from meeting": Peggy Mann, *Golda: The Life of Israel's Prime Minister* (New York: Coward, McCann & Geoghegan, 1971), p. 129.

172   Villa Harun al-Rashid: George Bisharat, "Golda Meir Lived in My Home," *Ha'aretz*, January 4, 2004.

172   "A woman stands in the kitchen five hours": Slater, *Golda*: p. 101.

172   the foundations were too high: Menachem Meir, *My Mother Golda Meir*, p. 140.

173   "a progressive country cannot": Golda was speaking to the Second National Convention of the Union of Engineers, Architects and Surveyors in Haifa, on August 22, 1949, as reported in the *Palestine Post* of that day.

173   "To make us a road is very kind": Golda Meir, *My Life* (New York: Putnam, 1975), pp. 280–81; and Peggy Mann, p. 130.

173   head lice and gynecology: Dvora Hacohen, *Immigrants in Turmoil*.

174   "Golda doesn't think in terms of hours of work": Peggy Mann, *Golda*, p. 127.

174   At Passover in 1950: Slater, *Golda*, p. 101.

174   "go to the *balabusta*": Ibid.

175   The birth of Meira Meyerson: The fullest background on Meira Meyerson is contained in Slater's *Golda*, pp. 107–17. This section was based on author's interview with Ari Rath, December 2004.

175   "Friends tried to mediate": Author interview with Ari Rath.

175   "She was never the grandmother": Author interview with Rolf Kneller, December 2004.

176   "It is a momentous occasion": Speech reproduced in Marie Syrkin, ed., *Golda Meir Speaks Out* (London: Weidenfeld & Nicolson, 1973), pp. 86–88.

177   "The State of Israel will not tolerate": Golda Meir, *My Life*, pp. 275–76.

178   Bill for Compulsory Service for Women: Syrkin, *Way of Valor*, pp. 291–97. For background on the dispute, see "National Service Law, 1953," in

Joseph Badi, ed., *Fundamental Laws of the State of Israel* (New York: Twayne Publishers, 1961), pp. 311–13.

178 "You will not force your way of life": Speech on the floor of the Knesset, quoted in Syrkin, *Way of Valor*, pp. 196–97.

CHAPTER ELEVEN

181 The torment of Moshe Sharett: There are many accounts of the deterioration in Ben-Gurion and Sharett's relationship. This section relies heavily on Medzini's *Ha-Yehudiyah*, chapter 11.

181 chief of operations of the Israel: Shabtai Teveth, *Moshe Dayan: The Soldier, the Man, the Legend* (Boston: Houghton Mifflin, 1973).

182 "But who will be foreign minister?": Golda Meir, *My Life* (New York: Putnam, 1975), p. 291.

182 "All a foreign minister does is talk": Robert Slater, *Golda: The Uncrowned Queen of Israel* (Middle Village, N.Y.: Jonathan David, 1981), p. 118.

182 asked her to jump out: Yosef Almogi, *Total Commitment* (New York: Herzl Press, 1982).

183 "think it's a cocktail party": Author interview with Yitzhak Navon, December 30, 2004.

183 "Golda didn't take a single note": Author interview with Ari Rath, December 23, 2004.

183 "hard woman": Anne Marie Lambert interview, March 4, 1981, Golda Meir Library Archives, University of Wisconsin–Milwaukee (UWM), tape 17, collection 21.

183 "queen bee": Author interview with Esther Herlitz, December 16, 2004.

183 "That makes two of us": Simcha Dinitz interview, Golda Meir Library Archives, UWM, tapes 8 and 9, collection 21.

184 Eleanor Roosevelt: From Roosevelt's foreward to Henry M. Chrisman, ed., *This Is Our Strength* (New York: Macmillan, 1962).

184 charmed the notoriously diffident: Golda Meir, *My Life*, p. 315.

184 After Argentina appealed: Speech delivered at the Security Council on June 22, 1960, and reprinted in Marie Syrkin, ed., *Golda Meir Speaks Out* (London: Weidenfeld & Nicolson, 1973), pp. 126–34.

184 "Former Milwaukee Schoolteacher Is": *Montreal Gazette*, May 28, 1958.

184 "Israeli FM 'Thinks Like a Man'": *Boston Globe*, November 18, 1956,

184 "Grandmother-Diplomat": *New York Times*, July 12, 1958.

185   "A 20-Hour Workday": *New York Tribune*, October 17, 1961.

185   "What is the use or realism": Address to the General Assembly, October 7, 1957.

185   "Like the Biblical Rachel": *The Telegram*, October 12, 1962.

185   The first hint of the role: Abba Solomon Eban, *Personal Witness: Israel Through My Eyes* (New York: G. P. Putnam's Sons, 1992).

186   Four days after Golda took office: Interview with Michael Bar-Zohar, December 21, 2004.

186   One of the first contacts Golda wanted: Matti Golan, *Road to Peace: A Biography of Shimon Peres*. Translated by Akiva Ron. (New York: Warner Books, 1989).

187   "I told G[olda] that I am worried": Diaries of David Ben-Gurion, July 12, 1957, Sde Boker.

187   floated on a boat down the Seine: Slater, *Golda*, p. 136.

188   skeptical about Peres' rosy reports: Golda's account of the meeting appears in Golda Meir, *My Life*, pp. 286–88, and Peres' in his *Battling for Peace: A Memoir* (New York: Random House, 1995), pp. 102–14.

189   The British and French delayed bombing: Avi Shlaim, *The Iron Wall* (New York: W. W. Norton, 2001), p. 179.

189   "restrain the aggressors": Quoted in Eban, *Personal Witness: Israel Through My Eyes*.

190   "Israel's people went into the desert,": Quote is from a speech delivered to the United Nations General Assembly on December 5, 1956, reproduced in part in Marie Syrkin, ed., *A Land of Our Own: An Oral Autobiography* (New York: G. P. Putnam's Sons, 1973).

191   "It wasn't just arrogance on Golda's": Author interview with Esther Herlitz, December 16, 2004.

191   When staff members showed up: Ibid.

191   "Had we waited": *Baltimore Sun*, December 11, 1956.

192   "Let's just have it printed": Author interview with Esther Herlitz.

192   "you will be a traitor": Ibid.

192   Golda called Ben-Gurion and begged for a delay: Slater, *Golda*, p. 126.

193   "washed my underwear": Ibid., p. 128.

194   Isser Harel, Israel's master spy: The story of the German scientists in Egypt is culled from "Nasser's Hired Germans," *Saturday Evening Post* 248, April 1976; Michael Bar-Zohar, *Ben-Gurion: The Armed Prophet* (New York: Prentice-Hall, 1968); and the coverage of the *Jerusalem Post* in 1963 and 1964.

194   Ben-Gurion and Peres were mollified: Golan, *Road to Peace*; and author interview with Bar-Zohar.

194   "There is no doubt that the motives": Knesset debate, November 20, 1963, covered in following day's *Jerusalem Post*.

195   in August: *Jerusalem Post*, August 15, 1965.

195   "If we're serious about a special": Author interview with Ari Rath, December 23, 2004.

196   Anastasio Somoza: A good explanation of the Peres and Somoza fiasco is found in Benjamin Beit-Hallahmi, *The Israeli Connection: Who Israel Arms and Why* (New York: Pantheon, 1987).

196   "a catastrophe, colonialism, imperialism": Ibid.

196   "Wait your turn": Author interview with Ari Rath, December 23, 2004.

197   "She was extremely vindictive": Author interview with Benny Morris, January 2005.

197   female members of the ministry: Author interview with Esther Herlitz, December 16, 2004.

197   less than exact with his per diem: Author interview with Yaakov Nitzan, December 2004.

197   "I don't care about people's behavior at home": Ibid.

197   "she went to Ben-Gurion": Author interview with Esther Herlitz, December 16, 2004.

197   long feud over Dimona: Peres' account of the Dimona plan appears in his autobiography *Battling for Peace*; pp. 115–24. For a more objective view, see Avner Cohen, *Israel and the Bomb* (New York: Columbia University Press, 1998).

198   installation of Léopold Senghor: Peres, *Battling for Peace*, pp. 199–20.

199   "Regarding Dimona, there is no need to stop the work": From minutes of meeting of senior Foreign Ministry staff, June 13, 1963, Israel State Archives.

200   during tough days of angry arguments: The trip was well covered on a daily basis by the *Jerusalem Post*, with key articles appearing on March 12, 24, and 27 and April 4.

200   "Independence came to us": Golda Meir, *My Life*, p. 318.

201   "tell them about the mistakes we made": Peggy Mann, *Golda*, p. 200.

201   "She mobilized top people": Interview with Hanan Aynor, August 1987, Golda Meir Library Archives, UWM, tape 16, collection 21.

201   With a low investment, Golda reasoned: Ibid.

201   the 500 delegates: Ehud Avriel, "Israel's Beginnings in Africa," in Michael

Curtis and Susan Gitelson, *Israel and the Third World* (New Brunswick, N.J.: Transaction Books, 1976) pp. 69–83.

202 "Back then, if you told a taxi driver": Author interview with Shlomo Hillel, December 21, 2004.

202 She sat for an interview: The list is culled from the archives of the *Jerusalem Post*.

203 "We don't want to hide the fact": Meir interview with Martin Levin, June 6, 1963, transcript, Golda Meir Library Archives, UWM.

203 President David Dacko: Mordechai E. Krenin, *Israel and Africa: A Study in Technical Cooperation* (New York: Praeger, 1964).

203 "Imagine that instead of coming to see us": Slater, *Golda*, p. 142.

204 "I can do without the Falls": *Jerusalem Post*, October 30, 1964, and Golda Meir, *My Life*, p. 336.

204 South Africa: On Israel's relationship with South Africa, see Michael Comay interview, OHD, 6(24).

204 "have the right—and justly so": *Jerusalem Post*, November 10, 1962.

204 "fight for liberty and freedom": *Jerusalem Post*, November 13, 1962.

204 "The danger lurks": "Israel and Africa," For the Record, Cairo: Arab League Press and Information Department, no. 175, July 31, 1963.

204 Taking the long view: On the diplomatic impact of Israeli-African ties, see Krenin, *Israel and Africa*; and Tibor Rodin, *Political Aspects of Israeli Foreign Aid in Africa*, PhD dissertation, University of Nebraska, 1969.

205 During a 1964 visit to Nigeria: *Jerusalem Post*, October 29, 1964.

205 "You'll forgive me—but": Speech to Socialist International Council, Haifa, April 1960.

206 "the African delegates would all line up": Author interview with Yaakov Nitzan, December 26, 2004.

CHAPTER TWELVE

207 leadership needed to be infused: All of the major biographies of Ben-Gurion discuss his sentiments on this regard. See, in particular, Dan Kurzman, *Ben-Gurion, Prophet of Fire* (New York: Simon & Schuster, 1983), pp. 426–28.

208 "It's not a question of age": *Jerusalem Post*, December 19, 1967.

208 "complete intolerance": Author interview with Uri Avnery, January 4, 2005.

209   When she was not displeased: Author interview with Miriam Eshkol, January 12, 2005.

209   At the Mapai convention: Shabtai Teveth, *Moshe Dayan: The Soldier, the Man, the Legend* (Boston: Houghton Mifflin, 1973), p. 287; and Naftali Lau-Lavie, *Moshe Dayan: A Biography* (London: Vallentine Mitchell, 1968), pp. 175–76.

209   At a meeting of the students' club: Ibid.

210   would not serve in the cabinet: Teveth, *Moshe Dayan*, p. 293, and Yosef Almogi, *Total Commitment* (New York: Herzl Press, 1982).

210   Pinhas Lavon: One of the most contentious political struggles in Israeli history, the Lavon Affair has been written about extensively. The most complete analysis is Shabtai Teveth, *Ben Gurion's Spy: The Story of the Political Scandal That Shaped Modern Israel* (New York: Columbia University Press, 1996).

212   But Golda had gamed out every move: Ibid., p. 305.

213   "Stating that no party member": Almogi, *Total Commitment*.

214   Ben-Gurion prepared himself carefully: For background on Ben-Gurion's mood, see Kurzman, *Ben-Gurion, Prophet of Fire*, pp. 446–48.

214   thrown down the gauntlet: Medzini, *Ha-Yehudiyah*, chapter 14.

214   "I don't know what the people": Speeches from the Night of the Long Knives are contained in the Central Zionist Archives, A245/139.

215   From a wheelchair: *Jerusalem Post*, February 18, 1965, and Almogi, *Total Commitment* p. 229.

215   kissed the dying man: Michael Bar-Zohar, *Ben-Gurion: The Armed Prophet* (New York: Prentice-Hall, 1968), p. 309.

215   the Night of the Long Knives: *Jerusalem Post*, February 17, 1965.

215   Wilted by the onslaught: Kurzman, *Ben-Gurion, Prophet of Fire*, p. 447.

215   "He made excuses for her": Author interview with Yitzhak Navon, December 30, 2004.

216   "It was sad to hear her speak": Ben-Gurion diaries, February 19, 1965.

216   The election campaign was the ugliest: *Jerusalem Post*, May 16, 1965.

216   "Tammany Hall": *Jerusalem Post*, July 18, 1965.

216   "LIAR!" shouted Golda: *Jerusalem Post*, July 29, 1965.

217   "It's hard to teach an old horse": *New York Times Magazine*, April 17, 1966.

218   "such an Amazon": Author interview with Miriam Eshkol, January 12, 2005.

218   The Soviets had informed: The impulse for the Soviet disinformation is well laid out in Benny Morris, "Provocations," *The New Republic*, July 23, 2007, pp. 47–51, in a review of *Foxbats over Dimona*, in which authors

Isabella Ginor and Gideon Remez argue against the conventional wisdom that the Soviet attempt to push the Egyptians into a show of strength went awry and led to war or, alternatively, that the Russians assumed that the well-armed Egyptians would destroy Israel, cementing the Egyptian-Soviet alliance. Rather, they suggest that the Russians wanted to provoke Israel into a preemptive strike to give them and the Egyptians license to destroy the Israeli nuclear facility at Dimona. Isabella Ginor and Gideon Remez, *Foxbats over Dimona: The Soviets' Nuclear Gamble in the Six-Day War* (New Haven, Conn.: Yale University Press, 2007).

219   "U Thant's ludicrous surrender": Golda Meir, *My Life* (New York: Putnam, 1975), p. 355.

219   "It was against all rhyme or reason for a force": Ibid., p. 354.

219   "She'll be tougher": David Kimche and Dan Bawley, *Sandstorm; the Arab-Israeli War of June, 1967: Prelude and Aftermath* (New York: Stein and Day, 1968).

220   "These two horses cannot be hitched": S. Ilan Troen and Zaki Shalom, "Ben Gurion's Diary for the 1967 Six Day War," *Israel Studies* 4, no. 2: 197.

221   protested that any change in the government: Kimche and Bawley, *Sandstorm*.

221   "A whispering campaign has been started": Ibid.

222   Realizing that the nation needed to hear: "Eshkol's Black Sunday," in Ahron Bregman and Jihan El-Tahri, *The Fifty Years War: Israel and the Arabs* (London: Penguin Books, BBC Books, 1998).

222   "A one-woman stumbling block": David Ben Gurion, *Israel: A Personal History* (New York: Funk & Wagnalls, 1971).

222   "As long as the Old Lady is there": Slater, *Golda*.

223   "I understand the Arabs wanting": Kimche and Bawley, *Sandstorm*.

223   Golda drove: Kaddar interview, (56)88, OHD.

224   "Again, we won a war": Peggy Mann, *Golda*, p. 227.

225   Quirky and intensely: There were numerous examples of Dayan's wild swings. The most vivid was the position he took on the Egyptian violation of the temporary cease-fire in 1970. In cabinet meetings, he was insistent that Israel force Egypt to remove the SAM batteries that the Egyptian army had moved forward, going so far as to suggest that military action was warranted if they refused. Yet having convinced the cabinet to press the United States on the matter and to pull out of the Jarring negotiations, he then publicly declared that Israel should continue talking with the UN emissary.

225   "She comes clumping along": *Newsweek*, March 17, 1969.

225   "Golda knew what power was": Interview with Abraham Harman, 1981, Golda Meir Library Archives, University of Wisconsin-Milwaukee.

226   "You people don't really want me": *Jerusalem Post*, February 3, 1968.

226   When she abruptly resigned again: The attempts to convince her to remain were covered by the *Jerusalem Post*, July 19–26, 1964.

226   "I want to be able to live": Interview with Golda, Kol Yisrael, July 12, 1968.

227   Israelis were stunned: *Christian Science Monitor*, July 11, 1968.

227   "No one is indispensable": *Jerusalem Post*, July 15, 1968.

227   "I am not going into a political wilderness": *Jerusalem Post*, July 16, 1968.

227   "I do not intend to retire to a political nunnery": Slater, *Golda*, p. 169.

CHAPTER THIRTEEN

231   "Golda is going to replace him": Author interview with Yossi Sarid, January 2, 2005.

231   Switzerland to sound out Golda: Meron Medzini, *Ha-Yehudiyah ha-geah: Goldah Meir ya-ḥazon Yiśra'el: Biyografyah poliṭit.* Tel Aviv: Edanim, 1990), chapter 15.

231   Sapir arrived at the prime minister's residence: Robert Slater, *Golda: The Uncrowned Queen of Israel* (Middle Village, N.Y.: Jonathan David, 1981), p. 175.

231   Miriam Eshkol so despised: Author interview with Miriam Eshkol, January 12, 2005.

232   "He will be buried in Jerusalem": Slater, *Golda*, p. 175.

233   the "shrew": Meron Medzini, *Ha-Yehudiyah*, chapter 15.

233   "She couldn't wait": Author interview with Rinna Samuel, December 18, 2004.

233   "Seventy is not a sin": *Time*, September 19, 1969.

233   "But to Senta Josephtal": Author interview with Senta Josephtal, January 9, 2005.

233   The latest poll showed: Poll from Public Opinion Research of Israel, Medzini, *Ha-Yehudiyah*, chapter 16

233   At Eshkol's funeral: Slater, *Golda*, p. 175.

234   *Ha'aretz*: a major Israeli daily newspaper: *Time*, March 7, 1969.

234   "milkman became an officer": Golda Meir, *My Life* (New York: Putnam, 1975), p. 379.

235   Golda gritted her teeth: Author interview with Eliezer Shmueli, October 16, 2007.

237 "I am opposed to a Jewish war": Transcript of interview with *Time* and *Life*, May 5, 1969, 12.5.1969, from Israel State Archives.

237 "Golda told them all to shut up": *New York Times*, June 5, 1969.

237 Ever inclined toward the puritanical: Author interview with Eliezer Shmueli, September 24, 2007.

237 Like a schoolmarm: Saadia Touval, *The Peacebrokers: Mediators in the Arab-Israeli Conflict, 1948–1979* (Princeton, N.J.: Princeton University Press, 1982).

237 "Conditional party membership is not": *Jerusalem Post*, March 29, 1969.

238 "If you could only watch her come": *Newsweek*, September 29, 1969.

238 use only 200 words: Author interview with Eitan Bentsur, December 28, 2004.

238 "There was never a mother": Dinitz interview, Golda Meir Library Archives, UWM, tapes 8 and 9, collection 21.

238 "never seen a more stupid schedule": Lou Kaddar interview, Golda Meir Library Archives, UWM, tapes 7 and 12, collection 21.

240 Greeting Katherine Graham: Medzini, *Ha-Yehudiyah*, chapter 16.

240 When Senator Frank Church came: Author interview with Marlin Levin, December 27, 2004.

240 How's your health: *Time*, October 10, 1969.

240 stylish Hermès bag in Paris: Author interview with Miriam Eshkol, January 12, 2005.

240 approval ratings had reached 61: Pori poll, *Jerusalem Post*, March 30, 1968.

240 By July, her approval: Pori poll, *Jerusalem Post*, July 19, 1969.

241 "She wasn't there to be": Author interview with Yossi Beilin, January 4, 2005.

242 "These aren't the lowest-paid workers": From discussion at Editors Committee luncheon, *Jerusalem Post*, December 2, 1972.

242 "I will not spend my last days": *Jerusalem Post*, September 7, 1971.

242 she threatened to send in volunteers: *Jerusalem Post*, August 8 and 9, 1971.

243 "I simply cannot stand by": *Jerusalem Post*, July 20, 1973.

243 "From where will the money": *Jerusalem Post*, April 26, 1973.

244 televising on Sabbath: Medzini, *Ha-Yehudiyah*, chapter 16.

244 Ephraim Kishon: Quoted in *New York Times*, June 8, 1969.

244 "For this they had to make": Medzini, *Ha-Yehudiyah*, chapter 16.

245 "If God has to be": *New York Times*, April 4, 1974.

245 "I cannot accept the statement": *Jerusalem Post*, April 5, 1971.

245 "Such a law will create discrimination": *Jerusalem Post*, November 9, 1971.

245 "I am not religious": Medzini, *Ha-Yehudiyah*, chapter 16.

246   tried to avoid the use of a chuppah: "Plain Talk with Golda," *Jerusalem Post*, December 7, 1972.

246   "If I had remained in the Soviet": Oriana Fallaci, *Interview with History*, (Boston: Houghtonn Miffin, 1976), p. 109.

246   "In Golda, you had": Author interview with Shulamit Aloni, December 23, 2004.

247   "But can she type?": Slater, *Golda*, p. 193.

247   a curfew on women: Kenneth Harris interview with Golda, *Midwest Magazine*, March 21, 1971.

247   "In a free, egalitarian": Golda's remark was made during a debate about establishing minimum quotas for female candidates for office. The controversy was covered extensively by the *Jerusalem Post*, especially in June 1973. The quote is from the *Jerusalem Post*, July 12, 1973.

247   Organized feminism: Fallaci interview with Golda, *Ms.*, April 1973.

247   "I've never tried to be a man": *New York Times*, March 18, 1969.

248   "I'm a realist": Slater, *Golda*.

248   "It's no accident": Ibid.

248   "the woman who gives birth": Fallaci interview, *Ms.*

249   "If you can't control the press": Author interview with Meron Medzini, January 11, 2005.

249   "There is one thing they won't say": *Jerusalem Post*, January 14, 1973.

249   "very thin-skinned": Author interview with Naftali Lavi, December 15, 2004.

249   "you felt yourself put in your place": Author interview with Rolf Kneller, December 28, 2004.

249   Bir'im and Ikrit: Author interview with Chaim Hefer, December 14, 2004.

250   Panthereim Schechorim: The best overview of the Mizrahi struggle from the point of view of the Panthers is Sami Shalom Chetrit, *The Mizrahi Struggle in Israel 1948–2003* (Tel Aviv: Am Oved Yisrael, 2004). See also the film *The Black Panthers Speak* (2003).

251   "Imagine how we felt as we watched": Author interview with Kochavi Shemesh, December 28, 2004.

252   "She was our special target": Ibid.

252   "We felt like we were meeting": Ibid.

252   "not nice boys": *Jerusalem Post*, May 19, 1971.

253   "brought deprivation": *Yediot Aharonot*, May 20, 1971.

253   "What has happened to us?": *New York Times*, August 1, 1971.

254   "A nation at war cannot": *Jerusalem Post*, November 20, 1971.

255   The next government, Golda promised: *Jerusalem Post*, March 28, 1973.

CHAPTER FOURTEEN

257   Golda had left orders: Interview with Simcha Dinitz, Golda Meir Library
        Archives UWM, tapes 8 and 9, collection 21.

258   jarring tones in her recurring nightmares: Amnon Barlizai, "Golda Meir's
        Nightmare," *Ha'aretz*, October 3, 2003.

258   Nasser decreed that: Gamal Abdel Nasser to the Egyptian parliament, No-
        vember 6, 1969.

258   Yasir Arafat, the chairman: Arafat interview in Oriana Fallaci, *Interview
        with History* (Boston: Houghton Mifflin, 1976), p. 131.

258   King Faisal of Saudi Arabia: *Beirut Daily Star*, November 17, 1972.

258   she was ready to go to Cairo: *Jersualem Post*, June 12, 1969.

259   mocked her offer: Golda Meir, *My Life* (New York: G. P. Putnam's Sons,
        1975), p. 384.

259   "by parcel post": *New York Times*, March 19, 1969.

260   "We refuse to be a protectorate": United Jewish Appeal dinner, September
        1, 1971, transcript at Israel State Archives.

261   "It isn't a question of, let's": *Meet the Press*, October 1969, transcript from
        Israel State Archives.

261   "The whole world is against us": Song by Yoram Tahar-Lev.

262   "Milwaukee lost the Braves": Tad Szulc, *The Illusion of Peace: Foreign Policy
        in the Nixon Years* (New York: Viking Press, 1978).

263   UN Security Council Resolution 242: Resolution 242 was passed on No-
        vember 22, 1967. Full text is available at http://www.un.org/documents/
        sc/res/1967/scres67.htm.resolution 242. Rogers' plan is well laid out in
        William Quandt, *Decade of Decisions: American Policy Toward the
        Arab-Israeli Conflict, 1967–1976* (Berkeley and Los Angeles: University
        of California Press, 1977).

263   Three Nos: The three nos were issued as the Khartoum Resolutions of
        September 1, 1967, passed at the Arab Summit Conference. They read:
        "no peace with Israel, no recognition of Israel, no negotiations with it,
        and insistence on the rights of the Palestinian people in their own
        country."

263   "back to the '67 borders": From unedited transcript of interview with Golda
        by Aftenposten, Oslo, which ran on February 19, 1972. Transcript un-
        numbered in ISA.

264   "We didn't survive three wars": Golda's remark from December 22, quoted
        in Douglas Little, *American Orientalism: The United States and the*

Middle East since 1945 (Chapel Hill: University of North Carolina Press, 2002).

264 "our Israel": UJA dinner, 1971.

264 firestorm of criticism: Isaiah L. Kenen, *Israel's Defense Line: Her Friends and Foes in Washington* (Buffalo, N.Y.: Prometheus Books, 1981).

265 "I wouldn't shed any tears": Avi Shlaim, *Collusion Across the Jordan: King Abdullah, the Zionist Movement, and the Partition of Palestine* (New York: Columbia University Press, 1988).

266 Rogers decided on one final gambit: Henry Kissinger, "From Turmoil to Hope," in *The White House Years* (Boston: Little, Brown, 1979).

266 Nixon appeared on national television: Richard Nixon, "Building for Peace," A Report to the Congress, February 25, 1971.

267 After the tenth time hearing the same story: Robert Slater, *Golda: The Uncrowned Queen of Israel* (Middle Village, N.Y.: Jonathan David, 1981), p. 185.

268 "Everybody tells us": Remarks delivered at Dorchester Hotel, November 5, 1971, while she was on a Keren Hayesod United Israel Appeal tour. Transcript in Israel State Archives.

269 "If the Arabs felt that": *U.S. News & World Report* interview with Golda, September 22, 1969.

269 Clara . . . asked her to try: Interview with Clara, *Midweek* (BBC), November 29, 1973.

269 "he's the gonif": *New York Times*, March 18, 1969.

269 "This is not the border": *New York Times*, March 21, 1971.

270 "It would be a fatal error": Speech to the Knesset, June 30, 1969, Israeli Government Printing Office.

270 pluck honor from his shame: *Jerusalem Post*, June 8, 1969.

270 "Do you think that an old lady like myself": Slater, *Golda*, p. 194.

271 "Do you still love me": Author interview with Medzini, January 11, 2005.

271 "Better Sharm al-Sheikh": Press conference with Dayan on Israel television, November 23, 1977.

272 Sadat's first statements: *New York Times*, January 3, 1971.

272 Sadat wasn't trying to forge: Sadat laid out his intentions clearly in his autobiography, *In Search of Identity: An Autobiography* (New York: Harper & Row, 1978), and in his *Those I Have Known* (London: Continuum International Publishing Group, 1984).

274 drew a sketchy map: *Times* of London, March 13, 1971.

274 When she asked Rogers to fly: Ibid.

275   "That interpretation made me happy": Sadat, *In Search of Identity*.

276   "better boundaries than King David": Avi Shlaim, "Interview with Abba Eban, 11 March 1976," *Israel Studies* 8, no. 1 (2003). pp. 153–77.

276   Standing on Masada: Howard Blum, *The Eve of Destruction: The Untold Story of the Yom Kippur War* (New York: HarperCollins, 2003), p. 22.

276   Four Wise Men: Saadia Touval, *Peace Brokers: Mediators in the Arab-Israeli Conflict, 1948–1979* (Princeton, N.J.: Princeton University Press, 1982), pp. 203–23.

277   Israel's emblematic informality: *Times*, November 15, 1971.

277   shortly after a trip by Ceaușescu: Golda Meir, *My Life*, pp. 400–402.

278   Mahmoud Riad: For his list of the efforts, see Mahmoud Riad's *The Struggle for Peace in the Middle East* (London: Quartet Books, 1982).

278   Pope Paul VI: *Ma'ariv*, January 19, 1973.

278   The most infamous private peace: For background on the Goldmann fiasco, see Tom Segev, "A Jew Without Borders," *Ha'Aretz*, December 28, 2002; Amnon Rubinstein, "And Now in Israel a Fluttering of Doves," *New York Times Magazine*, July 26, 1970; and *Le Monde*, April 8, 1972.

279   "Nasser, all of a sudden": Amnon Rubenstein, "And Now in Israel a Fluttering of Doves." Golda's quotes are from her interview with the *Jerusalem Post*, April 20, 1970.

280   The most dogged was Marek Halter: Information on the Halter initiatives is derived from his *The Jester and the Kings: A Political Autobiography* (New York: Arcade Publishing, 1989).

CHAPTER FIFTEEN

283   By 9 A.M.: The best account of the Munich story is Michael Bar-Zohar and Eitan Haber, *The Quest for the Red Prince* (New York: Morrow, 1983).

283   For three years: Lists of terrorist incidents are available at http://www .jewishvirtuallibrary.org/jsource/Terrorism/incidents.html and in the database of the Centre for Defence and International Security Studies, Lancaster, UK.

285   "infectious disease": Speech to the Knesset, May 31, 1972, Israeli Government Press Office.

285   "Golda believed that the European states": Yossi Melman, "Preventive Measures," *Ha'aretz*, February 17, 2006.

286   placed the question of terrorism: Hans Koechler, "The United Nations, the

International Rule of Law and Terrorism," Fourteenth Centennial Lecture, Supreme Court of the Philippines, 2002.

286   "Whoever looks on passively": *Jerusalem Post*, October 1972.

287   "to consciously plan": *Washington Post* interview with Yariv, November 24, 1993.

287   "Thanks to your inertia": *L'Europeo*, November 1972.

287   Pope Paul VI lectured her: *Ma'ariv*, January 19, 1973.

288   "Palestine is only a small drop": Fallaci interview with Arafat, *New Republic*, November 16, 1974.

288   "For nineteen years": Golda Meir speech, June, 27, 1972, Israel State Archives, C/788.

288   "Why did the West Bank Palestinians accept": Robert Slater, *Golda: The Uncrowned Queen of Israel* (Middle Village, N.Y.: Jonathan David, 1981), p. 197.

288   "There is no such thing as Palestinians": *New York Times*, January 14, 1976.

289   "Where was the holy principle": *New York Times*, August 27, 1972.

289   "By serenading?": Unedited transcript of Golda Meir interview with Cyrus Sulzberger, *New York Times*, January 29, 1972, Israel State Archives.

289   "We don't want peace": Fallaci interview with Arafat, *New Republic*, November 16, 1974.

289   "When Arafat speaks": *Aftenposten*, Oslo, February 19, 1972.

289   "That would solve the refugee problem": An excellent collection of documents about the UN and the refugee problem appears in Meron Medzini, *Israel's Foreign Relations: Selected Documents* (Jerusalem: Ministry for Foreign Affairs, 1976), pp. 365–465.

290   "practically the same number as that": Speech by Golda Meir at seventeenth session of the United Nations General Assembly, December 1962.

290   "How can you reconcile the outcry": Ibid.

290   "The Arab leaders exploit": Statement to the Knesset, November 12, 1962, Israeli Government Printing Office.

290   No single voice was better: The Eliav story is based on author's interview with Eliav in January 2005 and on Gershom Gorenberg, *The Accidental Empire* (New York: Times Books, Henry Holt, 2006).

291   Marlin Levin of *Time*: Author interview with Levin, December 27, 2005. "The Lion's Roar" ran in *Time* on January 26, 1970.

293   "accidental empire": Gorenberg, *The Accidental Empire*.

294   "Jewish state with a decisive": Golda Meir, Knesset speech, June 25, 1969, Israeli Government Printing Office, and *Jerusalem Post*, June 26, 1969.

294   "avoid gaining tens": *Jerusalem Post*, December 8, 1972.

294   at the request of the mayors: *Jerusalem Post*, June 19 and 21, 1969.

294   "It is better for the Arabs of Judea and Samaria": Answer given during a question-and-answer evening with members of Jerusalem Labour Council, *Jerusalem Post*, December 8, 1972.

295   "All the Jews will be getting PhDs": *Time*, September 19, 1969.

295   The spectacle of Israelis taking the law: The finest of account of the history of the settlements is Gorenberg, *The Accidental Empire*.

295   Yigal Allon: The development of the Allon plan is outlined by Gorenberg.

296   "We should see our presence": A good summary of Dayan's position is included in Rael Jean Isaac, *Israel Divided: Ideological Politics in the Jewish State* (Baltimore: Johns Hopkins University Press, 1976).

298   the Grand Debate: The final meeting was covered in the *Jerusalem Post*, April 12, 1973. See also Gershon Kievel, *Party Politics in Israel and the Occupied Territories* (Westport, Conn.: Greenwood Press, 1983), pp. 70–87.

298   "Does Zionism end at the Green Line?": Ibid.

299   "how many Jewish babies": Isaac, *Israel Divided*, p. 124.

299   "strengthened the foundations": *Jerusalem Post*, July 25, 1973.

299   Galili document: Kievel, *Party Politics*, pp. 76–78; and *Jerusalem Post*, September 3, 1973.

299   "the lash of time": *Jerusalem Post*, September 4, 1973.

300   at his apostasy: Ibid.

300   announced a proposal: The Hussein speech was made on March 15, 1972.

301   since their first meeting: Information on meetings between the Israelis and Hussein is based on Avi Shlaim; *Time*, November 23, 1970; and Moshe Zak, *Hussein the Peacemaker: The History of Israel-Jordan Secret Relations 1964–94* (Beersheva: Bar-Ilan University Press, 1998).

301   "talked about our dreams": Transcript of Avi Shlaim interview with King Hussein, *Ascot*, December 3, 1996.

302   Regular fighting had broken: For background on Hussein and the Palestinians, see Peter Snow, *Hussein: A Biography* (New York: Robert B. Luce, 1972).

303   Golda was in the midst of a forty-minute: Information from Tad Szulc, *The Illusion of Peace: Foreign Policy in the Nixon Years* (New York: Viking Press, 1974); Henry Kissinger, *The White House Years* (Boston: Little, Brown, 1979); William B. Quandt, *Decade of Decision* (Berkeley: University of California Press, 1977); Alan Dowty, *Middle East Crisis: U.S. Decision-Making in 1958, 1970 and 1973* (Berkeley: University of

California Press, 1984); Donald Neff, *Warriors Against Israel* (Brattle-boro, Vt.: Amana Books, 1988); and author's interview with Mordechai Gazit, January 2005.

305   exquisite strand: Author interview with Rinna Samuel, January 18, 2005.

306   "Jordan should have done this": *New York Times*, March 16, 1972.

306   "agent of Zionism and imperialism": *New York Times*, March 11, 1972.

306   "It's not a viable state": Golda Meir speech during Cleveland Mission, November 15, 1972, Israel State Archives.

306   Reuniting the peoples of the two banks: Unedited transcript of Golda Meir interview with Issues and Answers, submitted June 9, 1969, Israel State Archives.

306   she lashed out: Golda Meir speech to Knesset, March 16, 1972, *Jerusalem Post*, March 17, 1972.

CHAPTER SIXTEEN

309   seventy-fifth birthday: Golda's birthday celebration was covered extensively by the *Jerusalem Post*, May 3, 1973.

309   A group of high school students: *Jerusalem Post*, May 4, 1973.

309   73 percent of the Israeli: *Jerusalem Post*, May 10, 1973.

309   most admired: Golda was voted the most admired woman in Britain in a Gallup Poll published in the *Sunday Telegraph*, December 9, 1973. In 1972, she became only the second woman from outside the United States to be at the top of the list of the most admired women in America compiled by the Gallup Poll, an achievement she repeated in 1973 and 1974.

310   the French practically: *New York Times*, December 28, 1972; New York Post, January 2, 1973; and *Jerusalem Post*, various, December 26, 1972, through January 11, 1973.

310   loved her: Richard Nixon interview with Frank Gannon, May 27, 1983, University of Georgia Library.

310   "radiated force": Eli Mizrachi interview, Golda Meir Library Archives, UWM, tape 14, collection 21.

311   "sacking me": *Jerusalem Post*, December 12, 1972.

311   "I am old": Meron Medzini, *Ha-Yehudiyah ha-geah: Goldah Meir ya-ḥazon Yiśra'el: Biyografyah poliṭit.* (Tel Aviv: Edanim, 1990), chapter 16.

311   Senator Henry Jackson: See Stuart Altshuler, *From Exodus to Freedom: A History of the Soviet Jewry Movement* (Lanham, Md.: Rowman and Lit-

tlefield, 2005); and Noam Kochavi, "Insights Abandoned, Flexibility Lost," *Diplomatic History* 29, no. 3 (2005), pp. 503–30.

312 "one foot in war, one foot": Amos Elon, "Israelis Believe War Is Inevitable," *Life*, February 6, 1970.

312 "I want to get things moving": *Jerusalem Post*, July 5, 1973.

313 more money to domestic problems: *Jerusalem Post*, July 11 and 26, 1973.

313 The landing of the Bell 206: Details on Hussein's visit were stitched together from author interviews with Mordechai Gazit and Eitan Haber; Lov Kaddar interview, (56)88, OHD; Ze'ev Schiff, "Was There a Warning?" *Ha'aretz*, June 12, 1998; Amnon Barzilai, "Golda Meir's Nightmare," *Ha'aretz*, October 3, 2003; and Abraham Rabinovich, *The Yom Kippur War: The Epic Encounter That Transformed the Middle East* (New York: Schocken, 2004).

313 who'd served coffee and tea: Kaddar interview, (56)88, OHD.

314 Zusia Kniazer: Ze'ev Schiff, "Was There a Warning?" *Ha'aretz*, June 12, 1998, and Amnon Barzilai, "Golda Meir's Nightmare," *Ha'aretz*, October 3, 2003.

314 Austrian chancellor Bruno Kreisky: The Kreisky story was stitched together from reporting in the *Jerusalem Post*, October 1–5, 1973, from *Time*, October 15, 1973, and *Newsweek*, October 15, 1973.

314 "'a plague on both your houses'": *Jerusalem Post*, October 2, 1973.

315 In-Law, who was, in fact, Nasser's son-in-law: Rabinovich, *The Yom Kippur War*, p. 504; and Howard Blum, *The Eve of Destruction* (New York: HarperCollins, 2003).

316 "wanton conceit": London *Sunday Times* Team, *Yom Kippur War* (London: G. B. Deutsch, 1975).

316 "Is there anything": Rabinovich, *The Yom Kippur War*, p. 642.

316 Unbeknownst to Golda: Information on the lead-up to the war is derived from Golda Meir, *My Life* (New York: G. P. Putnam's Sons, 1975); Blum, *The Eve of Destruction*; Chaim Herzog, *War of Atonement: The Inside Story of the Yom Kippur War* (Boston: Little, Brown, 1975); Zeev Schiff, *October Earthquake: Yom Kippur 1973*, trans. Louis Williams (Tel Aviv: University Pub. Projects, 1974); Rabinovich, *The Yom Kippur War*; and author's interviews with Arye Shalev, Hanoch Bar-Tov, Meir Peil, Uri Bar-Yosef, Ya'akov Hasaid, and Yoav Gelber in December 2004 and January 2005.

318 "Maybe this should tell us something": Rabinovich, *The Yom Kippur War*, p. 81.

318 "Get the first flight out": Interview with Simcha Dinitz, Golda Meir Library Archives UWM.

319 "Golda, I will sleep well tonight": Author interview with Yaacov Hasdai, December 17, 2004.

319 "And what about tomorrow?": Blum, *The Eve of Destruction*, p. 145.

319 Golda agreed with Dayan about the dangers: Golda Meir, *My Life*, p. 426, and Rabinovich, *The Yom Kippur War*, 89–90.

319 But she sided with Dado: Robert Slater, *Golda: The Uncrowned Queen of Israel* (Middle Village, N.Y.: Jonathan David, 1981), p. 239.

320 "better to be in proper shape": Rabinovich, *The Yom Kippur War*, p. 90

320 "What will you do with all those": Ibid., p. 93.

320 Keating was waiting: Amnon Barzilai, "Golda Meir's Nightmare," *Ha'aretz*, October 3, 2003. Walter Isaacson, *Kissinger: A Biography* (New York: Simon & Schuster, 1992); and Rabinovich, *The Yom Kippur War*.

320 "I will not open fire": Rabinovich, *The Yom Kippur War*, p. 92.

322 "We have no doubt": Peter Allen, *The Yom Kippur War* (New York: Charles Scribner's Sons, 1982).

322 "The people of Tel Aviv will be able to sleep": *Time*, October 15, 1973.

322 "we've never actually done it": Rabinovich, *The Yom Kippur War*, p. 171.

324 "The destruction of the Third Temple": Blum, *The Eve of Destruction*, p. 195.

324 "Golda, I was wrong about everything": Allen, *The Yom Kippur War*.

324 "sort of grayish khaki": Interview with Kaddar, (56)88, OHD.

324 "generalissimo": Quoted in Zeev Schiff, *October Earthquake: Yom Kippur 1973*, trans. by Louis Williams (Tel Aviv: University Pub. Projects, 1974), p. 148.

325 "Today we hit bottom": Blum, *The Eve of Destruction*.

325 "The great Moshe Dayan!": Chaim Herzog, *The War of Atonement: The Inside Story of the Yom Kippur War* (Boston: Little, Brown, 1975).

325 "The Third Commonwealth might be destroyed": Slater, *Golda*, p. 241.

326 "We do not now have the strength": The full text of Dayan's remarks were printed in *Ha'aretz*, February 15, 1974.

326 "If you say on television": London *Sunday Times* Team, *Yom Kippur War*, p. 216.

326 Another editor called Golda: Rabinovich, *The Yom Kippur War*, p. 270.

327 "[Dayan] was beaten": Quoted in Robert Slater, *Warrior Statesman: The Life of Moshe Dayan* (New York: St. Martin's Press, 1991).

327 arming of Israeli nuclear weapons: The information for this section came

from Seymour Hersh, *The Samson Option: Israel's Nuclear Arsenal and American Foreign Policy* (New York: Random House, 1991); Avner Cohen, *Israel and the Bomb* (New York: Columbia University Press, 1998); and Blum, *Eve of Destruction*. The article from Stern was reported in the *Jerusalem Post*, March 12, 1980.

328 the new battlefront became Washington: The controversy about Kissinger and weapons for Israel has been laid out in Tad Szulc, *The Illusion of Peace: Foreign Policy in the Nixon Years* (New York: Viking Press, 1974), pp. 735–39; Walter Isaacson, *Kissinger: A Biography*, pp. 450–78; Edward Luttwak and W. Z. Laqueur, "Kissinger and the Yom Kippur War," *Commentary* 58, no. 3 (September 1974), pp. 33–62; Edward Sheehan, "How Henry Kissinger Did It," *Foreign Policy* (spring 1976), pp. 21–69; Elmo R. Zumwalt, *On Watch: A Memoir* (New York: Quadrangle/New York Times Book Co., 1976); Matti Golan, *The Secret Conversations of Henry Kissinger: Step-by-Step Diplomacy in the Middle East* (New York: Quadrangle/New York Times Book Co., 1976).

331 "get just as much blame": Isaacson, *Kissinger*, p. 522.

332 "back to being ourselves": Donald Neff, *Warriors against Israel* (Brattleboro. Vt.: Amana Books, 1988).

332 "given time to enjoy his defeat": Golda on *Face the Nation*, CBS, October 28, 1973.

334 "I presume she is wild," he remarked to Eban: Matti Golan, *The Secret Conversations of Henry Kissinger: Step-by-Step Diplomacy in the Midle East*, trans. by Ruth Geyra Stern and Sol Stern (New York: Quadrangle/New York Times Book Co., 1976) p. 88.

335 "Well, in Vietnam": Isaacson, *Kissinger*, p. 528.

335 "If the Egyptians fail": Michael Brecher, *Decisions in Israel's Foreign Policy* (New Haven, Conn.: Yale University Press, 1975), p. 223.

335 "I won't stand for the destruction of the Third Army": Golan, *Secret Conversations*.

335 Golda usually communicated: Golda Meir, *My Life*, p. 445.

336 "There is only one country": Ibid., pp. 371–72.

CHAPTER SEVENTEEN

337 Motti Ashkenazi took up: Author interview with Motti Ashkenazi, December 19, 2004; and Abraham Rabinovich, *The Yom Kippur War: The Epic*

Encounter That Transformed the Middle East (New York: Schocken, 2004).

338 "MY SON DIDN'T FALL IN BATTLE": Author interview with Meron Medzini, December 15, 2004.

338 "It doesn't matter what logic": Golda Meir, *My Life* (New York: Putnam, 1975), p. 425.

338 "never be the same Golda": Q and A on Israeli television, covered by the *Jerusalem Post*, April 26, 1974.

339 "Golda, is this all worth it?": Author interview with Yehuda Avner, January 4, 2005.

339 "Golda, be strong": Meron Medzini, *Ha-Yehudiyah ha-geah: Goldah Meir va-ḥazon Yisra'el: Biyografyah poliṭit* (Tel Aviv, Edanim, 1990), chapter 19.

339 "It's ridiculous": White House Memorandum of Conversation between Kissinger and Meir et al., Blair House, November 1, 1973, National Archives of the United States.

341 "There can be no greater mistake": *Wisconsin Jewish Chronicle*, June 1, 1973.

341 "They don't want us here": *New York Times*, October 26, 1969.

341 "We both have Jewish": Matti Golan, *The Secret Conversations of Henry Kissinger: Step-by-step Diplomacy in the Middle-East*, trans. by Ruth Geyra Stern and Sol Stern (New York: Quadrangle/New York Times Book Co., 1976).

341 "We have to squeeze": Walter Isaacson, *Kissinger: A Biography* (New York: Simon & Schuster, 1992), p. 522.

342 oozing empathy: White House Memorandum of Conversation between Nixon and Meir et al., Oval Office, November 1, 1973, National Archives of the United States.

342 "Where is the October 22 line?": White House Memorandum of Conversation between Kissinger and Meir et al., Blair House, November 2, 1973, National Archives of the United States.

343 As Kissinger emerged: Isaacson, *Kissinger*, p. 504.

343 "The nineteenth century is my specialty": Ibid., p. 552.

343 "We didn't begin the war": White House Memorandum of Conversation between Kissinger and Golda et al., Blair House, November 3, 1973, National Archives of the United States.

344 forcing his Orthodox parents: Isaacson, *Kissinger*, p. 505.

344 "But on Israeli policy, I have my PhD": Richard Valeriani, *Travels with Henry* (Boston: Houghton Mifflin Company, 1979), p. 200.

345   "You are not giving me anything": Golan, *Secret Conversations*, p. 109.

345   "all we have, really, is our spirit": Golda Meir, *My Life*, p. 447.

345   breakfast with fourteen: *Jerusalem Post*, November 4, 1973.

345   Her friends George Meany of the AFL-CIO: Golan, *Secret Conversations*, p. 109.

346   "For the first time in a quarter of a century": Golda Meir, *My Life*, p. 448.

346   "methods of Goebbels": Golan, *Secret Conversations*, p. 96.

347   "I am a very, very": London *Sunday Times*, p. 442.

347   "if we ever see Joe again": Marvin Kalb and Bernard Kalb, *Kissinger* (Boston: Little, Brown, 1974), p. 512.

348   "icing on the cake": For a view of Kissinger's thinking on guarantees, see Isaacson, *Kissinger*, p. 541, and Theodore Draper, "The United States and Israel," *Commentary* 59, no. 4 (April 1975).

349   WHERE IS MY FATHER: *Time*, November 12, 1973.

349   "My dear General": Kenneth W. Stein, "The Talks at Kilometer 101," in Richard B. Parker, ed., *The October War* (Gainesville: University Press of Florida Press, 2001), pp. 361–73.

349   off-kilter Star of David: Author interview with Rinna Samuel, December 18, 2004.

350   Twenty-one Socialist political leaders: Golda Meir, *My Life*, p. 446, and Harold Wilson, *The Chariot of Israel* (New York: W. W. Norton, 1981).

352   "Moshe, you created the feeling": Author interview with Michael Bar-Zohar, December 21, 2004.

352   the minister of justice screamed: Golda Meir, *My Life*, p. 450.

352   "strength of a warrior without fear": Medzini, *Ha-Yehudiyah*, chapter 17.

352   "45 percent of the voters": *Time*, December 10, 1973, poll commissioned by *Ha'aretz*.

352   "who is for whom and what if for what": *Jerusalem Post*, November 23, 1973.

353   "1 could offer excuses": Labor Party Central Committee, December 5, 1973, Israel State Archives, 106/810/2P, and *New York Times*, December 30, 1973.

355   Kissinger cajoled Golda: The information about Geneva is from Eban, *Abba Eban: An Autobiography* (New York: Random House, 1977), pp. 548–53; *Newsweek*, December 31, 1973; and the *Sunday Times* of London, pp. 483–84.

355   "One must not expect": *Jerusalem Post*, November 12, 1973.

356   Kissinger gave new meaning: Background on the first shuttle is from Peter

Allen, *The Yom Kippur War: The Epic Encounter That Transformed the Middle East* (New York: Charles Scribner's Sons, 1982); Isaacson, *Kissinger*; and Edward R. F. Sheehan, *The Arabs, Israelis, and Kissinger: A Secret History of American Diplomacy in the Middle East* (New York: Readers Digest Press: distributed by Crowell, 1976).

358 "The United States holds most": *Le Monde*, January 22, 1975.

358 nasty case of herpes: Richard Valeriani, *Travels with Henry* (Boston: Houghton Mifflin, 1979).

358 "All our sympathies": Donald Neff, *Warriors Against Israel* (Brattleboro, Vt.: Amana Books, 1988).

358 "He had a unique ability": Valeriani, *Travels*, p. 208.

358 A course described as: Ibid., pp. 210–11.

359 "played the role of an extremely talented": Ibid., p. 208.

359 Metternich's view: Isaacson, *Kissinger*, p. 554.

359 "It struck me as strange": Slater, *Golda*, p. 256.

359 probably mythic State Department story: Valeriani, *Travels*, pp. 197–98.

359 "benevolent aunt": Henry Kissinger, *The White House Years* (Boston: Little, Brown, 1979).

360 "You must take my word seriously": Ibid., p. 836.

360 "I am deeply conscious": Ibid., p. 844.

361 "The state has gone through a devaluation": Quoted in *Time*, March 14, 1974.

361 "Why are you protecting him?": Interview with Yossi Sarid, January 2, 2005.

361 approval rating fell: *Time*, March 14, 1974.

361 "The streets will not tell me": Medzini, *Ha-Yehudiyah*, chapter 19.

362 Shulamit Aloni, for defying: Author interview with Shulamit Aloni, December 23, 2004.

362 "Golda Meir had many great": Abba Solomon Eban, *Personal Witness: Israel Through My Eyes* (New York: G.P. Putnam's Sons, 1992), p. 554.

362 "I don't think they use sensible methods": Slater, *Golda*, p. 258.

363 "I have asked him": *Jerusalem Post*, February 21, 1974.

363 "Had I resigned last August": *Jerusalem Post*, March 4, 1974.

363 "Golda, come back": Author interview with Aharon Yadlin, December 30, 2004.

364 At the last minute: Eban, *Personal Witness*, p. 562, and *Jerusalem Post*, March 6 and 8, 1974.

364 Agranat Commission: For a thorough account of the commission and its

report, see Phina Lahav, *Judgment in Jerusalem* (Los Angeles and Berkeley: University of California Press, 1997).

365 "You realize that it won't end": *Time*, April 22, 1974.

365 On Thursday, April 10: *Jerusalem Post*, April 11, 1974.

366 "I am very tired": Menachem Meir, *My Mother Golda Meir: A Son's Evocation of Life with Golda Meir* (New York: Arbor House, 1983), p. 226.

367 "Two wars in seven": Isaacson, *Kissinger*, p. 576.

367 "I didn't just get up one day in 1967": Slater, *Golda*.

367 "like a rug merchant": Isaacson, *Kissinger*, p. 570.

368 "she had a grudge": Author interview with Ari Rath, December 23, 2004.

368 "Miss Israel" and "the beauty of Jerusalem": Isaacson, *Kissinger*, p. 553, and Valeriani, *Travels*, p. 295.

368 "Go get the best numbers you can": Valeriani, *Travels*, p. 200.

368 town of Ma'alot: A good recounting of the Ma'alot massacre ran in *Time*, March 27, 1974.

369 "better get back to peacemaking": Eban, *Personal Witness* p. 571.

CONCLUSION

372 slow down the immigration: *Jerusalem Post*, February 16, 1978.

372 Kolleck, the mayor of Jerusalem: Comments made in a television interview, reported in the *Jerusalem Post*, February 5, 1978.

372 "Doctrinaire and obsessed": Chaim Herzog, *Living History: The Memoirs of a Great Israeli Freedom-Fighter, Soldier, Diplomat and Statesman* (New York: Pantheon, 1996).

373 "So you're tired": *New York Post*, April 22, 1974.

373 compared her to Winston: Quoted in *Jerusalem Post*, April 23, 1974.

373 Kissinger threw her a black-tie: *Jerusalem Post*, December 20, 1974.

373 President Gerald Ford: *Jerusalem Post*, December 21, 1975, and May 20, 1976.

373 And when Jimmy Carter was running: *New York Times*, May 27, 1976.

374 declared Richard Nixon: Exchange of toasts between President Katzir and President Nixon in the Knesset, June 16, 1974, Israel Minister of Foreign Affairs, Historical Documents 1974–1977.

374 Even Sadat jumped: *Time*, January 2, 1978.

374 "I wouldn't want the West Bank": *Jerusalem Post*, May 22, 1978.

374 Her autobiography: Author interview with Rinna Samuel, December 18, 2004.

375   If I'd looked and sounded like Bancroft: Author interview with William
           Gibson, July 2004, and William Gibson, *Golda: Notes on How to Turn a
           Phoenix into Ashes* (New York: Atheneum, 1978), pp. 31–32.

375   "Israel will be astonished": *New York Times*, November 15, 1977.

375   "I'm not going to have people say": Author interview with Rinna Samuel,
           January 2, 2005.

375   "What's all the Messianic": *Jerusalem Post*, November 21, 1977.

376   "The time was not yet ripe": The repartee between Golda and Sadat was
           overheard by Yitzhak Rabin, who was standing nearby. From Rabin in-
           terview, Golda Meir Library Archives, UWM, tape 19, collection 21.

376   "I don't know about the Nobel": *Jerusalem Post*, November 25, 1977.

376   "Galili would be her agent": Author interview with Mordechai Gazit, Janu-
           ary 6, 2005.

377   It fell to Mordechai: Ibid.

377   spate of pamphlets and articles: That work included *Israeli Diplomacy and
           the Quest for Peace* (London: Routledge, 2002); "Peace between Egypt
           and Israel: A Missed Opportunity in 1971," *Zionism (Hatzionut)* 21 (1998);
           "Egypt and Israel—Was There a Peace Opportunity Missed in 1971?"
           Journal of Contemporary History 32, no. 1 pp. 97–115 (January 1997); *The
           Peace Process, 1969–1973: Efforts and Contacts* (Jerusalem: Magnes Press,
           1983); and "Golda and Peace," Jerusalem Post, January 1, 1980.

377   "the horrific Golda Syndrome": Doron Rosenblum, "Is He Another of the
           Goldas?" *Ha'aretz*, June 24, 2006.

378   "No Arab government offered": Simcha Dinitz interview, Golda Meir Li-
           brary Archives, UWM, tapes 8 and 9, collection 21.

378   his wife, Jihan: In interview with *Yediot Aharonot*, June 11, 1989.

378   "done better and gone farther than Menahem Begin": Sadat interview in
           October magazine, February 18, 1978.

380   " 'How dare you say Jews' ": Author interview with Shulamit Aloni, Decem-
           ber 23, 2004.

380   "It is true we do have a Masada complex": Golda was responding to Stewart
           Alsop's column in *Newsweek*, July 12, 1971.

380   "We also have a pogrom complex. We have a Hitler complex.": "Israel at
           40," *Time*, February 5, 2007.

380   "We are not Czechoslovakia": *L'Express*, July 10, 1970.

383   "Israeli society, for thirty years, a full generation": Author's interview with
           Yaakov Hasdai, December 17, 2004.

BIBLIOGRAPHY

Newspapers and Periodicals

Commentary
Commonweal
Current History
Foreign Policy
Fortune
Ha'aretz
Jerusalem Post
Life
Look
Middle East Record
The Nation
The New Republic
The New York Times
The New Yorker
Newsweek
The Palestine Post
Saturday Evening Post
Time
U.S. News and World Report
The Washington Post

Interviews

Achimer, Yossi. December 22, 2004
Adler, Haim. January 3, 2005
Aloni, Shulamit. December 23, 2004

Alpher, Yossi. December 22, 2004

Ashkenazi, Motti. December 19, 2004

Avner, Yehuda. January 6, 2005

Avnery, Uri. January 4, 2005

Bailey, Clinton. December 19, 2004

Bar-On, Dan. December 31, 2004

Bar-Yossef, Uri. January 2, 2005

Bar-Zohar, Michael. December 21, 2004

Bartov, Hanoch. December 27, 2004

Beilin, Yossi. January 4, 2005

Bentsur, Eitan. December 28, 2004

Biton, Charlie. December 23, 2004

Bloch, Danny. December 29, 2004

Bushinski, Jay. December 16, 2004

Cohen, Amnon. December 24, 2004

Dayan, Yael. December 16, 2004

Eliav, Aryeh (Lova). December 29, 2004

Elpeleg, Tzvi. December 20, 2004

Eshkol, Miriam. January 12, 2005

Gazit, Mordechai. January 6, 2005

Gelber, Yoav. January 5, 2005

Gibson, William. July 15, 2004

Goldman, Ralph. January 12, 2005

Golembo, Alice. August 3, 2004

Gruber, Ruth. July 2004

Haber, Eitan. December 20, 2004

Harish, Micah. January 13, 2005

Harmon, David. December 14, 2004

Hasdai, Yaakov. December 17, 2004

Hefer, Haim. December 14, 2004

Herlitz, Esther. December 16, 2004

Hillel, Shlomo. December 21, 2004

Josephtal, Senta. January 9, 2005

Jubran, Salem. January 5, 2005

Kneller, Rolf. December 28, 2004

Lapidot, Yehuda. January 3, 2005

Lavi, Naftali. December 15, 2004

Levin, Marlin. December 27, 2004

Levy, Baruch. January 2, 2005
Lossin, Yigal. December 19, 2004
Medzini, Meron. December 15, 2004; January 11, 2005
Meyerson, Menachim. January 13, 2005
Morris, Benny. January 12, 2005
Nakdimon, Shlomo. December 20, 2004
Namir, Ora. January 9, 2005
Navon, Yitzhak. December 30, 2004
Nitzan, Yaakov. December 26, 2004
Oded, Aryre. December 17, 2004
Oren, Michael. January 11, 2005
Pail, Meir. January 2, 2004
Pazner, Avi. January 11, 2005
Rachamin, Mordechai. December 22, 2004
Rath, Ari. December 23, 2004
Rehabi, Sara. January 13, 2005
Rifman, Shmuel. December 30, 2004
Rosolio, Danny. January 10, 2005
Rothstein, Raphael. August 2004
Samuel, Rinna. December 18, 2004; January 2, 2005
Sarid, Yossi. January 2, 2005
Segev, Tom. January 7, 2005
Shalev, Arye. January 13, 2005
Sharett, Yaakov. January 4, 2005
Shemesh, Kochavi. December 28, 2004
Shmueli, Eliezer. September 24; October 16, 2007
Yadlin, Aharon. December 30, 2004
Zuckerman, Nomi. December 15, 2004

Oral Histories

Golda Meir Library Archives, University of Wisconsin at Milwaukee
    Aloni, Shulamit: tape 3, collection 21
    Avidar, General Joseph: tape 4, collection 21
    Avner, Yehuda: tapes 5 and 6, collection 21
    Aynor, Hannan: tape 16, collection 21
    Dinitz, Simcha: tapes 8 and 9, collection 21

Eshel, Tamar: tape 10, collection 21

Hamburger-Medzini, Regina: tape 11, collection 21

Harman, Abraham: tapes 13 and 14, collection 21

Harman, David: tape 15, collection 21

Harman, Zena: tape 13, collection 21

Kaddar, Lou: tapes 7 and 12, collection 21

Lambert, Anne Marie: tape 17, collection 21

Medzini, Meron: tape 18, collection 21

Meir, Golda: tape 1, collection 21

Meir, Golda: tape 2, collection 21, speech CJF General Assembly 1977

Meir, Golda: tape 24, collection 21, political functions and press conferences

Meir, Golda: tape 25, collection 21, addresses to Knesset; speeches and interviews

Meir, Golda: tape 26, collection 21, reflections and interviews

Mizrachi, Eli: tape 14, collection 21

Rabin, Yitzhak: tape 19, collection 21

Rahabi, Sara: tapes 20 and 21, collection 21

Rath, Ari: tape 22, collection 21

Scherf, Zeev: tape 5, collection 21

Oral History Division, Institute for Contemporary Jewry, Hebrew University of
Jerusalem

Beale, Harry

Boyar, Lou

Brailov, Mathilda

Comay, Michael

Davis, Moshe

Goldstein, Israel

Kaddar, Lou

Livingston, Julius

Montor, Harry

New York Public Library, Golda Meir, interview with Mitchell Krausse

Archives

Central Zionist Archives

Golda Meir Library Archives, University of Wisconsin–Milwaukee

Israel State Archives

National Archives of the United States
Nelson Riddle Collection, University of Arizona School of Music
New York Public Library
Oral History Division, Institute for Contemporary Jewry, The Hebrew University

Books

Abadi, Jacob. *Israel's Leadership: From Utopia to Crisis.* Westport, Conn.: Greenwood Press, 1993.

Abdullah, King of Transjordan. *Memoirs.* Edited by Philip R. Graves. London: Jonathan Cape, 1950.

Allen, Peter. *The Yom Kippur War: The Epic Encounter That Transformed the Middle East.* New York: Charles Scribner's Sons, 1982.

Allon, Yigal. *The Making of Israel's Army.* Foreword by Michael Howard. New York: Universe Books, 1970.

———. *My Father's House: Israel's Foreign Minister Looks Back at the Heroic People, Heroic Country of His Youth.* New York: W. W. Norton, 1976.

Almogi, Yosef. *Total Commitment.* New York: Herzl Press, 1982.

AlRoy, Gil Carl. *The Kissinger Experience: American Policy in the Middle East.* New York: Horizon Press, 1975.

Altshuler, Stuart. *From Exodus to Freedom: A History of the Soviet Jewry Movement.* Lanham, Md.: Rowman and Littlefield, 2005.

Antonius, George. *The Arab Awakening: The Story of the Arab National Movement.* Safety Harbor, Fla.: Simon Publications, 2001.

Aronoff, Myron Joel. *Power and Ritual in the Israel Labor Party: A Study in Political Anthropology.* Armonk, N.Y.: M. E. Sharpe, 1993.

Aruri, Naseer H., ed. *Middle East Crucible: Studies on the Arab-Israeli War of October 1973.* AAUG monograph, series no. 6. Wilmette, Ill.: Medina University Press International, 1975.

Associated Press. *Lightning Out of Israel: The Six-day War in the Middle East.* New York: Associated Press, 1967.

Avallone, Michael. *A Woman Called Golda.* New York: Leisure Books, 1982.

Avineri, Shlomo. *The Making of Modern Zionism: Intellectual Origins of the Jewish State.* New York: Basic Books, 1981.

———, ed. *Israel and the Palestinians.* New York: St. Martin's Press, 1971.

Avishai, Bernard. *Tragedy of Zionism: How Its Revolutionary Past Haunts Israeli Democracy.* New York: Helios Press, 2002.

Avriel, Ehud. *Open the Gates! A Personal Story of "Illegal" Immigration to Israel*. New York: Atheneum, 1975.

Azcarate, Pablo. *Mission in Palestine, 1948–1952*. Washington, D.C.: Middle East Institute, 1966.

Bailey, Clinton. *Jordan's Palestinian Challenge 1948–1983: A Political History*. Boulder and London: Westview Press, 1984.

Bailey, Sydney Dawson. *The Making of Resolution 242*. The Hague and Boston: M. Nijhoff, 1985.

Barer, Shlomo. *The Magic Carpet*. New York: Harper, 1952.

Bar-Siman-Tov, Yaacov. *The Israeli-Egyptian War of Attrition, 1969–1970: A Case-study of Limited Local War*. New York: Columbia University Press, 1980.

Bartov, Hanoch. *Dado: 48 Years and 20 Days*. Tel Aviv: Sifriyat Ma'ariv, 1978.

Bar-Zohar, Michael. *Ben-Gurion: The Armed Prophet*. New York: Prentice-Hall, 1968.

———. *Facing a Cruel Mirror: Israel's Moment of Truth*. New York: Scribner, 1990.

———. *Spies in the Promised Land: Isser Harel and the Israeli Secret Service*. Translated from the French by Monroe Stearns. Boston: Houghton Mifflin, 1972.

Bar-Zohar, Michael, and Eitan Haber. *The Quest for the Red Prince*. New York: Morrow, 1983.

Bass, Warren. *Support Any Friend: Kennedy's Middle East and the Making of the U.S.-Israel Alliance*. New York: Oxford University Press, 2003.

Bauer, Yehuda. *Flight and Rescue: Brichah*. New York: Random House, 1970.

Becker, Abraham, S. *Israel and the Palestinian Occupied Territories: Military-Political Issues in the Debate*. Santa Monica, Calif.: RAND, 1971.

Begin, Menachem. *Revolt: Story of the Irgun*. Translated from the original Hebrew text by Shmuel Katz. New York: Schuman, 1951.

Ben-Aharon, Yitzhak. *The Courage to Change*. Jerusalem: Jerusalem Academic Press, 1971

Ben-Gurion, David. *Israel: A Personal History*. New York: Funk & Wagnalls, 1971.

———. *Israel: Years of Challenge*. New York: Holt, Rinehart and Winston, 1963.

———. *Letters to Paula*. London: Valentine, Mitchell, 1971.

———. *Memoirs: David Ben-Gurion*. New York: World, 1970.

———. *Rebirth and Destiny of Israel*. Edited and translated from the Hebrew under the supervision of Mordekhai Nurock. New York: Philosophical Library, 1954.

Berger, Earl. The *Covenant and the Sword: Arab-Israeli Relations, 1948–56*. London: Routledge & Kegan Paul, 1965.

Bernadotte, Folke. *To Jerusalem*. London: Hodder and Stoughton, 1951.

Bernstein, Deborah, ed. *Pioneers and Homemakers: Jewish Women in Pre-state Israel*. Albany: State University of New York, 1992.

Bernstein, Marver H. *Politics of Israel: The First Decade of Statehood*. Princeton, N.J.: Princeton University Press, 1957.

Bethell, Nicholas William. *The Palestine Triangle: The Struggle Between the British, the Jews, and the Arabs, 1935–1948*. London: A. Deutsch, 1979.

Bethmann, Erich W. *Decisive Years in Palestine, 1918–1948*. New York: American Friends of the Middle East, 1957.

Bingham, June. *U Thant: The Search for Peace*. New York: Alfred A. Knopf, 1966.

Black, Ian, and Benny Morris. *Israel's Secret Wars: The Untold History of Israeli Intelligence*. New York: Grove Weidenfeld, 1991.

Blum, Howard. *The Eve of Destruction: The Untold Story of the Yom Kippur War*. New York: HarperCollins, 2003.

Bober, Arie. *The Other Israel: The Radical Case Against Zionism*. Garden City, N.Y.: Anchor Books, 1972.

Brecher, Michael. *Decisions in Israel's Foreign Policy*. New Haven, Conn.: Yale University Press, 1975.

———. *Foreign Policy System of Israel: Setting, Images, Process*. New Haven, Conn.: Yale University Press, 1972.

Bregman, Ahron, and Jihan El-Tahri. *The Fifty Years War: Israel and the Arabs*. London: Penguin Books, BBC Books, 1998.

Brookings Middle East Study Group. *Toward Peace in the Middle East: Report of a Study Group*. Washington, D.C.: Brookings Institution, 1975.

Brown, Michael. *The Israeli-American Connection: Its Roots in the Yishuv, 1914–1945*. Detroit: Wayne State University Press, 1996.

Bullocks, John. *The Making of War. The Middle East from 1967 to 1973*. London: Longman, 1974.

Burdett, Winston. *Encounter with the Middle East: An Intimate Report on What Lies Behind the Arab-Israeli Conflict*. New York: Atheneum, 1969.

Burns, E. L. M. (Eedson Louis Millard). *Between Arab and Israeli*. New York: I. Obolensky, 1963.

Byford-Jones, W. *Lightning War*. London: Hale, 1967.

Caplan, Neil. *Futile Diplomacy*, volume 2, *Arab-Zionist Negotiations and the End of the Mandate*. London: Frank Cass, 1986.

Carnegie Endowment for International Peace. *Israel and the United Nations: Report of a Study Group Set Up by the Hebrew University of Jerusalem: National Studies on International Organization*. New York: Manhattan Publishing Company, 1956.

Chetrit, Sami Shalom. *The Mizrahi Struggle in Israel, 1948–2003* [Hebrew]. Tel Aviv: Am Oved, 2004.

Christman, Henry. *The State Papers of Levi Eshkol*. New York: Funk & Wagnalls, 1969.

———, ed. *This Is Our Strength: Selected Papers of Golda Meir*. New York: Macmillan, 1962.

Churchill, Randolph S., and Winston S. Churchill. *The Six Day War*. New York: Houghton Mifflin, 1967.

Cohen, Avner. *Israel and the Bomb*. New York: Columbia University Press, 1998.

Cohen, Michael Joseph. *Origins and Evolution of the Arab-Zionist Conflict*. Berkeley: University of California Press, 1987.

———. *Palestine and the Great Powers, 1945–1948*. Princeton, N.J.: Princeton University Press, 1982.

———. *Palestine, Retreat from the Mandate: The Making of British Policy, 1936–48*. New York: Holmes & Meier, 1978.

———. *Zion and State: Nation, Class and the Shaping of Modern Israel*. New York: Columbia University Press, 1992.

Collins, Larry, and Dominique Lapierre. *O Jerusalem*. New York: Simon & Schuster, 1972.

Cordier, Andrew W., and Wilder Foote. *Public Papers of the Secretaries-General of the United Nations*. New York: Columbia University Press, 1969–1977.

Crossman, R. H. S. *A Nation Reborn; A Personal Report on the Roles Played by Weizmann, Bevin and Ben-Gurion in the Story of Israel*. New York: Atheneum, 1960.

———. *Palestine Mission: A Personal Record*. New York: Arno Press, 1977.

Crum, Bartley Cavanaugh. *Behind the Silken Curtain: A Personal Account of Anglo-American Diplomacy in Palestine and the Middle East*. Jerusalem: Milah Press, 1996.

Curtis, Michael, and Susan Aurelia Gitelson, eds. *Israel in the Third World*. New Brunswick, N.J.: Transaction Books, 1976.

Dallas, Roland. *King Hussein: A Life on the Edge*. New York: Fromm International, 1999.

Dawidowicz, Lucy S. *War Against the Jews, 1933–1945*. Toronto and New York: Bantam Books, 1986.

Dayan, Moshe. *Breakthrough: A Personal Account of the Egypt-Israel Peace Negotiations*. New York: Alfred A. Knopf, 1981.

———. *Diary of the Sinai Campaign*. Westport, Conn.: Greenwood Press, 1979.

——. *Moshe Dayan: Story of My Life: An Autobiography*. New York: William Morrow, 1976.

Dayan, Yaël. *My Father, His Daughter*. New York: Farrar, Straus & Giroux, 1985.

Donovan, Robert J. *Six Days in June: Israel's Fight for Survival*. New York: New American Library, 1967.

Douglas-Home, Charles. *Arabs and Israel*. London, Sydney [etc.]: Bodley Head, 1968.

Dowty, Alan. *Middle East Crisis: U.S. Decision-Making in 1958, 1970 and 1973*. Berkeley: University of California Press, 1984.

Eban, Abba Solomon. *Abba Eban: An Autobiography*. New York: Random House, 1977.

——. *American Foreign Policy in the 80's*. Dallas, Tex.: Corporate Affairs Dept. of the LTV Corp., 1980.

——. *Personal Witness: Israel Through My Eyes*. New York: G. P. Putnam's Sons, 1992.

——. *The Political Legacy of Golda Meir*. Morris Fromkin Memorial Lecture. Milwaukee: Golda Meir Library, University of Wisconsin, Milwaukee, 1995.

Edelman, Maurice. *Ben Gurion, A Political Biography*. London: Hodder and Stoughton, 1964.

——. *David: The Story of Ben Gurion*. New York: Putnam, 1965.

Eisenhower, Dwight D. *Waging Peace, 1956–1961: The White House Years*. Garden City, N.Y.: Doubleday, 1965.

Eisenhower, Julie Nixon. *Special People*. New York: Simon & Schuster, 1977.

Elath, Eliahu. *Zionism at the UN: A Diary of the First Days*. Translated from Hebrew by Michael Ben-Yitzhak; foreword by Howard M. Sachar. Philadelphia: Jewish Publication Society of America, 1976.

Elazar, Daniel Judah. *Judea, Samaria, and Gaza: Views on the Present and Future*. Washington, D.C.: American Enterprise Institute, 1982.

Ellis, Harry B. *Israel and the Middle East*. New York: Ronald Press, 1957.

Elon, Amos. *The Israelis: Founders and Sons*. New York: Holt, Rinehart and Winston, 1971.

Elpeleg, Zvi. *The Grand Mufti: Haj Amin al-Hussaini, Founder of the Palestinian National Movement*. London: Frank Cass, 1993.

ESCO Foundation for Palestine, Inc. *Palestine: A Study of Jewish, Arab, and British Policies*. 2 vols. New Haven, Conn.: Yale University Press; London: G. Cumberlege and Oxford University Press, 1947.

Eshkol, Levi. *State Papers*. Ed., with an introduction, by Henry M. Christman. New York: Funk & Wagnalls, 1969.

Eytan, Walter. *First Ten Years: A Diplomatic History of Israel*. New York: Simón & Schuster, 1958.

Fallaci, Oriana. *Interview with History*. Boston: Houghton Mifflin, 1976.

Fein, Leonard J. *Israel: Politics and People*. Boston: Little, Brown, 1968.

Feuerwerger, Marvin C. *Congress and Israel: Foreign Aid Decision-Making in the House of Representatives*. Westport, Conn.: Greenwood Press, 1979.

Fischl, Viktor. *Moscow and Jerusalem: Twenty Years of Relations Between Israel and the Soviet Union*. London and New York: Abelard-Schuman, 1970.

Frankel, William. *Israel Observed: An Anatomy of the State*. London: Thames and Hudson, 1980.

Friedman, Thomas. *From Beirut to Jerusalem*. New York: Anchor Books, 1989.

Gabbay, Rony. *A Political Study of the Arab-Jewish Conflict: The Arab Refugee Problem, a Case Study*. Geneva, Switzerland: E. Droz, 1959.

Ganin, Zvi. *Truman, American Jewry, and Israel, 1945–1948*. New York: Holmes & Meier, 1979.

Garcia-Granados, Jorge. *The Birth of Israel: The Drama as I Saw It*. New York: Alfred A. Knopf, 1948.

Gazit, Mordechai. *The Peace Process, 1969–1973: Efforts and Contacts*. Jerusalem: Magnes Press, 1983.

———. *President Kennedy's Policy Toward the Arab States and Israel: Analysis and Documents*. Tel Aviv: Shiloah Center for Middle Eastern and African Studies, Tel Aviv University, 1983.

Gelber, Yoav. *Israeli-Jordanian Dialogue, 1948–1953: Cooperation, Conspiracy, or Collusion?* Portland, Ore.: Sussex Academic Press, 2004.

———. *Jewish-Transjordanian Relations, 1921–48*. Portland, Ore.: Frank Cass, 1997.

Gervasi, Frank. *Case for Israel*. New York: Viking Press, 1967.

Gibson, William. *Golda: Notes on How to Turn a Phoenix into Ashes*. New York: Atheneum, 1978.

Gilbert, Martin. *Exile and Return: The Struggle for a Jewish Homeland*. Philadelphia: Lippincott, 1978.

Glubb, John Bagot, Sir. *The Changing Scenes of Life: An Autobiography*. London: Quartet Books, 1983.

———. *Soldier with the Arabs*. New York: Harper, 1957.

Golan, Matti. *Road to Peace: A Biography of Shimon Peres*. Translated by Akiva Ron. New York: Warner Books, 1989.

———. *The Secret Conversations of Henry Kissinger: Step-by-Step Diplomacy in the Middle-East*. Translated by Ruth Geyra Stern and Sol Stern. New York: Quadrangle/New York Times Book Co., 1976.

Golani, Motti. *Israel in Search of a War: The Sinai Campaign, 1955–1956.* Portland, Ore.: Sussex Academic Press, 1998.

Goldberg, Anatol. *Ilya Ehrenburg: Writing, Politics and the Art of Survival.* London: Weidenfeld & Nicolson, 1984.

Goldberg, J. J. (Jonathan Jeremy). *Jewish Power: Inside the American Jewish Establishment.* Reading, Mass.: Addison-Wesley, 1996.

Goldmann, Nahum. *The Autobiography of Nahum Goldmann: Sixty Years of Jewish Life.* Translated by Helen Sebba. New York: Holt, Rinehart and Winston, 1969.

Goldstein, Israel. *My World as a Jew: The Memoirs of Israel Goldstein.* New York: Herzl Press, Cornwall Books, 1984.

Gorenberg, Gershom. *The Accidental Empire.* New York: Times Books, Henry Holt, 2006.

Gorny, Joseph. *British Labour Movement and Zionism, 1917–1948.* London: Frank Cass, 1983.

Granott, A. (Abraham). *Land and the Jewish Reconstruction of Palestine.* Authorized translation. Jerusalem: "Mischar w T'aasia" Pub. Co., 1931.

Gruber, Ruth. *Ahead of Time: My Early Years as a Foreign Correspondent.* New York: Carroll & Graf, 1991.

———. *Destination Palestine: The Story of the Haganah Ship Exodus 1947.* New York: Current Books, 1948.

———. *Inside of Time: My Journey from Alaska to Israel.* New York: Carroll & Graf, 2003.

Gvati, Haim. *A Hundred Years of Settlement: The Story of Jewish Settlement in the Land of Israel.* English translation edited by Fred Skolnik. Jerusalem: Keter, 1985.

Haber, Eitan. *Menahem Begin: The Legend and the Man.* Translated by Louis Williams. New York: Delacorte Press, 1978.

Hacohen, Devorah. *Immigrants in Turmoil: Mass Immigration to Israel and Its Repercussions in the 1950s and After.* Translated from the Hebrew by Gila Brand. Syracuse, N.Y.: Syracuse University Press, 2003.

Halabi, Rafik. *The West Bank Story.* Translated from the Hebrew by Ina Friedman. San Diego: Harcourt Brace Jovanovich, 1985.

Halpern, Ben. *The Idea of the Jewish State.* Cambridge, Mass.: Harvard University Press, 1969.

Halter, Marek. *The Jester and the Kings: A Political Autobiography.* New York: Arcade Publishing, Little, Brown, 1976.

Harkabi, Yehoshafat. *Arab Attitudes to Israel.* Translated by Misha Louvish. Jerusalem: Keter, 1972.

———. *Palestinians and Israel.* New York: Wiley, 1975.

Hart, Alan. *Arafat, Terrorist or Peacemaker?* London: Sidgwick & Jackson, 1984.

Hassan, Sana. *Enemy in the Promised Land: An Egyptian Woman's Journey into Israel.* New York: Schocken Books, 1986.

Haykal, Muḥammad Ḥasanayn. *The Road to Ramadan.* New York: Quadrangle/New York Times Book Co., 1975.

Hazleton, Lesley. *Israeli Women: The Reality Behind the Myth.* New York: Simon & Schuster, 1977.

Hazony, Yoram. *The Jewish State: The Struggle for Israel's Soul.* New York: Basic Books, 2000.

Hersh, Seymour M. *The Samson Option: Israel's Nuclear Arsenal and American Foreign Policy.* New York: Random House, 1991.

Hertzberg, Arthur. *The Zionist Idea: A Historical Analysis and Reader.* New York: Atheneum, 1972.

Herzog, Chaim. *The Arab-Israeli Wars: War and Peace in the Middle East.* New York: Random House, 1982.

———. *Living History: The Memoirs of a Great Israeli Freedom-Fighter, Soldier, Diplomat and Statesman.* New York: Pantheon, 1996.

———. *The War of Atonement: The Inside Story of the Yom Kippur War.* Boston: Little, Brown, 1975.

Higgins, Rosalyn. *United Nations Peacekeeping: Documents and Commentary.* Oxford and New York: Issued under the auspices of the Royal Institute of International Affairs by Oxford University Press, 1969–1981.

Hirst, David. The *Gun and the Olive Branch.* London and Boston: Faber and Faber, 1983.

Hirst, David, and Irene Beeson. *Sadat.* London: Faber and Faber, 1981.

Hohenberg, John. *Israel at 50: A Journalist's Perspective.* Syracuse, N.Y.: Syracuse University Press, 1998.

Horowitz, Dan, and Moshe Lissak. *Origins of the Israeli Polity: Palestine Under the Mandate.* Translated from the Hebrew by Charles Hoffman. Chicago: University of Chicago Press, 1978.

Horowitz, David. *State in the Making.* Translated from the Hebrew by Julian Meltzer. New York: Alfred A. Knopf, 1953.

Howard, Michael, and Robert Hunter. *Israel and the Arab World: The Crisis of 1967.* London: Institute for Strategic Studies, 1967.

Hurewitz, J. C. *Diplomacy in the Near and Middle East: A Documentary Record.* Princeton, N.J.: Van Nostrand, 1956.

———. *The Struggle for Palestine.* New York: Schocken Books, 1976.

Hussein, King of Jordan. My *War with Israel*. As told to and with additional material by Vick Vance and Pierre Lauer. Translated by June P. Wilson and Walter B. Michaels. New York: Morrow, 1969.

——. *Uneasy Lies the Head: The Autobiography of His Majesty King Hussein I of the Hashemite Kingdom of Jordan*. New York: B. Geis Associates; distributed by Random House, 1962.

Isaac, Rael Jean. *Israel Divided: Ideological Politics in the Jewish State*. Baltimore: Johns Hopkins University Press, 1976.

Isaacson, Walter. *Kissinger: A Biography*. New York: Simon & Schuster, 1992.

Israeli, Raphael. *Man of Defiance: A Political Biography of Anwar Sadat*. London: Weidenfeld & Nicolson, 1985.

Jamasï, Mohamed Abdel Ghani. *The October War: Memoirs of Field Marshal El-Gamasy of Egypt*. Translation by Gillian Potter, Nadra Morcos, and Rosette Frances. Cairo, Egypt: The American University in Cairo Press, 1993.

Jewish Agency for Israel. *Government Year Book, from 1950*. Jerusalem: Jewish Agency for Palestine, 1950.

Jewish Agency for Israel. *The Jewish Plan for Palestine: Memoranda and Statements Presented to the United Nations Special Committee on Palestine*. Jerusalem: Jewish Agency for Palestine, 1947.

Jiryis, Sabrï. *Arabs in Israel*. Translated from the Arabic by Inea Bushnaq. New York: Monthly Review Press, 1976.

Joseph, Bernard. *British Rule in Palestine*. Washington, D.C.: Public Affairs Press, 1948.

——. *The Faithful City; the Siege of Jerusalem, 1948*. New York: Simon & Schuster, 1960.

Jurman, Pinchas. *Moshe Dayan: A Portrait*. New York: Dodd, Mead, 1969.

Kagan, Benjamin. *The Secret Battle for Jerusalem*. Cleveland: World Publishing, 1966.

Kalb, Marvin, and Bernard Kalb. *Kissinger*. Boston: Little, Brown, 1974.

Kaplan, Deborah. *The Arab Refugees: An Abnormal Problem*. Translated from the Hebrew by Misha Louvish. Jerusalem: Rubin Mass, 1959.

Katz, Shmuel. *Battleground: Fact and Fantasy in Palestine*. Toronto and New York: Bantam Books, 1973.

——. *Days of Fire*. Garden City, N.Y.: Doubleday, 1968.

Katzman, Jacob. *Commitment: The Labor Zionist Life-Style in America: A Personal Memoir*. New York: Labor Zionist Letters, 1975.

Kedourie, Elie. *Palestine and Israel in the 19th and 20th Centuries*. Edited by Elie Kedourie and Sylvia G. Haim. London and Totowa, N.J.: Frank Cass, 1982.

——. *Zionism and Arabism in Palestine and Israel.* London: Frank Cass; Totowa, N.J.: Biblio Distribution Centre, 1982.

Kenen, Isaiah L. *Israel's Defense Line: Her Friends and Foes in Washington.* Buffalo, N.Y.: Prometheus Books, 1981.

Khouri, Fred J. (Fred John). *The Arab-Israeli Dilemma.* Syracuse, N.Y.: Syracuse University Press, 1968.

Kieval, Gershon R. *Party Politics in Israel and the Occupied Territories.* Series: Contributions in Political Science, no. 93. Westport, Conn.: Greenwood Press, 1983.

Kimche, David, and Dan Bawley. *Sandstorm: the Arab-Israeli War of June 1967: Prelude and Aftermath.* New York: Stein and Day, 1968.

Kimche, Jon. *The Second Arab Awakening.* New York: Holt, Rinehart and Winston, 1970.

Kimche, Jon, and David Kimche. *Both Sides of the Hill: Britain and the Palestine War.* London: Secker & Warburg, 1960.

——. *A Clash of Destinies: The Arab-Jewish War and the Founding of the State of Israel.* New York: Praeger, 1960.

Kimmerling, Baruch. *Zionism and Territory: The Socio-Territorial Dimensions of Zionist Politics.* Berkeley: Institute of International Studies, University of California, 1983.

Kirkbride, Alec Seath. *From the Wings: Amman Memoirs 1947–1951.* London: Frank Cass, 1976.

Kissinger, Henry. *Crisis: The Anatomy of Two Major Foreign Policy Crises.* New York: Simon & Schuster, 2003.

——. *The White House Years.* Boston: Little, Brown, 1979.

——. *Years of Upheaval.* Boston: Little, Brown, 1982.

Koestler, Arthur. *Promise and Fulfillment: Palestine 1917–1949.* London: Macmillan, 1949.

Korey, William. *The Soviet Cage: Anti-Semitism in Russia.* New York: Viking Press, 1973.

Korngold, Sheyna. *Zikhroynes.* Tel Aviv: Farlag Idpres, 1968.

Kraines, Oscar. *Government and Politics in Israel.* Boston: Houghton Mifflin, 1961.

Krenin, Mordechai Elihau. *Israel and Africa: A Study in Technical Cooperation.* New York: Praeger, 1964.

Kurzman, Dan. *Ben-Gurion, Prophet of Fire.* New York: Simon & Schuster, 1983.

——. *Genesis 1948: The First Arab-Israeli War.* New York: Da Capo Press, 1992.

Kutler, Stanley I. *Abuse of Power: The New Nixon Tapes.* Edited with an introduction and commentary by Stanley I. Kutler. New York: Free Press, 1997.

Lahav, Pnina. *Judgment in Jerusalem*. Berkeley: University of California Press, 1997.

Lall, Arthur S. *UN and the Middle East Crisis, 1967*. New York: Columbia University Press, 1970.

Lapidot, Yehuda. *Chapters in the History of the Irgun*. Tel Aviv: Jabotinsky Institute, 1999.

———. *The Flames of Revolt*. Tel Aviv: Ministry of Defense, 1996.

———. *The Hunting Season*. Israel: Jabotinsky Institute, 1994.

———. *Upon Thy Walls*. Tel Aviv: Ministry of Defense, 1992.

Laqueur, Walter. *A History of Zionism*. New York: Schocken Book, 1972.

Lash, Joseph P. *Dag Hammarskjold, Custodian of the Brushfire Peace*. Garden City, N.Y.: Doubleday, 1961.

Lau-Lavie, Naphtali. *Moshe Dayan: A Biography*. London: Vallentine, Mitchell, 1968.

Lauterpacht, Eli. *The United Nations Emergency Force: Basic Documents*. London: Stevens; New York: Praeger, 1960.

Lesch, Ann Mosely. *Arab Politics in Palestine, 1917–1939: The Frustration of a Nationalist Movement*. Ithaca, N.Y.: Cornell University Press, 1979.

Levenberg, S. *Jews and Palestine: A Study in Labour Zionism*. London: Poale Zion, Jewish Socialist Labour Party, 1945.

Levin, Harry. *I Saw the Battle of Jerusalem*. New York: Schocken Books, 1950.

Levin, Marlin. *Balm in Gilead: The Story of Hadassah*. Foreword by Golda Meir. New York: Schocken Books, 1973.

Levitats, Isaac. *The Story of the Milwaukee Jewish Community*. Milwaukee, Wis.: Bureau of Jewish Education, 1954.

Lie, Trygve. *In the Cause of Peace: Seven Years with the United Nations*. New York: Macmillan, 1954.

Lilienthal, Alfred M. *The Zionist Connection II: What Price Peace?* New Brunswick, N.J.: North American, 1982.

Little, Douglas. *American Orientalism: The United States and the Middle East since 1945*. Chapel Hill: University of North Carolina Press, 2002.

Litvinoff, Barnet. *The Essential Chaim Weizmann: The Man, the Statesman, the Scientist*. London: Weidenfeld & Nicolson, 1982.

Lockman, Zachary. *Comrades and Enemies: Arab and Jewish Workers in Palestine, 1906–1948*. Berkeley: University of California Press, 1996.

Loftus, John, and Mark Aarons. *The Secret War Against the Jews: How Western Espionage Betrayed the Jewish People*. New York: St. Martin's Press, 1994.

London, Louise. *Whitehall and the Jews, 1933–1948: British Immigration Policy,*

Jewish Refugees, and the Holocaust. Cambridge and New York: Cambridge University Press, 2000.

London Sunday Times Team. *The Yom Kippur War.* London: G. B. Deutsch, 1975.

Lorch, Netanel. *One Long War: Arab versus Jew since 1920.* Jerusalem: Keter Books, 1976.

Lossin. Yigal. *Pillar of Fire: A Television History of Israel's Rebirth.* Video recording. Jerusalem: Israel Broadcast Authority, 2005.

Louis, William Roger. *British Empire in the Middle East, 1945–1951: Arab Nationalism, the United States, and Postwar Imperialism.* Oxford: Clarendon Press, 1984.

Lukacs, Yehuda. *Documents on the Israeli-Palestinian Conflict, 1967–1983.* Cambridge: Cambridge University Press, 1984.

———. *Israel, Jordan, and the Peace Process.* Syracuse, N.Y.: Syracuse University Press, 1997.

MacLeish, Roderick. *The Sun Stood Still.* New York: Atheneum, 1967.

Magnus, Ralph H. *Documents on the Middle East.* Washington, D.C.: American Enterprise Institute for Public Policy Research, 1969.

Mahmoud, Amin Abdullah. "King Abdullah and Palestine: An Historical Study of His Role in the Palestine Problem from the Creation of Transjordan to the Annexation of the West Bank, 1921–1950." PhD thesis, Georgetown University, 1972.

Mandel, Neville J. *The Arabs and Zionism Before World War I.* Berkeley: University of California Press, 1976.

Mann, Barbara. *A Place in History: Modernism, Tel Aviv, and the Creation of Jewish Urban Space.* Palo Alto, Calif.: Stanford Studies in Jewish History and Culture, 2006.

Mann, Peggy. *Golda: The Life of Israel's Prime Minister.* New York: Coward, McCann & Geoghegan, 1971.

Marshall, S. L. A. (Samuel Lyman Atwood). *Swift Sword: The Historical Record of Israel's Victory, June, 1967.* New York: American Heritage Pub. Co., 1967.

Martin, Ralph G. *Golda: Golda Meir, the Romantic Years.* New York: Scribner, 1988.

McCrackan, William Denison. *The New Palestine.* New York: Arno Press, 1977.

Medding, Peter. *Mapai in Israel: Political Organisation and Government in a New Society.* Cambridge: Cambridge University Press, 1972.

Medzini, Meron. *Ha-Yehudiyah ha-geah: Goldah Meir ya-ḥazon Yiśra'el: Biyografyah politit.* Tel Aviv: Edanim, 1990.

——. *Israel's Foreign Relations: Selected Documents*. Jerusalem: Ministry for Foreign Affairs, 1976.

Meinertzhagen, Richard. *Middle East Diary, 1917–1956*. London: Cresset Press, 1959.

Meir, Golda. *My Life*. New York: Putnam, 1975.

Meir, Menachem. *My Mother Golda Meir: A Son's Evocation of Life with Golda Meir*. New York: Arbor House, 1983.

Mendez-Mendez, Serafin. "Towards an Understanding of Didactic Rhetoric: The Case of Golda Meir, Teacher and Politician." PhD dissertation, University of Massachusetts, 1989.

Milstein, Uri. *History of the War of Independence*. Translated and edited by Alan Sacks. Lanham, Md.: University Press of America, 1996.

Monroe, Elizabeth. *Britain's Moment in the Middle East, 1914–1971*. Foreword by Peter Mansfield. London: Chatto & Windus, 1981.

Monroe, Elizabeth, and A. H. Farrar-Hockley. *Arab-Israel War, October 1973: Background and Events*. London: International Institute for Strategic Studies, 1974.

Moore, John Norton. *The Arab-Israeli Conflict: Readings and Documents*. Edited by John Norton Moore. Sponsored by the American Society of International Law. Princeton, N.J.: Princeton University Press, 1977.

Morris, Benny. *The Birth of the Palestinian Refugee Problem, 1947–1949*. Cambridge: Cambridge University Press, 1987.

——. *The Birth of the Palestinian Refugee Problem Revisited*. Cambridge: Cambridge University Press, 2004.

——. *Israel's Border Wars, 1949–1956. Arab Infiltration, Israeli Retaliation, and the Countdown to the Suez War*. Oxford: Clarendon Press, 1997.

——. *1948 and After: Israel and the Palestinians*. Oxford: Clarendon Press; New York: Oxford University Press; 1990.

——. *Righteous Victims*. New York: Vintage, 2001.

Morris, Terry. *Shalom, Golda*. New York: Hawthorn Books, 1971.

Morse, Arthur D. *While Six Million Died: A Chronicle of American Apathy*. New York: Hart, 1968.

Namir, Mordechai. *The Israeli Mission to Moscow* [Hebrew]. Tel Aviv: Am Oved, 1971.

Neff, Donald. *Warriors Against Israel*. Brattleboro, Vt.: Amana Books, 1988.

——. *Warriors for Jerusalem: The Six Days That Changed the Middle East*. New York: Linden Press/Simon & Schuster, 1984.

Nisan, Mordechai. *Israel and the Territories: A Study in Control, 1967–1977*. Ramat Gan, Israel: Turtledove, 1978.

Nixon, Richard M. *RN: The Memories of Richard Nixon*. New York: Simon & Schuster, 1990.

O'Ballance, Edgar. *The Arab-Israeli War, 1948*. New York: Praeger, 1957.

——. *The Third Arab-Israeli War*. Hamden, Conn.: Archon Books, 1972.

O'Brien, Conor Cruise. *Siege: The Saga of Israel and Zionism*. New York: Simon & Schuster, 1986.

Oren, Michael. *Six Days of War: June 1967 and the Making of the Modern Middle East*. New York and Oxford: Oxford University Press, 2002.

Ostrovsky, Victor. *The Other Side of Deception: A Rogue Agent Exposes the Mossad's Secret Agenda*. New York: HarperCollins, 1994.

Ostrovsky, Victor, with Claire Hoy. *By Way of Deception: The Making of a Mossad Officer*. Scottsdale, Ariz.: Wilshire Press, 1990.

Palestine Conciliation Commission. *Progress Reports Submitted to the General Assembly*. New York, 1949.

Parker, Richard B., ed. *The October War*. Gainesville: University of Florida Press, 2001.

Parmet, Herbert S. *Richard Nixon and His America*. Boston: Little, Brown, 1990.

Pearlman, Moshe. *The Army of Israel*. New York: Philosophical Library, 1950.

——. *Ben Gurion Looks Back in Talks with Moshe Pearlman*. New York: Simon & Schuster, 1965.

——. *Mufti of Jerusalem; The Story of Haj Amin el Husseini*. London: V Gollancz, 1947.

Peres, Shimon. *Battling for Peace: A Memoir*. New York: Random House, 1995.

——. *David's Sling*. London: Weidenfeld & Nicolson, 1970.

Peretz, Don. *The Government and Politics of Israel*. Boulder, Colo.: Westview Press, 1983.

——. *Israel and the Palestine Arabs*. Foreword by Roger Baldwin. Washington, D.C.: Middle East Institute, 1958.

Peters, Joan. *From Time Immemorial: The Origins of the Arab-Jewish Conflict Over Palestine*. New York: Harper & Row, 1984.

Pfaff, Richard H. *Jerusalem: Keystone of an Arab-Israeli Settlement*. Series: Legislative and special analyses; 91st Cong. 1st sess., no. 13. Washington, D.C.: American Enterprise Institute for Public Policy Research, 1969.

Polk, William Roe, David Stamler, and Edmund Asfour. *Backdrop to Tragedy; The Struggle for Palestine*. Boston: Beacon Press, 1957.

Postal, Bernard, and Henry W. Levy. *And the Hills Shouted for Joy: The Day Israel Was Born*. New York, D. McKay, 1973.

Prittie, Terence. *Eshkol: The Man and the Nation*. New York: Pitman, 1969.

——. *Israel: Miracle in the Desert*. Baltimore, Md.: Penguin Books. 1968.

Quandt, William B. *Decade of Decision*. Berkeley and Los Angeles: University of California Press, 1977.

——. *Soviet Policy in the October 1973 War*. RAND Report I-1864-ISA. Santa Monica, CA: RAND Corporation, 1976.

Quandt, William B., Fuad Jabber, and Ann Mosely Lesch. *The Politics of Palestinian Nationalism*. Berkeley: University of California Press, 1973.

Rabin, Leah. *Rabin: Our Life, His Legacy*. New York: G. P. Putnam's Sons, 1997.

Rabin, Yitzhak. *Rabin Memoirs*. Boston: Little, Brown, 1979.

Rabinovich, Abraham. *The Battle for Jerusalem, June 5–7, 1967*. Philadelphia: Jewish Publication Society of America, 1972.

——. *The Yom Kippur War: The Epic Encounter That Transformed the Middle East*. New York: Schocken, 2004.

Rabinovits, Zeev, ed. *Pinsk: sefer edut ye-zikaron li-kehilat Pinsk-Karlin*. Tel Aviv: Irgun yotse Pinsk-Karlin bi-Medinat Yiśrael, 1966–1977.

Rafael, Gideon. *Destination Peace: Three Decades of Israeli Foreign Policy: A Personal Memoir*. New York: Stein and Day, 1981.

Reich, Bernard. *Quest for Peace: United States–Israel Relations and the Arab-Israeli Conflict*. New Brunswick, N.J.: Transaction Books, 1977.

Revusky, Abraham. *Jews in Palestine*. New York: Bloch, 1945.

Riad, Mahmoud. *The Struggle for Peace in the Middle East*. London: Quartet Books, 1982.

Roth, Stephen J., ed. *The Impact of the Six-Day War: A Twenty-Year Assessment*. Houndmills, Basingstoke, UK: Macmillan in association with the Institute of Jewish Affairs, 1988.

Royal Institute of International Affairs. *Documents on International Affairs, from 1945*. London: Oxford University Press, 1929–1973.

Sachar, Howard Morley. *A History of Israel*. New York: Alfred A. Knopf, 1976–1987.

Sacher, Harry. *Israel: The Establishment of a State*. New York: British Book Centre, 1952.

Sadat, Anwar. *In Search of Identity: An Autobiography*. New York: Harper & Row, 1978.

——. *Those I Have Known*. London: Continuum International Publishing Group, 1984.

Safran, Nadav. *From War to War: The Arab-Israeli Confrontation, 1948–1967, A Study of the Conflict from the Perspective of Coercion in the Context of Inter-Arab and Big Power Relations*. New York: Pegasus, 1969.

——. *Israel, the Embattled Ally*. New preface and postscript by the author. Cambridge, Mass.: Belknap Press, 1981.

——. *The United States and Israel*. Cambridge, Mass.: Harvard University Press, 1963.

Said, Edward W. *The Question of Palestine*. New York: Vintage Books, 1992.

Samuel, Maurice. *Harvest in the Desert*. Philadelphia: Jewish Publication Society of America, 1944.

Sanders, Ronald. *The High Walls of Jerusalem: A History of the Balfour Declaration and the Birth of the British Mandate for Palestine*. New York: Holt, Rinehart and Winston, 1984.

Sandler, Shmuel, and Hillel Frisch. *Israel, the Palestinians, and the West Bank: A Study in Intercommunal Conflict*. Lexington, Mass.: Lexington Books, 1984.

Satloff, Robert B. (Robert Barry). *From Abdullah to Hussein: Jordan in Transition*. New York: Oxford University Press, 1994.

Schechtman, Joseph B. *The Arab Refugee Problem*. New York: Philosophical Library, 1952.

Schiff, Zeev. *October Earthquake: Yom Kippur 1973*. Translated by Louis Williams. Tel Aviv: University Pub. Projects, 1974.

Schiff, Zeev, and Ehud Ya'ari. *Israel's Lebanon War*. Edited and translated by Ina Friedman. New York: Simon & Schuster, 1984.

Schwar, Harriet D. *Arab-Israeli Dispute, 1964–1967*. Washington, D.C.: Dept. of State: U.S. GPO, Supt. of Documents, 2000.

Segal, Ronald. *Whose Jerusalem? The Conflicts of Israel*. London: Cape, 1973.

Segev, Samuel. *Crossing the Jordan: Israel's Road to Peace*. New York: St. Martin's Press, 1998.

Segev, Tom. *1949: The First Israelis*. New York: Henry Holt, 1986.

——. *One Palestine, Complete: Jews and Arabs Under the British Mandate*. New York: Metropolitan Books, 2000.

Segre, Dan Vittorio. *Israel: A Society in Transition*. London and New York: Oxford University Press, 1971.

Sela, Amnon, and Yael Yishai. *Israel: The Peaceful Belligerent, 1967–79*. New York: St. Martin Press, 1986.

Shaked, Haim. *From June to October: The Middle East between 1967 and 1973*. Edited by Itamar Rabinovich and Haim Shaked. New Brunswick, N.J.: Transaction Books, 1978.

Shaked, Haim, and Itamar Rabinovich, eds. The *Middle East and the United States: Perceptions and Policies*. New Brunswick, N.J.: Transaction Books, 1980.

Shalev, Michael. *Labor and Political Economy in Israel*. New York: Oxford University Press, 1992.

Shapiro, Yonathan. *The Formative Years of the Israeli Labour Party: The Organization of Power, 1919–1930*. London and Beverly Hills, Calif.: Sage Publications, 1976.

Sharef, Zeev. *Three Days*. Translated by Julian Louis Meltzer from the Hebrew. London: W. H. Allen, 1962.

Sharett, Ya'akov. *Imprisoned with Paper and Pencil: The Letters of Moshe and Zipporah Sharett* [Hebrew]. Tel Aviv: Moshe Sharett Institute, 2000.

Shazar, Rachel Katznelson. *The Plough Woman: Records of the Pioneer Women of Palestine*. New York: Herzl Press, 1975.

Sheehan, Edward R. F. *The Arabs, Israelis, and Kissinger: A Secret History of American Diplomacy in the Middle East*. New York: Readers Digest Press: distributed by Crowell, 1976.

Sheffer, Gabriel. *Moshe Sharett: A Biography of a Political Moderate*. Oxford: Claredon Press; New York: Oxford University Press; 1996.

Shenker, Israel, and Mary Shenker. *As Good as Golda: The Warmth and Wisdom of Israel's Prime Minister*. New York: McCall, 1970.

Shepherd, Naomi. *Ploughing Sand: British Rule in Palestine, 1917–1948*. Brunswick, N.J.: Rutgers University Press, 2000.

Sherman, A. J. *Mandate Days: British Lives in Palestine, 1918–1948*. Baltimore: Johns Hopkins University Press, 2001.

Shimshoni, Daniel. *Israeli Democracy: The Middle of the Journey*. New York: Free Press, 1982.

Shipler, David K. *Arab and Jew: Wounded Spirits in a Promised Land*. New York: Penguin Books, 2002.

Shlaim, Avi. *Collusion Across the Jordan: King Abdullah, the Zionist Movement, and the Partition of Palestine*. New York: Columbia University Press, 1988.

———. *The Iron Wall*. New York: W. W. Norton, 2001.

Silver, Abba Hillel. *Vision and Victory: A Collection of Addresses, 1942–1948*. New York: Zionist Organization of America, 1949.

Silver, Eric. *Begin: A Biography*. London: Weidenfeld & Nicolson, 1984.

Slater, Robert. *Rabin of Israel*. New York: St. Martin's Press, 1993.

———. *Golda, the Uncrowned Queen of Israel*. Middle Village, N.Y.: Jonathan David, 1981.

———. *Warrior Statesman: The Life of Moshe Dayan*. New York: St. Martin's Press, 1991.

Smith, Walter Bedell. *My Three Years in Moscow*. Philadelphia: Lippincott, 1950.

Smooha, Sammy. *Israel, Pluralism and Conflict*. London: Routledge & Kegan Paul, 1978.

Snetsinger, John. *Truman, the Jewish Vote, and the Creation of Israel*. Series: Hoover Institution Studies; no. 39. Stanford, Calif.: Hoover Institution Press, 1974.

Snow, Peter (Peter John). *Hussein: A Biography*. New York: R. B. Luce, 1972.

Snow, Peter, and David Phillips. *The Arab Hijack War: The True Story of 25 Days in September, 1970*. New York: Ballantine Books, 1971.

Spiegel, Steven L. *The Other Arab-Israeli Conflict: Making America's Middle East Policy, from Truman to Reagan*. Series: Middle Eastern studies (Chicago, Ill.); monograph 1. Chicago: University of Chicago Press, 1985.

St. John, Robert. *Ben-Gurion: A Biography*. Garden City, N.Y.: Doubleday, 1971.

——. *Eban*. New York: Doubleday, 1972.

——. *Shalom Means Peace*. Garden City, N.Y.: Doubleday, 1949.

Sternhell, Zeev. *The Founding Myths of Israel: Nationalism, Socialism, and the Making of the Jewish State*. Translated by David Maisel. Princeton, N.J.: Princeton University Press, 1998.

Stevenson, William. *Strike Zion!* New York: Bantam Books, 1967.

Stock, Ernest. *Israel on the Road to Sinai, 1949–1956. With a Sequel on the Six-Day War, 1967*. Ithaca, N.Y.: Cornell University Press, 1967.

Stoessinger, John George. *Henry Kissinger: The Anguish of Power*. New York: W. W. Norton, 1976.

Susser, Asher. *Jordan: Case Study of a Pivotal State*. Washington, D.C.: Washington Institute for Near East Policy, 2000.

Swichkow, Louis J., and Lloyd P. Gartner. *The History of the Jews of Milwaukee*. The Jacob R. Schiff Library of Jewish Contributions to American Democracy, no. 16. Philadelphia: Jewish Publication Society of America, 1963.

Sykes, Christopher. *Crossroads to Israel*. Bloomington: Indiana University Press, 1973.

Syrkin, Marie. *Blessed Is the Match: The Story of Jewish Resistance*. Philadelphia: Jewish Publication Society of America, 1947.

——, ed. *Golda Meir Speaks Out*. London: Weidenfeld & Nicolson, 1973.

——, ed. *A Land of Our Own: An Oral Autobiography by Golda Meir*. New York: G. P. Putnam's Sons, 1973.

——. *Way of Valor: A Biography of Golda Myerson*. New York: Sharon Books, 1955.

Szulc, Tad. *The Illusion of Peace: Foreign Policy in the Nixon Years*. New York: Viking Press, 1973.

Teveth, Shabtai. *Ben Gurion's Spy: The Story of the Political Scandal That Shaped Modern Israel*. New York: Columbia University Press, 1996.

———. *The Cursed Blessing: The Story of Israel's Occupation of the West Bank*. Translated from the Hebrew by Myra Bank. London: Weidenfeld & Nicolson, 1970.

———. *Moshe Dayan: The Soldier, the Man, the Legend*. Boston: Houghton Mifflin, 1973.

Thant, U. *View from the UN*. Garden City, N.Y.: Doubleday, 1978.

Touval, Saadia. *The Peacebrokers: Mediators in the Arab-Israeli Conflict, 1948–1979*. Princeton, N.J.: Princeton University Press, 1982.

Trevor, Daphne. *Under the White Paper; Some Aspects of British Administration in Palestine from 1939 to 1947*. Munich: Kraus International Publications, 1980.

Truman, Harry S. *Memoirs*. Garden City, N.Y.: Doubleday, 1955–56.

Uchill, Ida Libert. *Pioneers, Peddlers, and Tsadikim: The Story of Jews in Colorado*. Boulder: University Press of Colorado, 2000.

United Nations. *Year Book of the United Nations, New York, from 1947*.

United States Policy in the Middle East. September 1956–June 1957, Department of State Publication 6505, Washington, D.C., 1957.

United Nations Relief and Works Agency. *Annual Reports of the Commissioner General to the General Assembly*. New York: United Nations Relief and Works Agency, 1949.

Urofsky, Melvin I. *We Are One! American Jewry and Israel*. Garden City, N.Y.: Anchor Press, 1978.

Urquhart, Brian. *Life in Peace and War*. New York: Harper & Row, 1987.

Valeriani, Richard. *Travels with Henry*. Boston: Houghton Mifflin, 1979.

Van der Linden, Frank. *Nixon's Quest for Peace*. Washington, D.C.: R. B. Luce, 1972.

Viorst, Milton. *Sands of Sorrow: Israel's Journey from Independence*. New York: Harper & Row, 1987.

Wallach, Janet, and John Wallach. *Arafat: In the Eyes of the Beholder*. New York: A Lyle Stuart Book, Carol Publishing Group, 1990.

Ward, Richard J., Don Peretz, and Evan M. Wilson. *The Palestine State: A Rational Approach*. Port Washington, N.Y.: Kennikat Press, 1977.

Weidenfeld, George. *Remembering My Good Friends: An Autobiography*. New York: HarperCollins, 1995.

Wiesel, Elie. *And the Sea Is Never Full: Memoirs*. Translated from the French by Marion Wiesel. New York: Alfred A. Knopf, 1999.

Weisgal, Meyer Wolfe. *Chaim Weizmann; A Biography by Several Hands*. Preface by David Ben-Gurion. Edited by Meyer W. Weisgal and Joel Carmichael. New York: Atheneum, 1963.

Weizman, Ezer. *The Battle for Peace.* Toronto and New York: Bantam Books, 1981.

Weizmann, Chaim. *Trial and Error: The Autobiography of Chaim Weizmann.* New York: Harper, 1949.

Weizmann, Vera. *The Impossible Takes Longer; The Memoirs of Vera Weizmann, Wife of Israel's First President, as Told to David Tutaev.* New York: Harper & Row, 1967.

Wilson, Evan M. *Decision on Palestine: How the U.S. Came to Recognize Israel.* Stanford, Calif.: Hoover Institution Press, Stanford University, 1979.

———. *Jerusalem, Key to Peace.* Washington, D.C.: Middle East Institute, 1970.

Wilson, Harold. *The Chariot of Israel: Britain, America and the State of Israel.* New York W. W. Norton, 1981.

Wilson, Mary C. (Mary Christina). *King Abdullah, Britain, and the Making of Jordan.* New York: Cambridge University Press, 1987.

Yaari, Ehud. *Strike Terror: The Story of Fatah.* Translated from the Hebrew by Esther Yaari. New York: Sabra Books, 1970.

Yoram, Peri. *Between Ballots and Battles: Israeli Military in Politics.* New York: Cambridge University Press, 1983.

Zak, Moshe. *Hussein the Peacemaker (Ohseh Shalom): The History of Israel-Jordan Secret Relations 1964–94* [Hebrew]. Ramat Gam: Bar Ilan University Press, 1998.

Zuckerman, Baruch. *Excerpts from Memoirs.* Jerusalem: Mini Dfus, 2000.

Articles

Abramov, S. Z. "Lavon Affair," *Commentary* 31 (February 1961): 100–105.

Allon, Yigal. "Israel: The Case of Defensible Borders," *Foreign Affairs* 55, no. 1 (1976): 38–53.

AlRoy, G. C. "Do the Arabs Want Peace?" *Commentary* 57 (February 1974): 56–61.

———. "Do the Arabs Want Peace?" discussion, *Commentary* 57 (May 1974): 58 4ff; (July 1974): 18.

Aronson, Geoffrey. "Israel's Policy of Military Occupation," *Journal of Palestine Studies* 7, no. 4 (1978): 79–98.

Avneri, Shlomo. "Beyond Camp David," *Foreign Policy* 46 (1982): 19–36.

——. "Rethinking Israel's Position; Views of Eliezer Livneh, Gad Yaacobi and Arie Eliav," *Commentary* 55 (June 1973): 51–55.

——. "Rethinking Israel's Position; Views of Eliezer Livneh, Gad Yaacobi and Arie Eliav," discussion, *Commentary* 56 (November 1973): 22ff.

Ball, George W. "The Coming Crisis in Israeli-American Relations," *Foreign Affairs* 58, no. 2 (winter 1979–80): 31–56.

Battle, D. "Arabs: Why Now?" *New York Times Magazine*, October 21, 1973, 33.

——. "First Israeli Revolution," *New York Times Magazine* May 6, 1973, 12–14.

Ben-Gurion, David. "Israel's First Decade and the Future," *New York Times Magazine*, April 20, 1958, 9–11ff.

——. "Peace Is More Important Than Real Estate; Interview," ed. by J.M. Roots, *Saturday Review* 54 (April 3, 1971): 14–16.

——. "Traveler, Consider My Israel," ed. R. Joseph, *Esquire* 68 (July 1967): 94–95.

——. "Why I Retired to the Desert," *New York Times Magazine*, March 28, 1954, 17ff.

Ben-Gurion, David, and S. A. Dolgin. "Can We Stay Jews Outside the Land?" exchange of letters, *Commentary* 16 (September 1953): 233–40.

Brown, Michael. "The American Element in the Rise of Golda Meir," *Jewish History* 6:1–2 (March 1992): 16–50.

Brown, Neville. "Jordanian Civil War," *Military Review* 51 (September 1971): 38–48.

Bruzonsky, M. A. "U.S., the P.L.O. and Israel," *Commonweal* 103 (May 21, 1976): 326–30.

Buckley, W. F. "Viva Eban," *National Review* 22 (October 20, 1970): 1124–25.

Cline, Ray. "Policy Without Intelligence," *Foreign Policy* 17 (winter 1974–75): 121–35.

Clugston, M. "Golda Years," *Macleans* 91 (December 1978): 27–30.

Cohen, C. "Democracy in Israel," *Nation* 219 (July 20, 1974): 43–47.

Coughlin, R. "Modern Prophet of Israel," *Life* 43 (November 18, 1957): 154–56.

Crossman, R. H. S. "War in Palestine?" *New Republic* 114 (May 20, 1946): 718–19.

Currivan, G. "Two Worlds Meet in Tense Palestine," *New York Times Magazine*, November 18, 1945, 9ff.

Dawidowicz, Lucy. "Toward a History of the Holocaust," *Commentary* 47, no. 4 (1969): 51–58.

Dayan, Moshe. "Hopeful Truths of the New Reality," *Life* 63 (September 29, 1967): 120–120B.

——. "Israel's Border and Security Problems," *Foreign Affairs* 23, no. 2 (January 1955).

Dayan, Yael. "Father and Hero," *Look* 31 (August 22, 1967): 15–19.

Doctorow, E. L. "After the Nightmare," *Sports Illustrated* 44 (June 28, 1976): 72–76.

Draper, T. "From 1967 to 1973; The Arab-Israeli Wars," *Commentary* 56 (December 1973): 21–45.

——. "United States and Israel: Tilt in the Middle East," *Commentary* 59 (April 1975): 29–45.

——. "United States and Israel: Tilt in the Middle East," discussion, *Commentary* 60 (September 1975): 18ff.

Eban, Abba. "Camp David: The Unfinished Business," *Foreign Affairs* 57, no. 2 (1978–79): 343–54.

——. "Israel's Road to Peace," *Nation* 171 (December 16, 1950): 617–18.

——. "Reaching for Peace and Partnership," *Saturday Evening Post* 245 (November 1973): 8ff.

Elon, Amos. "Israelis Believe War Is Inevitable," *Life* 68 (February 6, 1970): 46–48.

——. "Letter from Tel Aviv," *Commentary* 53 (February 1972): 75–78.

——. "Mood: Self Confidence and a Subdued Sadness," *New York Times Magazine*, May 6, 1973, 56–57.

——. "Governing Is Harder than Conquering," *New York Times Magazine*, May 4, 1975, 11ff.

Elozur, Y. "Heart of Golda," *New Leader* 62 (January 1, 1979): 7–8.

Fallaci, O., ed. "Yasir Arafat: Interview," excerpt from *Interview with History. New Republic*, 171:10–12, November 16, 1974.

Feron, J. "Eshkol Sticks to His Guns," *New York Times Magazine*, 40+, April 17, 1966.

——. "German Ambassador to Israel," *New York Times Magazine*, 102+, October 31, 1965.

——. "Israel Has Found a Replacement for Golda Meir. It's Golda Meir," *New York Times Magazine*, 52–53+, October 26, 1969.

——. "That New Boy in Israel's Foreign Office," *New York Times Magazine*, 40–1+, April 17, 1966.

——. "Yigal Allon Has Supporters, Moshe Dayan Has Disciples," *New York Times Magazine*, April 27, 1969.

Forster, Arnold. "Women: Golda Meir," *Playboy*, October 1988.

Gary, D. "Israel: Jews versus Jews," *Atlantic*, 245:18+, June, 1980.

Gelber, Yoav. "The Negotiations between the Jewish Agency and Transjordan, 1946–1948," *Studies in Zionism* 6/1, 53–83, 1985.

Gellhorn, M. "Israeli Secret Weapon," *Vogue*, 150–192–3, October 1, 1967.

Gersh, G. "Israel's Aid to Africa," *Commonweal*, 85:226–8, November 25, 1966.

Gervasi, F. "Myths and Realities: The Rights of the Palestinians," *The Nation*, 222:530–2, May 1, 1976.

Goldman, Nahum. "Future of Israel," *Foreign Affairs*, 48, 443–59, April, 1970.

———. "Zionist Ideology and the Reality of Israel," *Foreign Affairs*, 57, No. 1, 70–82, 1978.

Gramont, S. "Nasser's Hired Germans; Israelis Fight to Stop Missiles Being Built," *Saturday Evening Post* 236 (July 13, 1963): 60–61.

Halevi, Yossi. "Tough Love," *New Republic* 231, no. 6 (August 9, 2004): 16.

Halkin, H. "Dayan as a Politician," *Commentary* 55 (January, 1973): 51–55.

Hassan, S. "Egyptian's Vision of Peace," *New York Times Magazine*, February 10, 1974, 8–9ff.

Herzog, Chaim, "Talk with Chaim Herzog: Interview," ed. G. Samuels, *New Leader* 61 (June 19, 1978): 4–9.

Holden, D. "Hero of the Crossing. They Shout, Where Is Our Breakfast? Sadat's Role in Peace Negotiations," *New York Times Magazine*, June 1, 1975, 12–13ff.

Izraeli, Dafna. "The Zionist Women's Movement in Palestine, 1911–1927: A Sociological Analysis," *Signs* 7:1 (autumn 1981), pp. 97–114.

Jiryas, Sabri. "Secrets of State: An Analysis of the Diaries of Moshe Sharett," *Journal of Palestine Studies* 10, no. 1 (1980): 35–57.

Kalb, M., and Bernard Kalb. "Notes and Comment: Henry Kissinger's Strategy," *New Yorker* 50 (July 8, 1974): 25.

Kaplan, J. P., ed. "Interview with Abba Eban," *New Republic* 170 (March 23, 1974): 17–20.

Karnow, S., ed. "Interview with King Hussein," *New Republic* 172 (February 22, 1975): 19–21.

Kimche, J. "Dayan, Meir and the Libyan Plane," *The Nation* 216 (March 12, 1973): 325–26.

Koestler, Arthur, "The Great Dilemma That Is Palestine," *New York Times Magazine*, September 1, 1946, 5.

Kogan, N. S. "Israel Moves Right," *National Review* 22 (January 27, 1970): 83.

Kostyrchenko, Gennadii. "Golda at the Metropol Hotel," *Russian Studies in History* 43, no. 4 (fall 2004): 77–85.

Krantz, J. "At Home in Jerusalem," *Good Housekeeping* 145 (July 1957): 68–71.

Krosney, Herbert. "In Israel, Warriors Sick of War," *The Nation* 217 (November 12, 1973): 487–89.

———. "Promises in the Promised Land," *Nation* 214, no. (March 13, 1974): 336–38.

———. "Talk with Yigal Allon," *Nation* 212 (May 3, 1971): 550–52.

Laqueur, W. Z. "Israel After the War; Peace with Egypt?" *Commentary* 57 (March 1974): 34–40.

———. "Israel's Great Foreign Policy Debate," *Commentary* 20 (August 1955): 109–15.

Lewis, Bernard. "The Arab-Israeli War. The Consequences of Defeat," *Foreign Affairs* 46, no. 2 (1968): 321–35.

———. "The Emergence of Modern Israel," *Middle Eastern Studies* [London] 8, no. 3 (1972): 421–27.

Liebling, A. J. "Letter from Tel Aviv," *New Yorker* 33 (March 30, 1957): 115–20.

Lomax, L. E. "Memo from Amman: The Death of Hussein's Peace Mission," *Look* 32 (May 14, 1968): 95.

Luttwak, Edward, and W. Z. Laqueur. "Kissinger and the Yom Kippur War," *Commentary* 58 (September 1974): 33–40.

McCormick, A. O. "There Is No Present Tense in Israel," *New York Times Magazine*, February 13, 1949, 7ff.

Meir, Golda. "Golda Meir on the Palestinians," *Saturday Evening Post* 248 (April 1976): 10ff.

———. "Israel in Search of Lasting Peace," *Foreign Affairs* 51 (April 1973): 147–61.

———. "My First Days at Kibbutz Merhavia," *Jewish Affairs* (December 1970).

Middleton, D. "Who Lost the Yom Kippur War?" *Atlantic Monthly* 208 (November 1961): 83–149.

Morris, Benny. "The Crystallization of Israeli Policy Against a Return of the Arab Refugees, April–December 1948," *Studies in Zionism* 6, no. 1 (1985): 85–118.

Moskin, J. R. "Golda," *Look* 33 (October 7, 1969): 94ff.

———. "Israel; Twenty Years of Siege and Struggle: The Next Twenty Years," *Look* 32 (April 30, 1968): 28–40.

Orshefsky, M. "Bulldozers Carve Out Israel's New Frontiers," *Life* 70 (March 12, 1971): 32–34.

Pepper, C. G. "Hawk of Israel: Moshe Dayan," *New York Times Magazine*, July 9, 1967, 5ff.

Peretz, Don, and Sammy Smooha. "Israel's 10th Knesset Elections: Ethnic Upsurgence and Decline of Ideology," *Middle East Journal* 35, no. 4 (1981): 506–26.

Peretz, M. "The American Left and Israel," *Commentary* 44 (November 1967): 27–34.

———. "Woman of Valor," *New Republic* 179 (December 23, 1978): 42.

Perlmutter, Amos. "Crises Management," *International Studies Quarterly* 19 (September 1975): 316–43.

———. "A Race Against Time: The Egyptian-Israeli Negotiations over the Future of Palestine," *Foreign Affairs* 57, no. 5 (1979): 987–1004.

Raab, E. "Is Israel Losing Popular Support? The Evidence of the Polls," *Commentary* 57 (January 1974): 26–29.

Raider, Mark A. "Emissaries in the Promised Land," *Judaism* 49 (winter 2000).

Reed, D. "Golda Meir: Israel's Tough Grandmother—Prime Minister," *Reader's Digest* 99 (July 1971): 109–13.

Reich, Bernard. "Israel Between War and Peace," *Current History* 66, no. 390 (1974): 49–52.

———. "Israel's Policy in Africa," *Middle East Journal* 18 (winter 1964): 14–26.

Reichley, A. J. "Sign of the Dove in Israel," *Fortune* 89 (February 1974): 66–69.

Remba, O. "Can Israel Support Herself?" *Commentary* 22 (November 1956): 433–41.

Richardson, Elliot. "Nixon Behind the Scenes," *Newsweek* (February 1, 1971): 16–17.

Rostow, Eugene. "America, Europe and the Middle East," *Commentary* 57 (February 1974): 40–55.

Rubenstein, A. "Damn Everybody Sums Up the Angry Mood of Israel," *New York Times Magazine*, February 9, 1969, 24–27.

———. "Now in Israel, a Fluttering of Doves," *New York Times Magazine*, July 26, 1970, 8–9.

———. "Why the Israelis Are Being Difficult," *New York Times Magazine*, April 18, 1971, 32–33.

Safran, N. "Israeli Politics Since the 1967 War," *Current History* 60 (January 1971): 19–25.

Sagan, Scott. "Lessons of the Yom Kippur Alert," *Foreign Policy* 36 (fall 1979): 160–77.

Samuel, M. "If the Arabs Had Won," *Look* 31 (July 25, 1967): 80–82.

Samuels, G. "Israel at Thirteen, B-G at Seventy-five," *New York Times Magazine*, September 24, 1961, 34ff.

———. "Man of Sde Boker," *New York Times Magazine*, October 16, 1966, 42ff.

——. "Refugees to Israel, Five Years Later," *New York Times Magazine*, October 3, 1954, 12ff.

Schmidt, D. A. "Israel's Little War of the Borders," *New York Times Magazine*, November 18, 1945, 10ff.

Schoenbaum, David, and Henry Brandon. "Jordan: The Forgotten Crises," *Foreign Policy* 10 (spring 1973): 171–81.

Shapiro, A. E. "Israel as Sparta," *New York Times Magazine*, December 22, 1974, 9ff.

Shaw, J. "Death of Israel's Prime Minister Complicates Middle East Crisis," *Life* 66 (March 7, 1969): 54–56.

Sheehan, E. R. F. "Talk with Golda Meir," *New York Times Magazine*, August, 27, 1973, 78ff.

——. "Visit with Hussein, the Palestinians and Golda Meir," *New York Times Magazine*, August 27, 1972, 10–11.

Sheehy, Gail. "Riddle of Sadat," *Esquire* 91 (January 30, 1979): 25–34.

Shlaim, Avi. "Conflicting Approaches to Israel's Relations with the Arabs: Ben-Gurion and Sharett, 1953–1956," *Middle East Journal* 37, no. 2 (1983): 180–201.

Syrkin, M. "Arab Refugees; A Zionist View," *Commentary* 41 (January 1966): 23–30.

Szulc, T. "Seeing and Not Believing: Misjudging Arab Intentions," *New Republic* 169 (December 22, 1973): 13–14.

Tucker, R. W. "Israel and the United States: From Dependence to Nuclear Weapons?" *Commentary* 60 (November 1975): 29–43.

Vital, D. "Israel After the War: The Need for Political Change," *Commentary* 57 (March 1974): 46–50.

Welles, S. "New Hope for the Jewish People," *The Nation* 160 (May 5, 1945): 511–13.

Welser, B. "Israel. Faith, Courage and Taxes," *Commentary* 22 (September 1956): 211–18.

Wheeler, K. "Egypt's Premier Reveals How He Made Red Arms Deal," *Life* 39 (November 14, 1955): 127–28.

Winocour, J. "London and Palestine," *New Republic* 115 (July 22, 1946): 78.

Wren, Christopher. "Confronting the PLO," *New York Times Magazine*, September 9, 1979, 15.

Yoffle, H. "Dayan as a Politician; Reply with Rejoinder," *Commentary* 55 (April 1973): 24ff.

Yost, C. "Last Chance for Peace." *Life* 70 (April 9, 1971): 4.

INDEX

BOOKS BY ELINOR BURKETT

GOLDA

ISBN 978-0-06-078666-3 (paperback)

To produce this definitive account of Meir's life, Burkett mined historical records never before examined by any researcher. The result is an astounding portrait of a woman whose uncompromising commitment to the creation and preservation of a Jewish state fueled and framed the ideological conflicts that still define Middle Eastern relations today.

SO MANY ENEMIES, SO LITTLE TIME
An American Woman in All the Wrong Places

ISBN 978-0-06-052443-2 (paperback)

Whether she's writing about being served goat's head in a Kyrgyz yurt, checking out bowling alleys in Baghdad, or trying to cook a chicken in a crumbling apartment, Burkett offers an eclectic series of adventures that are alternately comical, poignant, and discomfiting.

ANOTHER PLANET
A Year in the Life of a Suburban High School

ISBN: 978-0-06-050585-1 (paperback)

In the wake of school shootings across the country, one question haunted America: What is going wrong inside our nation's schools? To find out, award-winning journalist Elinor Burkett spent nine months—from the opening pep rally to graduation day—in a suburban Minneapolis high school. She attended classes, hung out with students, listened to parents, and joined teachers on the front lines.